Advanced Sciences and Technologies for Security Applications

Indexed by SCOPUS

The series Advanced Sciences and Technologies for Security Applications comprises interdisciplinary research covering the theory, foundations and domain-specific topics pertaining to security. Publications within the series are peer-reviewed monographs and edited works in the areas of:

- biological and chemical threat recognition and detection (e.g., biosensors, aerosols, forensics)
- crisis and disaster management
- terrorism
- cyber security and secure information systems (e.g., encryption, optical and photonic systems)
- traditional and non-traditional security
- energy, food and resource security
- economic security and securitization (including associated infrastructures)
- transnational crime
- human security and health security
- social, political and psychological aspects of security
- recognition and identification (e.g., optical imaging, biometrics, authentication and verification)
- smart surveillance systems
- applications of theoretical frameworks and methodologies (e.g., grounded theory, complexity, network sciences, modelling and simulation)

Together, the high-quality contributions to this series provide a cross-disciplinary overview of forefront research endeavours aiming to make the world a safer place.

The editors encourage prospective authors to correspond with them in advance of submitting a manuscript. Submission of manuscripts should be made to the Editor-in-Chief or one of the Editors.

Reza Montasari

Cyberspace, Cyberterrorism and the International Security in the Fourth Industrial Revolution

Threats, Assessment and Responses

 Springer

Reza Montasari
Department of Criminology, Sociology
and Social Policy, School of Social
Sciences
Swansea University
Swansea, UK

ISSN 1613-5113   ISSN 2363-9466 (electronic)
Advanced Sciences and Technologies for Security Applications
ISBN 978-3-031-50453-2   ISBN 978-3-031-50454-9 (eBook)
https://doi.org/10.1007/978-3-031-50454-9

This Springer imprint is published by the registered company Springer Nature Switzerland AG
The registered company address is: Gewerbestrasse 11, 6330 Cham, Switzerland

Paper in this product is recyclable.

To the Heavenly Father
and to my wife:
Anna

Acknowledgments

I would like to extend my sincere gratitude to those who have supported me in the creation of this book:

First and foremost, I am thankful to my Heavenly Father for his boundless love and blessings and for being a source of inspiration throughout this journey.

In loving memory of my dear aunt, Azi Meri, who remains in our hearts, your support in spirit has been a guiding light.

I offer my deepest appreciation to my wife, Anna, and my family for their unwavering support during the extensive research and writing process.

I would also like to acknowledge my church community for their prayers and encouragement, which provided invaluable moral and spiritual support.

Furthermore, I express my profound thanks to the following individuals who played instrumental roles in various aspects of this work:

Aime Sullivan, Megan Wilmot McIntyre, Charlotte Warner, Kaycee Jacka, Zaryab Baig, Georgina Butler, Molly Chapman, Vanessa Serranheira Montinho, and Jessica Bloxham.

Their contributions have significantly enriched the content and overall quality of this book.

Contents

Abbreviations

3D-GAN	Three-Dimensional Generative Adversarial Network
ACF	Active Change Foundation
AI	Artificial Intelligence
AIVD	Netherlands General Intelligence and Security Service
APTs	Advanced Persistent Threats
BD	Big Data
BDPA	Big Data Predictive Analytics
BZK	Ministry of the Interior and Kingdom Relations
CCTVs	Closed-Circuit Televisions
CISA	Cybersecurity and Infrastructure Security Agency
CNII	Critical National Information Infrastructure
CPS	Crown Prosecution Service
CRISP-DM	Cross-Industry Standard Process for Data Mining
CT	Counter-Terrorism
CT&S	Counter-Terrorism and Security
CTCO	Counter-Terrorism Case Officer
CTED	Terrorism Committee Executive Directorate
CTSA 2015	Counter-Terrorism and Security Act 2015
CVE	Countering Violent Extremism
DDoS	Distributed Denial of Service
DF	Digital Forensics
DL	Deep Learning
DML	Deep Machine Learning
DoS	Denial of Service
ECHR	European Convention on Human Rights
ECM	Enabling Cyber Militancy
EU	European Union
FBI	Federal Bureau of Investigation
FR	Facial Recognition
FRT	Facial Recognition Technology
GAN	Generative Adversarial Network

GCTF	Global Counter-Terrorism Forum
GDPR	General Data Protection Regulations
GIFCT	Global Internet Forum to Counter-Terrorism
GTD	Global Terrorism Database
GTO	Global Terrorism Overview
ICO	Information Commissioner's Office
ICRC	International Committee of the Red Cross
ICSR	International Centre for the Study of Radicalisation
ICSVE	International Center for the Study of Violent Extremism
ICT	Information and Communications Technology
IDF	Inverse Document Frequency
IDS	Intrusion Detection Systems
IHL	International Humanitarian Law
IoT	Internet of Things
IP	Internet Protocol
IRA	Irish Republican Army
IS	Islamic State
ISD	Institute for Strategic Dialogue
ISIS	Islamic State of Iraq and Syria
JTAC	Joint Terrorism Analysis Centre
KKK	Ku Klux Klan
K-NN	K-Nearest Neighbor
LEA	Law Enforcement Agency
LIWC	Linguistic Inquiry and Word Count
LTTE	Liberation Tigers of Tamil Eelam
ML	Machine Learning
NATO	North Atlantic Treaty Organization
NCMEC	National Center for Missing and Exploited Children
NGO	Non-governmental Organization
NHS	National Health Service
NLP	Natural Language Processing
NSA	National Security Agency
OIW	Offensive Information Warfare
OS	Operating System
OSINT	Open-Source Intelligence
PBDP	Predictive Big Data Policing
PCA	Principal Component Analysis
PMAP	Prevent Multi-agency Panel
PPV	Positive Predictive Value
PRC	Precision-Recall
PS	Prevention Systems
RFE	Recursive Feature Elimination
RFID	Radio-Frequency Identification
ROC	Receiver Operating Characteristic
RSICC	Rome Statute of the International Criminal Court

SAT	Situational Action Theory
SMOTE	Synthetic Minority Over-sampling TEchnique
SMP	Social Media Platform
START	Study of Terrorism and Responses to Terrorism
SVM	Support Vector Machine
TF	Term Frequency
TF-IDF	Term Frequency-Inverse Document Frequency
t-SNE	t-Distributed Stochastic Neighbor Embedding
UAVs	Unmanned Aerial Vehicles
UK	United Kingdom
UKSA	UK Statistics Authority
UML	Unsupervised Machine Learning
UN	United Nations
UNCCT	United Nations Counter-Terrorism Centre
UNICRI	United Nations Interregional Crime and Justice Research Institute
UNODC	United Nations Office on Drugs and Crime
US	United States
WMDs	Weapons of Mass Destruction
WMV	Weighted Majority Voting
WPS	Women, Peace, and Security Program

Chapter 1
Introduction

Cyberspace, Cyberterrorism and the International Security in the Fourth Industrial Revolution: Threats, Assessment and Responses

Abstract This scholarly book critically examines the complex relationship between cyberspace, cyberterrorism, national and international security, and artificial intelligence (AI) technology. To this end, it delves into the dual nature of these elements, analysing their potential for both constructive and malicious applications within the dynamic landscape of contemporary cyber threats. In pursuit of the research aim, the book provides a deeper understanding of the motives, methods, and consequences of cyberterrorism. Additionally, it conducts a comprehensive assessment of international efforts to counter cyberterrorism, highlighting progress in establishing a unified response and addressing persistent challenges and disagreements. The book also evaluates current practical strategies aimed at deterring, identifying, and responding to terrorist attacks. Therefore, it focuses on analysing the effectiveness, strengths, weaknesses, and opportunities for enhancement of these strategies. Furthermore, the text critically examines the ethical, legal, technical, and operational challenges that arise when harnessing AI and associated technologies to combat both traditional terrorism and cyberterrorism. Following these analyses, the book presents a series of recommendations that serve as a framework for understanding and mitigating the multifaceted challenges arising from the intersection of cyberspace, cyberterrorism, and AI in the context of national and international security. Additionally, it offers actionable insights and practical solutions by providing a wide range of strategies and best practices for countering cyberterrorism and enhancing cyber resilience.

Keywords Cyberspace · Cyber threats · Cyberterrorism · Artificial Intelligence · National security · International security · Cyberattacks · Technical challenges · Ethical challenges · Legal challenges · Recommendations · Practical strategies · Cyber resilience

© The Author(s), under exclusive license to Springer Nature Switzerland AG 2024
R. Montasari, *Cyberspace, Cyberterrorism and the International Security in the Fourth Industrial Revolution*, Advanced Sciences and Technologies for Security Applications, https://doi.org/10.1007/978-3-031-50454-9_1

1.1   About the Book

In recent years, the pervasive presence of cyberspace has fundamentally transformed individuals' lives, revolutionising the way they live, communicate, trade, socialise, access news, conduct banking transactions, entertain themselves, publish content, and consume media created by others [1]. This transformation coincides with an era marked by unparalleled technological progress and global interconnectedness, where the boundaries of human experience have surpassed the physical domain into the complex and ever-evolving landscape of cyberspace (Ibid.). The Fourth Industrial Revolution, commonly referred to as Industry 4.0 or 4IR, signifies a profound and rapid transformation of technologies, industries, and societal structures in the 21st century. This transformation is primarily driven by the increasing interconnectedness of systems and the widespread adoption of intelligent automation [2]. In its technological dimension, 4IR encompasses various facets, including automation and data exchange, within a diverse array of technologies and processes. These technologies comprise artificial intelligence (AI); advanced robotics; cyber-physical systems; online social and gaming platforms; the internet of things (IoT) including the internet of military things and the industrial internet of things; cloud computing; cognitive computing (Ibid.); the Surface Web, the Deep Web and the Dark Web; and numerous other innovations.[1]

While these technologies hold substantial potential to elevate global income levels and improve the quality of life for populations worldwide (Ibid.), they concurrently pose a myriad of challenges that impact businesses, governments, and broader societies. Equally significant is the potential impact of this technological revolution on the national and international security landscape, influencing the likelihood and nature of both cyber and conventional warfare (Ibid.). Furthermore, as increasingly powerful technologies emerge and become more accessible, individuals and groups, such as terrorists and organised criminals, gain enhanced capabilities to inflict widespread devastation and chaos (Ibid.).

Consequently, as cyberspace becomes more complex and ubiquitous, so too do the tactics of modern cybercriminals and terrorist groups. As a result, this evolving digital landscape has become a fertile breeding ground for various threats, with cyberterrorism emerging as a formidable adversary that transcend geopolitical boundaries. Terrorists and their sympathisers exploit cyberspace both as a tool and as a means to execute various acts of terrorism, including the glorification of terrorist acts, incitement, propagation of ideology, recruitment, radicalisation, financing, training, planning, and the commission of terrorist attacks [9]. These activities are facilitated by the aforementioned technologies, which, when combined with sophisticated anti-forensic tools and methods, contribute to the formidable online arsenal of terrorist networks. Therefore, nations now stand at a crossroads of this digital age, where the virtual and physical domains are inextricably intertwined, and the consequences of

---

[1] Refer to the following publications for more detailed information:

Montasari [1, 3, 5], Montasari et al. [4, 6, 7] and Montasari and Jahankhani [8].

cyber threats reverberate across nations, affecting governments, organisations, and individuals alike.

Against this backdrop, there has been a growing recognition of the need to understand how terrorist groups adapt to and exploit both technologies and cyberspace, with the aim of more effectively countering their activities. Acknowledging that cyberspace and technology can be harnessed for both positive and malicious purposes, this book delves into their dual use in the context of cybercrime and cyberterrorism. It is therefore within the context of this evolving landscape of cyberterrorism and its implications that this book aims to examine the complex relationship between cyberspace, cyberterrorism, national and international security, and, AI technology.

1.1.1 Aim and Objectives

Building upon the understanding of the evolving landscape of cyberterrorism and its implications discussed above, this book aims

to investigate the multifaceted relationship between cyberspace, cyberterrorism, national and international security, and AI technology, exploring their dual utility for both constructive and malicious purposes within the dynamic context of contemporary cyberthreats.

Therefore, to achieve the research aim, the following objectives will be pursued:

- Analyse a wide range of challenges posed by the convergence of cyberspace, cyberterrorism, and AI.
- Provide a deeper understanding of the motives, methods, and consequences of cyberterrorism, recognising that the impacts of cyberterrorism transcend national boundaries.
- Conduct a comprehensive assessment of international efforts to counter cyberterrorism, highlighting both progress in establishing a unified response and persistent challenges and disagreements.
- Evaluate current practical strategies aimed at deterring, identifying, and responding to terrorist attacks, with a focus on analysing their effectiveness, strengths, weaknesses, and opportunities for enhancement.
- Investigate the dual use of both cyberspace and AI as powerful tools that enable criminal activities and serve as crucial assets in countering them.
- Examine the ethical, legal, technical, and operational challenges that arise when harnessing these cutting-edge technologies to combat both traditional terrorism and cyberterrorism.
- Present a series of recommendations that serve as a framework for understanding and mitigating the multifaceted challenges arising from the intersection of cyberspace, cyberterrorism and AI in the context of national and international security.

- Emphasise actionable insights and practical solutions by providing a wide range of strategies and best practices for countering cyberterrorism and enhancing cyber resilience.

1.1.2 Contributions

This book undertakes a comprehensive and interdisciplinary approach, covering a wide range of topics, from traditional counter-terrorism strategies to cyberterrorism, machine learning, ethics, gender, and international security. Drawing insights from fields such as computer science, social science, political science, cybersecurity, ethics, gender studies, and international relations, it offers practical recommendations for a diverse audience, including LEAs, governments, policymakers, security professionals, academics, and researchers. Furthermore, the book recognises and addresses ethical, legal, technical, and operational challenges in counterterrorism efforts, providing a holistic understanding of the evolving security landscape. Furthermore, to navigate this rapidly changing world, this book serves as a roadmap for a more secure digital future, contributing to the existing body of knowledge on AI, cybersecurity, and cyberterrorism while seamlessly integrating with broader discourse on these critical topics. Specifically, the book advances the discourse on national and global security in the following ways:

- Conduct the first systematic, interdisciplinary investigation into the intersection of cyberspace, cyberterrorism, and AI.
- Equip policymakers, scholars, and concerned citizens with insights for navigating the complex landscape of cyberspace through historical analysis, contemporary incidents, and future scenarios.
- Explore effective applications of AI and cyberspace in countering terrorism.
- Provide a diverse range of cybersecurity solutions and best practices to assist with global peace and security.
- Offer practical recommendations for enhancing cyber defences for governments, LEAs, policymakers, and organisations.
- Equip policymakers, investigators, and prosecutors with the necessary tools to combat cyberterrorism effectively.
- Extract insights from real-world cases to inform technological and criminal justice responses to cyberterrorism.
- Offer a synthesis of current issues and potential solutions, navigating the intricate nexus between technology, crime, and security.
- Assist LEAs and other stakeholders in combating online terrorist content by presenting practical measures to protect online users.
- Deliver a comprehensive analysis of cyberterrorism, integrating insights from various academic disciplines to provide a holistic perspective.

1.1.3 Target Audience

This book draws upon a wide range of fields, including computer science, cyber-security, law, social science, criminology, psychology, politics, and international relations. Considering its multidisciplinary approach, it caters to a diverse audience. Therefore, it serves as an indispensable resource for a wide range of professionals and scholars, including law enforcement officers, government officials, security professionals, military and industry leaders, technology strategists, political scientists, policymakers, digital forensic investigators, cybersecurity experts, researchers, academics, and graduate and advanced undergraduate students. Furthermore, legal professionals, criminologists, and technology enthusiasts will also find its contents engaging and insightful. This broad appeal arises from the book's comprehensive examination at the intersection of technology, security studies, and international relations, positioning it as an essential reference for anyone interested in navigating the intricate landscape of cyberterrorism in the Fourth Industrial Revolution.

1.1.4 Structure of the Book

This book adopts a structured organisation comprising four distinct themes, with each theme represented by a dedicated Part. Within this organisational framework, each overarching theme is systematically explored through a series of chapters, providing readers with a clear and thematic roadmap for their journey through the content. The subsequent discourse offers a description regarding the structural composition of the book. For a detailed description of each specific chapter, refer to Sect. 1.1.2.

Part I: Understanding Terrorism and Counter-Terrorism Strategies

Part I of this book explores complexities surrounding contemporary global security challenges. It serves as the foundational segment of the book, encompassing three chapters that critically analyse various dimensions of terrorism and the strategies implemented to combat it. Within this section, the book delves into the complex relationship between state actions and the rise of extremist ideologies. This examination begins with a critical analysis of the Iraq War and its profound global consequences, shedding light on the factors that contribute to the emergence of terrorism. Moving forward, the book examines the effectiveness of counter-terrorism strategies, with a particular focus on the United Kingdom (U.K.). These strategies are critically assessed, and alternative approaches are considered to identify methods that could enhance national and international security. Additionally, Part I explores a less-explored aspect of terrorism, namely the roles played by women in extremist activities. This examination challenges conventional assumptions and strives to comprehend the multifaceted nature of their involvement, ultimately contributing to the refinement of counter-terrorism strategies. Part I consists of the following chapters:

- Chapter 2: Unravelling State Crimes: A Critical Analysis of the Iraq War and Its Global Ramifications
- Chapter 3: Assessing the Effectiveness of UK Counter-Terrorism Strategies and Alternative Approaches
- Chapter 4: Understanding and Assessing the Role of Women in Terrorism

Collectively, the chapters within Part I provide readers with a solid foundation for comprehending the intricacies of terrorism in the modern world. Furthermore, the section serves as a platform for a more informed and nuanced approach to addressing these complex global threats in the subsequent portions of the book.

Part II: Cyberterrorism Landscape

Part II of this book offers an in-depth assessment of the current cyberterrorism landscape, a domain that is continually evolving. This section comprises two critical chapters, each contributing to a comprehensive understanding of the contemporary threats posed by cyberterrorism and their implications for national security. In pursuit of this objective, the section begins by exploring the impact of the COVID-19 pandemic on online terrorism. It sheds light on how global crises have transformed the digital realm into a breeding ground for new threats. Additionally, this section analyses the tactics and strategies employed by cyberterrorists in response to this global crisis, providing valuable insights into the evolving nature of this threat. Subsequently, Part II shifts its focus to the imminent cyberterrorism threats facing national security. To this end, it delves deeper to uncover the potential risks that cyberterrorism poses to the security of nations. Through a critical assessment of vulnerabilities and emerging threats, this section offers readers a comprehensive view of the challenges that governments and organisations encounter as they strive to safeguard their national interests in an increasingly interconnected world. Part II comprises the following chapters:

- Chapter 5: Exploring the Current Landscape of Cyberterrorism: Insights, Strategies, and the Impact of COVID-19
- Chapter 6: Exploring the Imminence of Cyberterrorism Threat to National Security

Together, the chapters within Part II illuminate the dynamic landscape of cyberterrorism, offering readers a comprehensive understanding of current threats and their implications for national security. In addition, this section serves as a valuable resource for individuals seeking to navigate the complex landscape of cybersecurity and its intersection with contemporary global security concerns.

Part III: Countering Cyberterrorism with Technology

Part III of this book forms the core of the book's exploration into leveraging technology to mitigate the threats of cyberterrorism. This section encompasses four critical chapters, collectively providing an in-depth understanding of the intersection between technology and counterterrorism strategies. To commence, it examines how technology impacts the radicalisation to violent extremism and terrorism in the contemporary security landscape. Furthermore, it explores the digital tools and platforms that facilitate the spread of extremist ideologies and provides insights into the challenges of countering radicalisation in the digital age. The focus then shifts to the role of machine learning (ML) and deep learning (DL) techniques in countering cyberterrorism. To this end, this section delves into the potential of these advanced technologies to enhance cybersecurity and offers valuable perspectives on their application in safeguarding against cyber threats. Moving forward, the section critically examines the ethical, legal, technical, and operational challenges associated with applying ML in countering cyberterrorism. It also highlights the complexities involved in deploying these technologies effectively and responsibly. Part III concludes by addressing the ethical, legal, technical, and operational challenges faced in counterterrorism efforts that utilise AI. Furthermore, it offers recommendations and strategies for navigating these challenges, aiming to ensure technology's effectiveness in the fight against cyberterrorism. Part III comprises the following chapters:

- Chapter 7: The Impact of Technology on Radicalisation to Violent Extremism and Terrorism in the Contemporary Security Landscape
- Chapter 8: Machine Learning and Deep Learning Techniques in Countering Cyberterrorism
- Chapter 9: Analysing Ethical, Legal, Technical and Operational Challenges of the Application of Machine Learning in Countering Cyber Terrorism
- Chapter 10: Addressing Ethical, Legal, Technical, and Operational Challenges in Counterterrorism with Machine Learning: Recommendations and Strategies

Together, the chapters within Part III provide readers with a comprehensive overview of the dynamic relationship between AI technology and counterterrorism. Furthermore, they equip readers with insights into the potential benefits and challenges associated with deploying advanced technologies in the pursuit of national and global security.

Part IV: Artificial Intelligence and National and International Security

Part IV of this book delves into the intricate relationship between AI technology and the broader security landscape. Comprising three pivotal chapters, this section provides an in-depth understanding of AI's transformative role in shaping the future of national and international security. This part begins by critically analysing the dual role of AI in online disinformation, illuminating how AI technology can be both a tool for spreading disinformation and a tool for countering it. Furthermore, it provides valuable insights into the complexities surrounding the use of AI in the information warfare domain. Moving forward, Part IV explores the impact of facial recognition technology (FRT) on the fundamental right to privacy. To this end, it delves into the ethical and legal considerations surrounding the use of FRT and offers recommendations to strike a balance between security and individual privacy rights. Part IV concludes by analysing the specific challenges posed by deepfake technology within the context of the UK. Subsequently, it offers both legal and technical insights into the deepfake landscape and provides recommendations to address these emerging threats effectively. Part IV comprises the following chapters:

- Chapter 11: The Dual Role of Artificial Intelligence in Online Disinformation: A Critical Analysis
- Chapter 12: Responding to Deepfake Challenges in the United Kingdom: Legal and Technical Insights with Recommendations
- Chapter 13: The Impact of Facial Recognition Technology on the Fundamental Right to Privacy and Associated Recommendations

Collectively, the chapters within Part IV offer readers a comprehensive view of the multifaceted relationship between AI technology and national and international security. They equip readers with insights into the ethical, legal, and technical dimensions of AI's impact on security matters, serving as a valuable resource for those navigating this rapidly evolving landscape.

1.2   Chapter Previews

Chapter 2: *Unravelling State Crimes: A Critical Analysis of the Iraq War and Its Global Ramifications*

Chapter 2 delves into the complexities of international politics, state accountability, and the lasting repercussions of military interventions. At its core, this chapter undertakes a meticulous examination of the Iraq War of 2003, a pivotal moment in recent history that remains steeped in controversy. To this end, the chapter examines the state's violation during this tumultuous period, shedding light on actions that violated both domestic and international law. The Iraq War, a subject of intense global scrutiny, raised doubts and reservations among not only its allies but also international bodies such as the United Nations (UN). This chapter seeks to understand the motivations behind this conflict, shaped in part by the spectre of terrorism and the perceived presence of weapons of mass destruction in Iraq, which drove the push for humanitarian intervention. The chapter also delves into the historical backdrop, examining international tensions and state sovereignty, especially after the events of September 11, 2001. This context is crucial for understanding the Iraq War' complexity. Furthermore, the chapter analyses the events during and after the war, bringing into focus its global and local impacts. This consolidates the findings of the chapter, tracing the threads that weave the Iraq War's overall outcomes and enduring implications, offering a more profound understanding of the far-reaching consequences that have altered history's course. Finally, the research presented in the chapter concludes that the Iraq War was an act of aggression, violating the principles in the UN Charter.

Chapter 3: *Assessing the Effectiveness of UK Counter-Terrorism Strategies and Alternative Approaches*

This chapter critically assesses the U.K.'s counter-terrorism strategies, specifically focusing on the CONTEST mechanism and its constituent components, including Prevent, Pursue, Protect, and Prepare. In pursuit of this objective, the chapter performs a critical analysis of the effectiveness of these strategies, illuminating their strengths and shortcomings. As part of this analysis, the chapter uncovers the intricacies of the UK's counter-terrorism efforts and carefully scrutinise the identified problematic aspects and the negative consequences that have emerged as a result of these strategies. This analysis forms the basis for recommendations and alternative approaches put forth in the subsequent sections of the chapter. These proposals are designed not only to assist with rectifying the highlighted issues but also to enhance the overall effectiveness of the counter-terrorism practices. To this end, an emphasis is placed on the need for a balanced, inclusive, and adaptable approach, one that harmonises security measures with the safeguarding of individual rights and societal values. In addition, this chapter underscores the significance of collaboration and knowledge exchange among diverse stakeholders, drawing on the experiences of other nations to bolster the UK's counter-terrorism endeavours. The chapter underscores the importance of ongoing research into the root causes of terrorism to develop a targeted response while upholding democratic principles and human rights. In summary, this chapter serves as a valuable resource for policymakers and practitioners, shaping the future of counter-terrorism strategies to effectively counter evolving threats while preserving democratic core principles.

Chapter 4: Understanding and Assessing the Role of Women in Terrorism

This chapter investigates gender-biased assumptions regarding the reasons and methods behind women's participation in terrorism. To this end, the chapter conducts a comprehensive analysis of the multifaceted roles that women play in the context of terrorism, emphasising the need to move beyond rigid dualistic paradigms and to capture the complexities inherent in their involvement. The analysis begins by critically examining the prevalent notion that women predominantly assume victim roles within terrorism, occupying subservient positions such as sympathisers, spouses, and maternal figures. However, drawing upon the expanding body of scholarship, the chapter repositions women not only as perpetrators of violence but also as active participants within the intricate network of terrorist organisations. Central to this analysis is the nuanced concept of agency, that is, an individual's capacity to exercise personal choices and actions, while also acknowledging the constraints that might curtail such agency. Throughout the discourse, the emphasis remains on cultivating a nuanced understanding that integrates agency with the complexities of choice, thereby enabling a more accurate portrayal of the diverse roles that women assume in the context of terrorism. In line with this objective, the chapter argues that the nature of terrorist organisations themselves can significantly influence these roles. To support the argument, the chapter delves deeper into the multifaceted roles played by women, examining how various types of terrorist organisations shape women's involvement in terrorism. This examination sheds light on women's roles as passive victims in the Islamic State and as active agents in the Tamil Tigers. By offering insights into the operational dynamics of different organisations and the specific contexts in which they operate, the chapter contributes towards a better understanding of the wide spectrum of roles that women might occupy within the complex landscape of terrorism. In summary, the chapter calls into question the proposition that tends to homogenise women into a facile dichotomy, alternating solely between the poles of vulnerable victims and empowered agents.

Chapter 5: Exploring the Current Landscape of Cyberterrorism: Insights, Strategies, and the Impact of COVID-19

This chapter delves into the dynamic realm of cyberterrorism, illuminating the intricate relationship between malicious actors, including terrorist groups and their sympathisers, and the digital landscape. Its aim is to provide valuable insights for policymakers, researchers, and security professionals, offering proactive strategies to combat the evolving cyber threats faced by governments worldwide. Beginning with a comprehensive exploration of the current cyberterrorism landscape, the chapter examines the impact of the COVID-19 pandemic on online terrorism. It also investigates how terrorist entities harness the internet's global reach and interactivity for various nefarious activities, ranging from propaganda dissemination and recruitment to training, planning, funding, and executing attacks. Furthermore, the chapter delves into the nuanced nature of cyberterrorism, unravelling the characteristics of cyberattacks, their most prevalent forms, and their underlying objectives. This understanding serves as a foundation for the development of effective countermeasures and the enhancement of cybersecurity on both national and international scales. In doing so, the chapter contributes to the existing body of knowledge on cyberterrorism by offering a nuanced perspective on its current state.

Chapter 6: Exploring the Imminence of Cyberterrorism Threat to National Security

As the title suggests, this chapter explores the imminence of the cyberterrorism threat to national security, shedding light on its multifaceted aspects. Beginning with essential background information, the chapter lays the foundation for comprehending the intricacies of cyberterrorism by elucidating the significance of the technological landscape in today's interconnected society. It then delves into the reasons behind growing concerns about cyberterrorism, offering insights into the current technological arsenal employed by cyberterrorists and why this mode of terrorism might be more attractive to them than conventional methods. Additionally, the chapter uncovers the evolving capabilities fuelling apprehensions about the threat of cyberterrorism to national security while also addressing counterarguments that contest its severity. Similarly, it provides readers with a well-rounded understanding of the complex relationship between technological progress, vulnerabilities, and the perceived cyberterrorism threat. Ultimately, this chapter equips readers with the knowledge needed to navigate the intricate intersection of technology and security in the digital age, enabling them to gain an informed perspective on the cyberterrorism threat to national security.

Chapter 7: The Impact of Technology on Radicalisation to Violent Extremism and Terrorism in the Contemporary Security Landscape

This chapter explores the profound changes brought about by digital technologies in today's interconnected world and investigates the intricate relationship between technology and violent ideologies. In doing so, it sheds light on the convergence that has given rise to unprecedented challenges and opportunities for security experts and society at large. Additionally, the chapter delves into the complex interplay between offline and online expressions of extremism, revealing their symbiotic relationship and their influence on extremist narratives and ideologies. Moreover, it assesses how the internet and associated technologies facilitate radicalisation to violent extremism, drawing insights from real-world cases. Building on these insights, the chapter also explores the realm of the term 'new terrorism', uncovering its evolving strategies and its reliance on technology as a powerful tool for recruitment and organisation. Shifting focus to the IoT, the chapter similarly examines its role in enabling and complicating terrorist activities. Ultimately, this chapter provides readers with a deeper understanding of how technology has reshaped the landscape of radicalisation, extremism, and terrorism in the contemporary security context. It also contributes significantly to the development of strategies that enhance global security in an era of unprecedented technological influence.

Interconnected Chapters: 8, 9, *and* 10

To establish the context for the descriptions of the following three chapters, it is first crucial to acknowledge their close interconnection in the unfolding narrative of this book. Each chapter serves as a distinct segment of the broader discourse, building upon the foundations laid by its predecessor. Chapter 8 begins the exploration, providing a comprehensive overview of the core ML methods deployed in the battle against cyber threats. In contrast, Chap. 9 delves deeper into the complexities that emerge when applying ML techniques in this context. It examines ethical dilemmas, legal intricacies, and technical and operational challenges, establishing the context for a more profound understanding of the multifaceted challenges at hand. Finally, in Chap. 10, the three chapters culminate their journey with a strategic approach, offering insights, recommendations, and strategies aimed at not only mitigating the challenges identified but also charting a path toward responsible and effective use of ML in bolstering cybersecurity efforts. It is also important to note that, while each of these chapters stand independently, readers are encouraged to explore them within the context of one another. This interconnected triad forms a cohesive narrative that progressively elucidate the intricacies of countering cyberterrorism through the lens of ML, providing a comprehensive and holistic understanding of this critical subject matter. Similarly, they provide valuable insights and guidance for policymakers, law enforcement agencies, and cybersecurity professionals. Furthermore, upon concluding the journey through these chapters, readers will gain not only a profound understanding of the challenges at hand but also the necessary tools and strategies to navigate and overcome these challenges, ultimately contributing to a safer and more secure cyber landscape.

Chapter 8: Machine Learning and Deep Learning Techniques in Countering Cyberterrorism

This chapter explores the integration of AI in national security efforts, with a particular focus on the roles of ML and DL. Within this context, it provides a comprehensive examination of ML techniques in countering cyberterrorism, emphasising their central role in detecting extremist content and online radical behaviours. The chapter also investigates the critical topic of feature extraction and selection, essential for optimising the accuracy of ML algorithms in countering cyberterrorism. Similarly, the chapter investigates techniques used to identify relevant data points for analysis. Furthermore, it critically examines the intersection of DL and intrusion detection systems as tools to strengthen cybersecurity measures. Additionally, the chapter then delves into counter-terrorism policing, offering insights into recruitment tactics used by cyber terrorists. This chapter contributes to a deeper understanding of how AI bolsters cybersecurity measures and combats the growing threat of cyberterrorism. By exploring the technical dimensions of AI in cybersecurity, it equips readers with a nuanced perspective on the roles of ML and DL in addressing evolving challenges posed by cyberterrorism.

Chapter 9: *Analysing Ethical, Legal, Technical and Operational Challenges of the Application of Machine Learning in Countering Cyber Terrorism*

This chapter builds upon the foundation established in the previous chapter, which examined the core ML methods in the context of counterterrorism. To this end, the chapter delves deeper into the complexities that arise when utilising ML techniques for countering cyber terrorism, exploring challenges across technical, operational, ethical, and legal dimensions. With regards to the ethical and legal considerations, the chapter investigates issues related to individual autonomy, human rights preservation, algorithmic bias, privacy protection, and the consequences of mass surveillance. Within the same context, the chapter also examines topics such as hybrid classifiers, accountability assignment, and jurisdictional complexities that surface in this domain. The chapter then shifts its focus to the technical and operational challenges associated with the application of AI and ML in counterterrorism. Here, the chapter explores the intricacies of managing vast amounts of data, the difficulties posed by class imbalances, the curse of dimensionality, identification of spurious correlations, detection of lone-wolf terrorists, implications of mass surveillance, and the application of predictive analytics. Furthermore, transparency and explainability in these technical and operational aspects are emphasised, along with potential unintended consequences stemming from the integration of AI and ML into counterterrorism practices. By systematically analysing technical, operational, ethical, and legal obstacles, this chapter offers a comprehensive understanding of the complexities involved in applying AI and ML to counter cyber terrorism. It contributes significantly to the existing knowledge base, laying the groundwork for informed strategies and solutions that enhance security while safeguarding fundamental human rights.

Chapter 10: *Addressing Ethical, Legal, Technical, and Operational Challenges in Counterterrorism with Machine Learning: Recommendations and Strategies*

As previously discussed, Chap. 8 explored the technical aspects of using ML techniques to combat cyberterrorism, while Chap. 9 highlighted the challenges inherent in their application. This chapter, an essential component of the trilogy, builds upon the foundations laid by its predecessors. To this end, the chapter provides a roadmap for addressing the multifaceted challenges identified in particular within Chap. 9. Within this context, the chapter serves a dual purpose. Firstly, it addresses the challenges articulated in Chap. 9 by offering insightful recommendations and strategic guidance. Secondly, it outlines a responsible and effective trajectory for the integration of ML in bolstering cybersecurity efforts. To this end, the chapter begins by presenting a comprehensive set of technical and operational recommendations aligned with the challenges elucidated in Chap. 9. These recommendations pertain to the application of ML in counterterrorism initiatives, offering practical guidance for practitioners. Moving forward, the chapter also offers a diverse array of ethical and legal recommendations that correspond to the ethical and legal challenges intertwined with the use of ML within counterterrorism efforts, as discussed in Chap. 9. By doing so, the chapter provides readers with a comprehensive and holistic understanding of the critical subject matter of countering cyberterrorism through the lens of ML.

Chapter 11: The Dual Role of Artificial Intelligence in Online Disinformation: A Critical Analysis

This chapter critically analyses the intricate role of AI in the context of online disinformation. To this end, it delves into AI's capability to craft highly convincing disinformation and rapidly disseminate it across large audiences on social media platforms (SMPs). Furthermore, this study examines AI's potential as a countermeasure to combat this harmful phenomenon while also scrutinising the accompanying challenges and ethical dilemmas. To address these multifaceted issues, the chapter proceeds to provide a comprehensive set of recommendations aimed at mitigating these concerns. Throughout the chapter, ethical considerations are highlighted as essential elements in the development and deployment of AI tools. This is due to the unintentional negative consequences that might arise from their use, including inherent limitations that give rise to ethical concerns and counterproductive outcomes.

Chapter 12: Responding Deepfake Challenges in the United Kingdom: Legal and Technical Insights with Recommendations

This chapter explores the multifaceted realm of deepfakes, examining the complex interplay between legal and technical challenges posed by this evolving technology. With a specific focus on the UK, the chapter scrutinises the absence of comprehensive legislation explicitly tailored to deepfakes, highlighting the implications of this regulatory gap. To address this challenge, it offers insights into the potential emergence of legislation and its ramifications for stakeholders. Furthermore, the chapter illuminates the limitations of existing detection mechanisms, revealing their struggle to keep pace with the rapidly evolving landscape of deepfake creation. Similarly, it investigates pornography as a critical focal point and explores available legal provisions aimed at protecting individuals ensnared in the realm of image-based abuse. Finally, the chapter explores the profound implications of disinformation, emphasising how deepfake technology can amplify the perilous landscape of mis- and disinformation, thereby raising concerns about the dissemination of false narratives and their societal impact. The chapter contributes to the existing body of knowledge by deepening understanding of the legal, technical, and societal aspects of deepfake technology, particularly in the context of the UK. It also underscores the need for regulatory measures and the challenges associated with combating the misuse of deepfakes.

Chapter 13***: The Impact of Facial Recognition Technology on the Fundamental Right to Privacy and Associated Recommendations***

This chapter critically examines the multifaceted dimensions of FRT, considering its advantages and inherent limitations. While recognising its potential to bolster security measures, the chapter also underscores the need for a balanced perspective that considers its potential infringement and impact on privacy rights. Within this context, the chapter offers a series of recommendations aimed at safeguarding the fundamental right to privacy concerning the deployment of FRT, emphasising the critical convergence of technology and ethics in upholding individual privacy. This chapter contributes to the existing body of knowledge by offering valuable insights and guiding informed decision-making in the application of FRT.

1.3 Disclaimer

This book provides a comprehensive examination of critical issues at the intersection of cyberspace, cyberterrorism, and international security. Its primary aim is to offer a comprehensive and unbiased assessment of various viewpoints, theories, and analyses put forth by scholars, experts, and individuals from diverse backgrounds. To this end, it is imperative to clarify that the perspectives presented within these pages are not reflective of the author's personal views or biases. Instead, the perspectives presented within these pages are intended to be impartial and devoid of personal bias. The author's aim here is to present a neutral and objective analysis by impartially showcasing the arguments, criticisms, and viewpoints of various scholars and experts. Furthermore, it is vital to emphasise that this book is not intended to advocate for any particular perspective or stance. Rather, it serves as a platform for the exploration and evaluation of differing viewpoints, allowing readers to form their own informed judgments based on the array of analyses presented. To ensure a well-rounded understanding of the topics discussed, a wide spectrum of perspectives is included, even those with which the author might personally disagree. It is the author's belief that fostering a diverse and inclusive dialogue is essential for addressing complex issues such as the ones explored in this book. By presenting opposing viewpoints and critical analyses side by side, the author seeks to empower readers to engage critically with the subject matter and arrive at their own conclusions. Finally, this book is an academic endeavour to provide readers with a thorough, balanced, and nuanced exploration of the intricate issues surrounding cyberspace, cyberterrorism, and international security. The author remains committed to the principles of academic rigor, objectivity, and open discourse, and he invites readers to approach the contents of this book with the same commitment to intellectual inquiry and thoughtful evaluation.

References

1. Montasari R (2023) Countering cyberterrorism: the confluence of artificial intelligence, cyber forensics and digital policing in US and UK national cybersecurity. Springer, Cham
2. Schwab K (2018) The fourth industrial revolution: what it means, how to respond. World Economic Forum. https://www.weforum.org/agenda/2016/01/the-fourth-industrial-revolution-what-it-means-and-how-to-respond/. Accessed 26 Sept 2023
3. Montasari R (ed) (2023) Applications for artificial intelligence and digital forensics in national security. Springer, Cham
4. Montasari R, Carpenter V, Masys A (eds) (2023) Digital transformation in policing: the promise, perils and solutions. Springer, Cham
5. Montasari R (ed) (2022) Artificial intelligence and national security. Springer, Cham
6. Montasari R, Carroll F, Mitchell I, Hara S, Bolton-King R (eds) (2022) Privacy, security and forensics in the internet of things (IoT). Springer, Cham
7. Montasari R, Jahankhani H, Hill R, Parkinson S (eds) (2021) Digital forensic investigation of internet of things (IoT) devices. Springer, Cham
8. Montasari R, Jahankhani H (eds) (2021) Artificial intelligence in cyber security: impact and implications. Springer, Cham
9. United Nations Office on Drugs and Crime (2012) The use of the Internet for terrorist purposes. https://www.unodc.org/documents/frontpage/Use_of_Internet_for_Terrorist_Purposes.pdf. Accessed 26 Sept 2023

Part I
Understanding Terrorism and Counter-Terrorism Strategies

Chapter 2
Unravelling State Crimes: A Critical Analysis of the Iraq War and Its Global Ramifications

Abstract The Iraq War serves as a crucial focal point for research, illuminating the intricacies of international politics, state responsibility, and the enduring consequences of military interventions. This case study delves into the Iraq War of 2003, one of the most contentious political decisions in recent decades, to comprehend the specific state crimes committed during the conflict. The war's controversial nature garnered scepticism from allies, the United Nations (UN), and much of the non-Western population. The chapter defines state crime as acts or omissions by the state that violate domestic and international law, human rights, or systematically harm its own or another state's population. The conclusion establishes the Iraq War as a war of aggression, violating the UN Charter. The study also examines the challenges surrounding humanitarian intervention and its motive, largely driven by the perceived threat of terrorism and weapons of mass destruction (WMDs) in Iraq. The chapter presents a historical context and explores the events at both macro and micro levels, including the aftermath of the war. The overall outcomes and lasting implications of this significant event are also discussed.

Keywords The Iraq War · Weapons of mass destruction · United Nations · International politics · Terrorism · International security · Military interventions · Controversy

2.1 Introduction

The Iraq War serves as a pivotal focal point for research, shedding light on the intricacies of international politics, state responsibility, and the enduring consequences of military interventions. This case study critically analyses the 2003 Iraq War, a conflict of significant global significance, which drew scepticism from allies, the UN, and a substantial portion of the non-Western world [8]. The chapter investigates specific state crimes committed during the war, using the definition of state crime as acts or omissions by a state that result in "violations of domestic and international law, human rights, or systematic institutionalized harm of its or another state's population … regardless of whether there is or is not self-motivation" [26, p. 102]. The Iraq War of 2003 stands as a notable and divisive political decision of recent times, necessitating a thorough exploration of its implications. Therefore, this study seeks to evaluate the conflict in light of the state crimes perpetrated during the war, while also addressing the complexities surrounding humanitarian intervention. Notably, the Iraq War has been labelled as a war of aggression [9], namely a war that involves the exertion of control over a sovereign state, which inherently contravenes the principles outlined in the UN Charter (Article 8 bis, Rome Statute of the International Criminal Court (RSICC)). This case study also highlights the complexities associated with humanitarian intervention, a concept defined by Hehir [13] as state-led military action within the territory of another state without their consent, with the intent to protect the welfare of the other state's citizens. The motive for humanitarian intervention in this instance was largely fuelled by concerns over terrorism and WMDs purportedly present in Iraq, justifying it as a proportionate self-defence measure [19].

Subsequent sections of this study are structured as follows. Section 2.2 delves into the historical context surrounding the 2003 Iraq war, shedding light on relevant international issues prevalent during that period. This contextualisation of the Iraq War through an examination of state sovereignty and the impact of preceding events, particularly on September 11, 2001 (9/11), is vital for a comprehensive analysis of the conflict and its consequences. Section 2.3 comprehensively evaluates the macro and micro-level events, providing a detailed analysis of the aftermath of the Iraq War. Section 2.4 consolidates the study's findings, outlining the overall outcomes and lasting implications that emanated from this significant event. Finally, the chapter is concluded in Sect. 2.5.

2.2   Context

To gain a comprehensive understanding of the Iraq War, it is essential to delve into relevant international issues and key concepts. Central to this analysis is the notion of state sovereignty, which encompasses state authority and autonomy, granting the state the right to govern its territory and citizens [2]. Ayoob [2] underscores the significance of recognising and accepting the principle of state sovereignty within an international system characterised by significant disparities in power between nations. This recognition is essential to safeguard the interests of weaker states and institutions. This principle is notably emphasised in the UN Charter [27], particularly in Article 2(4), which mandates, "all members shall refrain … from the threat or use of force against the territorial integrity or political independence of any state." An in-depth analysis of events leading up to the Iraq War is also crucial to gain a comprehensive understanding of the context of the case. To this end, particular emphasis must be placed on the pivotal event of the terrorist attack on the United States (US) on 9/11. During this historic moment, the terrorist group Al-Qaeda, comprising Islamic extremists, hijacked four commercial airplanes and orchestrated attacks on the World Trade Center in New York and the Pentagon in Washington DC, resulting in the tragic loss of approximately 3000 innocent lives [14]. Subsequently, The National Security Strategy of the USA [28] was launched, outlining the US President at the time, George W. Bush's commitment to thwart and eliminate terrorist organisations, targeting their leadership, command, control, communications, material support, and finances. The aftermath of 9/11 eventually led to the United Kingdom (UK) and US launching an attack on Iraq on March 20, 2003, resulting in an estimated death toll of up to 1.2 million Iraqis, according to expert studies [3].

2.3 Evaluation

Despite the UN Charter's clear assertion in Article 2(7) that the UN should avoid intervention in matters within a state's jurisdiction, arguments advocating forceful intervention were put forth by coalition states (UK, US, and Australia) to support the invasion of Iraq. The first argument hinged on an exception to Article 2(4) as outlined in Article 51 of the UN Charter, which upholds the inherent right of individual or collective self-defence in the event of an armed attack against a UN member state. The coalition contended that Iraq's alleged links to terrorism, specifically the 9/11 attack by Al-Qaeda, warranted the invasion. However, this claim was consistently propagated by the Bush Administration without any substantiated factual evidence [11]. Furthermore, the exclusion of pre-emptive self-defence in Article 51 rendered this argument void, as Iraq had not carried out an armed attack, negating any grounds for self-defence [5]. Another argument used to justify the attack on Iraq emanated from the National Security Strategy implemented by the Bush Administration. This strategy asserted the US's right to act pre-emptively against perceived threats, including those with access to WMDs [28]. The Bush Administration adamantly claimed on numerous occasions that Iraq possessed or was actively seeking WMDs [21]. However, no evidence of WMDs was ever discovered in Iraq, thereby rendering the basis for this argument invalid [6]. Moreover, the US and UK believed they had a responsibility to employ reasonable force through humanitarian intervention to halt the human rights abuses perpetrated by the Saddam Hussein regime against Iraqi citizens [17]. Under Articles 39 and 42 of the UN Charter (2019), the Security Council is authorised to take action to "maintain or restore international peace and security." Nevertheless, the coalition's invasion of Iraq lacked approval or authorisation from the UN Security Council, significantly undermining the humanitarian intentions of the war and its overall legitimacy [25].

The military campaign known as "Operation Iraqi Freedom," led by the US, commenced on 19th March 2003, with military support from the UK, Australia, and Poland [22]. Following the expiry of the 48-h ultimatum for Saddam Hussein and his sons to leave Iraq, the US launched their initial missiles at Baghdad on 20th March 2003, initiating the war that ultimately led to the overthrow of Hussein's regime and the establishment of US control over Iraq [3]. The attack predominantly consisted of targeted missiles and bombs directed at key military and government installations across Iraqi cities, based on suspected locations of the Iraqi leadership [10]. Despite President Bush declaring victory on 1st May 2003, with the statement that "major combat operations in Iraq have ended" [23]. Garden [10] notes that the figurative display of victory occurred on 9th April 2003 when US troops toppled the statue of Saddam Hussein in Baghdad. However, while the invasion succeeded in toppling the Iraqi government, it fell short in establishing lasting peace and conducting effective humanitarian intervention [20].

McGoldrick [22] clarifies how some of the US actions taken in response to terrorism contradicted their human rights and humanitarian responsibilities. Kramer et al. [18] further expound on how the invasion violated International Humanitarian Law (IHL) on multiple fronts. Firstly, there was a failure to protect innocent civilians or a disregard for their safety, evidenced by the use of weapons such as napalm, cluster bombs, missiles, and depleted uranium shells, indicating a greater focus on offensive measures than protective ones. Additionally, Amnesty International [1] emphasises that IHL was breached by the failure to ensure public safety and security through efficient policing and manpower, leading to extensive looting of public buildings, an increase in crime, and heightened fear among citizens. Bacon [4] observes that the Bush Administration exceeded its duty of deposing the Iraqi regime by exploiting its power to transform the Iraqi economy into a capitalist nation with unrestricted trade and widespread privatisation, primarily benefiting other nations and resulting in deteriorating human rights conditions for Iraqi citizens. Another violation of IHL pertains to the inadequate US responses to Iraqi resistance actions, resulting in unnecessary civilian deaths and casualties among US and British military personnel, for which the US must bear responsibility as the insurgent crimes were an outcome of their coalition invasion [18]. The International Committee of the Red Cross (ICRC) [16] reported severe violations of IHL concerning the treatment and conditions of Iraqi prisoners in the Abu Ghraib prison, which was under the control of coalition forces. Iraqi prisoners were subjected to heinous forms of torture, humiliation, and deprivation of basic human rights, including access to food, water, and sanitary facilities [16].

The Iraq War's exact death toll remains unrecorded officially; however, the Iraq Body Count [15] employs data verification from multiple sources to ascertain the number of documented violent deaths. The findings indicate that the documented civilian death count ranges from 185,632 to 208,716, while the total death count, including combatants, is estimated at 288,000. Moreover, approximately 250,000 additional civilians were injured, and around 200,000 individuals sought refuge as refugees [3]. The repercussions of the regime's overthrow in Iraq were not as liberating as anticipated, particularly for vulnerable groups. Green and Ward [12] emphasise that conditions worsened for certain groups of individuals. Another significant consequence of the Iraq War was the surge in "dual-purpose" criminality, encompassing crimes such as murder, robbery, rape, and smuggling, which simultaneously served individual and organisational objectives [12].

2.4 Discussion

The Iraq War, with its profound consequences, has caused significant loss of life and injury to civilians, attracting substantial scholarly attention. In his work, Crawford [7] emphasises the importance of acknowledging the indirect harm resulting from the breakdown of Iraq's infrastructure, rendering "internally displaced persons, refugees and children" vulnerable to disease, malnutrition, and sustained injury (p. 13). Considering the repercussions of the Iraq War, coupled with the evident breaches of international law, there exists a firm conviction that the US invasion and occupation of Iraq unequivocally qualify as a state crime [17]. The invasion of Iraq fails to meet the criteria of humanitarian intervention. The aggressive force employed by the coalition forces was not warranted, as the regime in Iraq did not pose an immediate threat [25]. Furthermore, the primary motive behind the intervention was not solely to provide humanitarian aid (Ibid.). The actions taken did not strive to comply with humanitarian law, and they lacked the approval of the UN Security Council. Consequently, it becomes evident that the interests of Iraqi citizens were not at the core of the intervention (Ibid.). The actions of the Bush Administration following the Iraq War have significantly hindered the US's capacity to undertake new humanitarian interventions [20]. The lack of legitimacy behind the justifications for intervention, as highlighted by Kurth [20], has cast a shadow on the credibility of future interventions. This catastrophic failure underscores the importance of adhering to legal and ethical considerations in any prospective humanitarian intervention [20]. Thus, the attainment of liberation from the scourge of the supreme international crime, specifically unlawful intervention, necessitates universal adherence to the principles of the rule of law by all nations, irrespective of their power dynamics [18]. This underscores the urgency of fostering a collective understanding of the justifiability of military intervention as a humanitarian intervention. Therefore, a more unified framework is required to prevent the abuse of humanitarian intervention as a tool by stronger and more powerful states, thereby safeguarding its effectiveness in addressing the needs of civilians facing predicaments [24].

2.5 Conclusion

The arguments presented by the coalition states to justify the intervention in Iraq exhibited various flaws and lacked sufficient support from international law and the UN Charter, raising concerns about the legitimacy and legality of the invasion. The Operation Iraqi Freedom campaign, while arguably achieving its military objectives, faced significant criticism for violations of human rights, humanitarian principles, and IHL, raising concerns about the ethical and legal implications of the intervention [1, 18, 22]. As a result, the Iraq War of 2003 remains a subject of intense scholarly debate due to its controversial nature. This case study has shed light on the commission of state crimes during the conflict and illuminated the complexities surrounding the concept of humanitarian intervention. As researchers continue to examine the consequences of Iraq War, it becomes evident that further research is necessary to understand its broader implications on international relations and human rights. To this end, the Iraq War stands as a critical case study in assessing the consequences and complexities of humanitarian interventions. The catastrophic impact on civilian lives due to the breakdown of infrastructure and the blatant violations of international law highlights the urgency of reforming the concept of humanitarian intervention. Building a more comprehensive and unified understanding of justifiability is paramount to ensure that future interventions genuinely prioritise the well-being of civilians in vulnerable situations [24]. By striving for international accountability and adherence to the rule of law, the international community can move towards more ethical and effective humanitarian interventions.

References

1. Amnesty International (2004) Amnesty international report 2004—Iraq, May 26. Available at: https://www.refworld.org/docid/40b5a1f710.html. Accessed 06 Sept 2023
2. Ayoob M (2002) Humanitarian intervention and state sovereignty. Int J Hum Rights 6(1):82–102. https://doi.org/10.1080/714003751
3. Bassil Y (2012) The 2003 Iraq War: operations, causes and consequences. J Hum Soc Sci 4(5):29–47
4. Bacon D (2004) Privatized Iraq: imposed economic and social policies raise human rights questions. Race Poverty Environ 11(1):46–49. http://www.jstor.org/stable/41554428
5. Bellamy AJ (2003) International law and the war with Iraq. Melb J Int Law 4(3):479–520
6. Cohn M (2013) Iraq: a war of aggression. No WMDs no connection to Al Qaeda. Global Research. http://pricewebss.pbworks.com/w/file/fetch/107163474/11A.%20Iraq-%20War%20of%20Aggression.pdf
7. Crawford NC (2013) Civilian death and injury in the Iraq War, 2003–2013. Costs of War. https://watson.brown.edu/costsofwar/files/cow/imce/papers/2013/Civilian%20Death%20and%20Injury%20in%20the%20Iraq%20War,%202003-2013.pdf
8. Danchev A, MacMillan J (eds) (2004) The Iraq War and democratic politics. ProQuest Ebook Central. https://ebookcentral.proquest.com
9. Falk R (2005) The world speaks on Iraq. Guild Pract 62:91–94
10. Garden T (2003) Iraq: the military campaign. Int Aff (Roy Inst Int Aff) 79(4):701–718
11. Gershkoff A, Kushner S (2005) Shaping public opinion: the 9/11-Iraq connection in the bush administration's rhetoric. Perspect Polit 3(3):525–537
12. Green P, Ward T (2009) The transformation of violence in Iraq. Br J Criminol 49(5):609–627. https://doi.org/10.1093/bjc/azp022
13. Hehir A (2013) Humanitarian intervention: an introduction, 2nd edn. Red Globe Press
14. Holloway D (2008) 9/11 and the War on Terror. Edinburgh University Press
15. Iraq Body Count (2021) https://www.iraqbodycount.org/
16. International Committee of the Red Cross (ICRC) (2004) Report of the ICRC on the treatment by the coalition forces of prisoners of war and other protected persons by the Geneva Conventions in Iraq during arrest, internment and interrogation, February. https://cryptome.org/icrc-report.htm
17. Kramer RC, Michalowski RJ (2005) War, aggression and state crime: a criminological analysis of the invasion and occupation of Iraq. Br J Criminol 45(4):446–469. https://doi.org/10.1093/bjc/azi032

18. Kramer R, Michalowski R, Rothe D (2005) The Supreme International Crime: how the U.S. War in Iraq threatens the rule of law. Soc Just 32(2):52–81. http://www.jstor.org/stable/297 68307

19. Kumar R (2014) Iraq War 2003 and the issue of pre-emptive and preventative self-defence: implications for the United Nations. India Q 70(2):123–137. https://www.jstor.org/stable/450 72829

20. Kurth J (2005) Humanitarian intervention after Iraq: legal ideals vs military realities. Orbis 50(1):87–101. https://doi.org/10.1016/j.orbis.2005.10.007

21. Lewis C, Reading-Smith M (2014) False pretenses. The Center for Public Integrity, June 30. https://publicintegrity.org/2008/01/23/5641/false-pretenses

22. McGoldrick D (2004) From '9–11' to the 'Iraq War 2003': international law in an age of complexity. Hart Publishing

23. McSmith A (2016) The Chilcot Inquiry—a timeline of the Iraq War. Independent, July 6. https://www.independent.co.uk/news/uk/politics/chilcot-report-a-timeline-of-the-iraq-war-and-the-disasters-that-ensued-a7119731.html

24. Roth K (2004) War in Iraq: not a humanitarian intervention. Human rights in the War on Terror, pp 143–157

25. Roth K (2006) Was the Iraq War a humanitarian intervention? J Milit Ethics 5(2):84–92. https://doi.org/10.1080/15027570600711864

26. Rothe DL, Kauzlarich D (2016) Crimes of the powerful; an introduction. Routledge

27. The United Nations (2019) United Nations charter. https://www.un.org/securitycouncil/content/purposes-and-principles-un-chapter-i-un-charter

28. The White House (2002) The national security strategy of the United States of America. https://2009-2017.state.gov/documents/organization/63562.pdf

Chapter 3
Assessing the Effectiveness of UK Counter-Terrorism Strategies and Alternative Approaches

Abstract This chapter critically assesses the effectiveness of the United Kingdom's (UK) counter-terrorism strategies, with a specific focus on the CONTEST mechanism and its key components: Prevent, Pursue, Protect, and Prepare. Through a comprehensive analysis of the current practices, the chapter identifies key challenges and proposes targeted recommendations aimed at achieving a delicate balance between robust security measures and the protection of individual rights. Emphasising the need for a comprehensive approach, the study advocates for the integration of both hard and soft strategies to effectively counter terrorism while upholding essential human rights and societal values. Furthermore, the chapter underscores the significance of engaging diverse stakeholders and learning from the experiences of other nations to enhance the adaptability and efficacy of UK counter-terrorism efforts. Collaboration and knowledge exchange are highlighted as indispensable tools in combating the ever-evolving nature of terrorism while preserving democratic principles and human rights. While contributing to the fight against terrorism, the research acknowledges the necessity of more in-depth investigations into the underlying causes of violent extremism for continuous improvement of the counter-terrorism strategy. In this context, the study sheds light on the importance of ongoing research to gain a deeper understanding of the root causes of terrorism, facilitating the development of more targeted and effective responses. The chapter contributes significantly to the field by providing valuable insights for policymakers and practitioners alike. By advocating for a balanced, inclusive, and adaptable approach, the study seeks to support ongoing efforts to combat terrorism while preserving the core principles that underpin a democratic society.

Keywords Countering violent extremism · UK · Prevent strategy · National security · CONTEST · International politics · Terrorism · International security · Human rights · Controversy

3.1 Introduction

In Britain, the nation upholds principles of freedom, allowing individuals to practice their chosen religion or opt for no religion at all, while emphasising the importance of understanding and respecting diverse religious beliefs [105]. This approach is rooted in democratic and egalitarian values, fostering a diverse and inclusive society that respects diverse religious beliefs (Ibid., [15]). However, despite these principles, the UK faces the ongoing challenge of countering terrorism. The Joint Terrorism Analysis Centre (JTAC), under the auspices of the Security Service MI5, has assigned the current national threat level from terrorism as "substantial", signifying a credible likelihood of a terrorist attack [83]. The notion of terrorism, as defined by the Terrorism Act of 2000, encompasses acts or threats of violence intended to exert influence over the government or intimidate the general public, with the aim of promoting specific political, religious, or ideological objectives. Such acts might involve endangering or inflicting serious violence upon individuals, causing substantial property damage, or disrupting electronic systems. Scholars have also provided an academic definition of terrorism, characterising it as a tactic that instils fear and employs "coercive political violence". It is often seen as a calculated and demonstrative practice involving "direct violent actions", devoid of "legal or moral constraints" [80, p. 158]. The primary targets of terrorism are civilians and non-combatants, and its objectives often revolve around propaganda dissemination and the psychological impact on adversaries [80].

In order to develop effective counterterrorism strategies, it becomes imperative to comprehend the underlying factors that contribute to terrorism [15]. By acknowledging and addressing these root causes, the UK can implement targeted and proactive measures to prevent acts of terrorism and promote societal harmony. Furthermore, embracing an approach that incorporates a comprehensive understanding of the drivers of terrorism can aid in formulating more nuanced and effective strategies for countering extremism and ensuring the safety and security of the nation. This chapter aims to explore the effectiveness of the UK counter-terrorism strategies and offer evidence-based alternative approaches. To this end, the chapter will undertake a critical evaluation of the current counter-terrorism practices implemented by the UK, considering the contentious debates surrounding their effectiveness. Specifically, the focus will be on the UK's counter-terrorism mechanism known as CONTEST Strategy.

The subsequent sections of the chapter are structured as follows. Section 3.2 provides background information which contextualises the study. Section 3.3 presents a detailed examination of CONTEST along with its key components: Prevent, Pursue, Protect, and Prepare. Following this, Sect. 3.4 critically analyses identified problematic elements and negative consequences. The subsequent Section of the chapter, namely Sect. 3.5, offers recommendations and alternative approaches that can be adopted to enhance the existing counter-terrorism practices and help to address the highlighted issues effectively. These suggestions seek to improve the overall approach and mitigate potential negative outcomes. The final section, i.e. Section 3.6, concludes the evaluation by summarising the key points of the UK's current counter-terrorism practices and reiterating the main recommendations proposed.

3.2 Background

The process of radicalisation encompasses three distinct stages consisting of Macro causes, Meso causes and Micro causes [32]. Macro causes constitute the first stage and involve broader societal factors, such as governmental and foreign policies, socioeconomic conditions, democratic influences, and governmental inadequacy in ensuring a secure environment. The second stage, Meso causes, pertains to contextual factors that emerge within physical, sociocultural, and local environmental settings. The third stage, Micro causes, delves into individual differences, encompassing stressors and mental health issues that might contribute to radicalisation (Ibid.). While these categories provide a comprehensive framework, it is also vital to acknowledge the role of ideology, which might play a crucial and independent role in the genesis of terrorism, as suggested by [19]. In countering terrorism, two distinct approaches can be identified: the "hard" (coercive) and "soft" (non-coercive) responses [39], which will be expounded upon in the subsequent two subsections.

3.2.1 Hard Approach

Hard approaches to counterterrorism, favoured and employed by prominent global leaders such as the United States (US), relies on kinetic methods, direct force, and coercive state power to combat terrorism. This encompasses the use of military action, intelligence gathering, drone attacks, law enforcement investigations, security measures, and persuasive measures [2, 37, 89]. The hard approach strategy is centred on disrupting and dismantling terrorist organisations through covert operations [74] and targeted killings of suspected terrorists, often involving drone strikes. One significant example of the efficacy of hard tactics was the 2011 operation in which Osama Bin Laden, the leader of Al-Qaeda and mastermind behind the 9/11 attacks on the World Trade Center, was eliminated during a military operation conducted by the US [16]. The deployment of hard tactics can deter terrorist acts by instilling fear in potential attackers regarding the swift and severe consequences they may face [20]. However, these methods have generated controversy due to their ethical implications, particularly in terms of disregarding the right to life and a fair trial for the targeted individuals [28]. The use of lethal force against terrorists in this manner has been argued to potentially contravene their inherent human rights [23] and may inadvertently bestow upon them the status of martyr figures [21]. Additionally, the media's portrayal of fear often leads to a preference for hard interventions, such as military action, as the primary response among the public [3].

3.2.2 Soft Approach

In contrast to hard approaches, the soft response involves employing strategic measures to engage with terrorists. One key component of this approach is the concept of Countering Violent Extremism (CVE) [37]. CVE programmes aim to prevent radicalisation and extremism by providing support, education, and rehabilitation for at-risk individuals, seeking to address the root causes that contribute to their susceptibility to extremist ideologies. These soft interventions are often perceived as more inclusive and just, recognising the importance of safeguarding human rights and offering opportunities for disengagement and reintegration into society. To this end, the UK adopts soft power strategies, emphasising educational initiatives and safeguarding efforts to address the underlying factors contributing to extremism (Ibid.). These soft interventions focus on protecting individuals and promoting respect and inclusion [57]. Furthermore, they aim to foster resilient communities, and implementing CVE programmes, which target prevention, deradicalisation, disengagement, and awareness-raising efforts [5]. This is exemplified by strategic communications that aim to curtail the proliferation of extremism [78].

In addition to its emphasis on soft strategies, the UK also adopts a proactive approach to ensure security during major events held in densely populated areas, such as the coronation of King Charles held on 6th May 2023. In this regard, Counter Terrorism Policing's Senior National Coordinator for Protect and Prepare, Laurence Taylor, highlighted the "extensive measures in place to mitigate risk and monitor the terrorist threat against the backdrop of a historic moment, both nationally and internationally" [30]. London, where the coronation was taking place, was heavily guarded with a strong police presence, surveillance, snipers, and emergency services on standby. However, some critics argue that the high level of protection is primarily focused on safeguarding high-profile individuals, such as the royal family, while other regions within the UK, such as Northern Ireland, might face substantial terror threats with potentially less attention and protection. In Northern Ireland, the threat from the IRA remained severe, with the possibility of attempting an attack during the coronation events, especially given their recent claims of responsibility for pipe bombs and an alleged plan to target Joe Biden's visit [84] from 11 to 14th April, 2023. This raises concerns about the differential treatment and risk exposure faced by the general public across the country. Furthermore, media coverage of terrorist incidents committed by Muslims tends to receive significantly greater attention, approximately 357% more extensive than that of other types of attacks. A more accurate and balanced reporting approach could potentially help align public perception and awareness with the full extent of terrorism threats, fostering a more comprehensive understanding of the issue [52].

3.3 Contest

During the peak of the War on Terror in 2003, the UK Labour Government initiated its comprehensive counter-terrorism strategy known as CONTEST, in response to the escalating terrorist threat following the 9/11 attacks in New York and Washington DC in 2001 [45]. Subsequently, the strategy underwent several reviews and revisions in 2006, 2009, 2011, 2018, and most recently in July 2023 [93, 96, 98, 100, 101] to adapt to the evolving nature of terrorism and align with new governmental policies on counter-terrorism. Following the 2011 review, a three-pronged strategy was devised to tackle the ideological challenge, prevent individuals from being attracted to terrorism, and provide support. It also aimed to collaborate with sectors and institutions at risk of extremist influences [70]. Subsequently, the Counter-Terrorism and Security (CT&S) Act 2015 incorporated the Prevent Duty (discussed subsequently) within the broader framework of CONTEST. This legislative inclusion imposes a mandate on specific organisations to undertake proactive measures in thwarting individuals from engaging in terrorism while simultaneously upholding principles of equity and approach [96].

The most recent iterations of CONTEST Strategy, carried out in June 2018 and July 2023 under the Conservative Government, emphasise enhanced structural coordination across the public sector, comprising intelligence agencies, local authorities, healthcare services, and other entities. This approach seeks to impede terrorists and their supporters from planning and executing attacks more effectively [93, 96]. The primary focus of CONTEST Strategy is to address the root causes of terrorism through early intervention and rehabilitation strategies (Ibid.). Specifically, the strategy aims "to reduce the risk to the UK and its citizens and interests overseas from terrorism, so that people can go about their lives freely and with confidence" [96]. To achieve this goal, CONTEST addresses several strategic factors that terrorists seek to exploit. Among these are extremism, which has the potential to sow divisions within communities, conflict and instability, which create conducive environments for terrorists to thrive; and advancements in technology, which facilitate terrorists' quest for anonymity and global reach (Ibid.). The CONTEST framework consists of four interrelated elements, commonly known as the 4 P's or 4 P work strands: Prevent, Pursue, Protect, and Prepare. Together, these components constitute CONTEST's Risk Reduction Model, collectively working in tandem to achieve the overarching goal of reducing the risk of terrorist attacks (Ibid.).

The following subsections will critically examine these four distinct strands with the main emphasis being placed on the Prevent component.

3.3.1 Prevent

The first component, referred to as "Prevent", assumes critical significance within the broader context of the CONTEST framework. The primary objective of the Prevent element is to proactively safeguard individuals considered susceptible to radicalisation and potential involvement in terrorism by means of early intervention and accessible support mechanisms [96]. By identifying and providing support to individuals vulnerable to radicalisation, the Prevent Strategy [99] aims to curtail the spread of terrorism at its roots and reduce the likelihood that young individuals might be lured into making regrettable decisions [67]. One approach employed in this endeavour is the implementation of CVE programmes, which takes a comprehensive approach to counter the proliferation of extremist ideologies by fostering resilient communities [61] and addressing the underlying macro, meso, and micro factors that contribute to the process of radicalisation and its connection to terrorism [44]. Moreover, Prevent endeavours to rehabilitate individuals with existing terrorist connections [96]. Another key aspect of Prevent is its focus on addressing the root causes of radicalisation and formulating responses to the ideological challenges posed by terrorism [96]. This strand holds the Prevent duty as a pivotal element, which was introduced in 2011 and constitutes a significant aspect of the overall CONTEST strategy. The fundamental objective of the Prevent duty can be articulated as the imperative to "prevent people from being drawn into terrorism" [94].

The preventive strategy is founded on three specific objectives: firstly, it aims to address the ideological challenges presented by terrorism and the inherent risks posed by those who lend support to such activities. Secondly, it strives to prevent individuals from being enticed into terrorism by providing relevant advice and support. Finally, it seeks to foster collaborative efforts with various sectors and institutions where the potential for radicalisation necessitates intervention [94]. The Prevent delivery model exemplifies how risk factors and root causes of radicalisation can be effectively mitigated, while simultaneously enabling interventions with individuals already involved in terrorism [96]. It endeavours to address the underlying causes of radicalisation through comprehensive initiatives both in the digital and physical realms, aimed at empowering individuals and communities as a cohesive whole (Ibid.). This approach acknowledges the significance of early intervention, applying safeguarding principles, and enlisting support from diverse agencies to assist those identified as being most vulnerable to radicalisation. Furthermore, the model utilises rehabilitation efforts to provide support and disengage individuals already associated with terrorist activities (Ibid.).

Section 26 of the [29] (CTSA) imposes a duty of care upon specific bodies, compelling them to exercise "due regard to the need to prevent people from being drawn into terrorism" [94]. This obligation, known as the Prevent duty, encompasses a range of authorities in various sectors, including the health sector, police forces, prisons, probation services, as well as educational institutions such as schools and universities (Ibid.). To fulfil this duty effectively, relevant frontline staff are expected to undergo standardised training, equipping them with a comprehensive understanding of the concepts of radicalisation and extremism, their interconnections with terrorism, and the reasons why certain individuals might be susceptible to being drawn into terrorist activities (Ibid.). Frontline workers, upon perceiving that an individual under their care might be at risk of involvement in terrorism, initiate a Prevent referral, subject to review by specialised police officers who conduct a 'gateway assessment' [95]. This assessment process entails the use of police databases and other pertinent resources to evaluate whether the level of risk and vulnerability meets the criteria for further engagement within the Prevent program. If deemed appropriate, the individual proceeds to the next stage, involving the multi-agency and local authority-led Channel panel [95].

Section 37 of the CTSA 2015 stipulates that the Channel panel must be chaired by the responsible local authority and comprise multiple panel members drawn from relevant safeguarding sectors, including representatives from the police, health, and education services. The Channel panel employs the Channel Vulnerability Assessment Framework [97] to ascertain the extent of vulnerability, considering three key elements: engagement with a particular group or ideology, the intent to cause harm, and the capability to cause harm [95]. When no risk is identified, cases will be concluded. However, in situations where risks are recognised, personalised and targeted support packages will be devised to address individual needs and identified vulnerabilities. This process entails the involvement of all relevant agencies and their resources (Ibid.). It is pertinent to emphasise that all support programs offered are entirely voluntary. In instances where individuals opt not to accept the provided support, they might be referred to alternative services for assistance. If, despite the refusal of support, the threat of terrorism persists, the cases will be referred to the Counter Terrorism Case Officer (CTCO) for police monitoring (Ibid.).

3.3.2 Pursue

The Pursue element is aimed at the prevention of terrorist attacks, both within the UK and against allied nations overseas. It aims to take proactive measures to disrupt and thwart potential terrorist plots before they can materialise [96]. The Pursue element comprises of several identified objectives. Its primary focus is on the detection and understanding of any terrorist activity, along with the capacity to investigate and disrupt such activities, which may involve pursuing prosecutions (Ibid.). To achieve these objectives, Pursue relies on various organisations, including the UK counter-terrorism police, national and international intelligence agencies, and the military. One of the measures employed under the Pursue strategy involves deporting foreign nationals involved in terrorist-related activities. A notable example of this approach is the case of Shemima Begum, who left the UK to join ISIS and subsequently had her UK nationality revoked. This decision sparked controversy, with public discourse emphasising her young age of 15 at the time of her decision to join ISIS, leading to arguments that she might not have fully comprehended the gravity of her actions [67]. Another case relevant to the Pursue strategy is that of Salman Abedi, the perpetrator of the 2017 Manchester Arena bombing. Prior to the tragic incident, UK intelligence agencies were already aware of Abedi and had previously conducted investigations into his activities [42]. This culminated in inquiries into whether the attack could have been prevented. Subsequently, UK law enforcement authorities carried out raids and arrests in the aftermath of the attack, leading to the apprehension of several individuals connected to the incident. However, the timing and scope of these actions have been subject to criticism, with some arguing they might have been insufficient and delayed, highlighting the limitations of the CONTEST framework [96]. Furthermore, the implementation of the 'pursue' strategy might inadvertently contribute to increased terrorism within communities labelled as suspect, as this labelling can align with the principles of labelling theory [6].

Therefore, in light of the aforementioned concerns, the UK Government should carefully consider the scope and application of powers under Schedule 7 and address challenges related to stop and search authority for suspected terrorists, as exemplified by the controversies surrounding Section 44 of the Terrorism Act 2000 [98]. Overall, the Pursue element of the CONTEST framework plays a crucial role in proactively detecting and preventing terrorist activities, employing various measures and collaborations among law enforcement and intelligence agencies. Nevertheless, notable cases such as those of Shemima Begum and Salman Abedi as exemplified underscore the complexities and challenges inherent in counterterrorism efforts, prompting ongoing discussions on optimizing the effectiveness of the CONTEST strategy [42, 67, 96].

3.3.3 Protect

As for the Protect strand of the UK's counterterrorism strategy, its main purpose is to enhance security against terrorist attacks in the UK and safeguard its interests abroad, thus reducing vulnerabilities. This objective is to be accomplished through the establishment of a robust, multi-layered defence system, facilitating the detection and handling of potential terrorists and dangerous materials at the borders. Additionally, other goals include mitigating risks faced by global aviation and other transportation sectors, bolstering the security of crowded public places, specific groups, and individuals, and thwarting terrorists' access to harmful materials and crucial knowledge [96]. The intelligence capabilities within the UK have played a crucial role in preventing several terrorist attacks. For instance, in 2017, a plot to bomb a plane flying from Australia to the UK on Etihad Airways was foiled through collaborative efforts between the UK and Australian authorities. The plot was successfully discovered through intelligence gathering and robust aviation security measures [107]. Border security measures in the UK have proven effective in reducing the risk of international terrorist attacks, as evidenced by the detection and prevention of over 100 potential terrorists by the UK's border force in 2017 [96].

Furthermore, the Protect strand encompasses a range of measures aimed at safeguarding against cyber-attacks. However, it is essential to critically assess the effectiveness of these measures in the face of evolving cyber threats as detecting and preventing such attacks is indeed a formidable challenge [77]. Despite having seemingly robust protection measures in place, the consequences of a successful cyber-attack can still be significant [59], with implications for national security, economic stability, and individual privacy. Moreover, the international nature of many cyber-attacks underscores the imperative for robust international cooperation in detecting and preventing such threats [40]. A critical examination of this cooperation reveals potential impediments such as geopolitical tensions, differing legal frameworks, and issues of sovereignty. Consequently, the effectiveness of the Protect component might be constrained if other nations do not have equivalent measures in place. For instance, in 2018, reports emerged that ISIS effectively exploited the secure messaging app Telegram for coordinating attacks and propagating their ideology [85]. In response, the UK government collaborated with technology companies to remove terrorist content from their platforms and enacted legislation aimed at criminalising the viewing of terrorist material online. While these measures demonstrate a proactive response, they have sparked public opposition due to concerns about potential limitations on free speech and government overreach [63].

In light of these critical considerations, it becomes evident that a more nuanced approach to the Protect strand is necessary. This approach should involve continuous evaluation and adaptation of cybersecurity measures to address evolving threats while carefully balancing security imperatives with individual rights [11]. Furthermore, these issues are closely linked with the Tor browser and the digital policing of terrorism on the Dark Web [12]. Some argue that if individuals with extreme views are banned from using public social media platforms (SMPs), they might seek to express themselves anonymously on the Dark Web, citing freedom of speech as a human right [46]. However, leaving the Dark Web unpoliced is considered unethical due to concerns about activities such as weapons and drug trafficking, as well as illegal practices such as paedophilia, which cannot be allowed to continue unchecked [66].

3.3.4 Prepare

The Prepare branch of the CONTEST strategy aims to minimise harm and save lives by ensuring a swift and efficient recovery in the aftermath of a terrorist attack. Stated objectives involve the implementation of a comprehensive multi-agency response to any form of terrorist attack, cultivating the UK's proficiency to respond effectively to both existing and potential future attacks, and reducing the impact of such attacks on civilians, essential services, and communities. To accomplish this, Prepare utilises all emergency services, including fire and ambulance personnel, as well as armed officers [96]. The Prepare component has also contributed to raising public awareness about the terrorism threat and educating individuals on how to prepare for such events, improving communication between different agencies. However, implementing the Prepare strand entails significant financial, material, and labour investments, which might not be readily available in smaller and less-affluent areas [10]. As a result, disparities in effectiveness might arise across regions perceived as less important, potentially leading to public discontent with the government's focus on wealthier areas.

3.4   Evaluation of CONTEST

CONTEST has demonstrated effectiveness in preventing terrorism. At its core, the strategy is built on the principle of shared responsibility, necessitating cooperation and coordination among key stakeholders in the public and private sectors, as well as various government authorities [18]. This collaborative approach has led to improvements in the accuracy and timeliness of threat assessments through the exchange of intelligence and information among diverse agencies, organisations, and international partners, thereby enhancing the quality of decision-making [79]. One notable success of CONTEST lies in the establishment of the Joint Terrorism Analysis Centre (JTAC), which assumes the responsibility of evaluating the terrorism threat level and devising a strategic response to potential terrorist incidents [88]. This multi-agency collaboration, involving experts from entities such as MI5, the police, and others, has resulted in enhanced response capabilities and reduced response times. Nevertheless, while international data sharing fosters global cooperation, concerns arise regarding the varying data protection and human rights standards beyond the EU [14]. Such disparities might lead to potential data leaks, particularly when sharing intelligence with non-EU partners. Managing intelligence presence for counterterrorism and anti-organised crime efforts outside the EU proves challenging and might extend beyond the scope of intelligence oversight (Ibid.). Therefore, careful attention to data protection standards and human rights principles is necessary to maintain the integrity and efficacy of intelligence cooperation on an international scale.

Furthermore, despite the non-specific and faith-neutral language employed in the Prevent strategy, there are substantial concerns regarding potential discrimination and the unintended targeting of innocent Muslim individuals through its implementation. Some scholars have contended that the UK's counter-terrorism strategy has, whether inadvertently or not, stigmatised the Muslim community, effectively designating them as a 'suspect' community [8]. The notion of a suspect community refers to a group within the population that has been singled out by the state as problematic and subject to specific policing measures based on their presumed affiliation with a sub-group [71]. An illustrative instance of the discriminatory application of Prevent was evident in the funding allocation of the strategy prior to the 2011 Review, where funding aimed at addressing extremism was apportioned based on the number of Muslims residing in a given area [91]. Additionally, in educational settings, there has been a disproportionate number of Prevent referrals involving Muslim children, a trend that contributes to the exacerbation of Islamophobia among young children (Ibid.). Similarly, the National Union of Teachers has expressed apprehensions regarding the potential implications of certain measures, which could result in the stigmatisation of Muslim students and teachers, potentially impeding free speech and academic freedom [65]. These concerns extend to the erosion of trust between teachers and students, thereby impacting the overall learning environment. Such measures might prompt some students to conceal their faith (Taqqiyah) in order to avoid discrimination, for instance, refraining from wearing a hijab, a practice that is deemed unethical and might inadvertently promote covert expressions of Islam, a tactic exploited by jihadists to elude law enforcement [22]. In light of these concerns, the National Union of Teachers advocates for a more nuanced approach to addressing the issue of radicalisation, one that takes into account broader social and political factors contributing to extremism [65]. Moreover, the media play a significant role in perpetuating these stereotypes, often portraying terrorists in a manner that is explicitly linked to their ethnicity and culture [68]. This portrayal arguably grants the public an ideological and moral justification for racist anti-Muslim hate crimes [75, 76], consequently exacerbating the marginalisation of the Muslim community and fostering a sense of distrust towards the police. Such breakdowns in vital relationships hinder the flow of information from these communities, which could otherwise assist in preventing acts of terrorism [49]. Consequently, the very measures intended to enhance security might inadvertently undermine it, while Muslim communities continue to endure the burden of being considered suspects [71].

Another contentious aspect of the Prevent strategy pertains to the definition of extremism, which, though not legally binding, serves as the sole definition that frontline staff must utilise to assess whether an individual's behaviour or ideologies qualify as extremist, subsequently leading to potential Prevent referrals [60]. The provided definition of extremism reads as follows: "vocal or active opposition to fundamental British values, including democracy, the rule of law, individual liberty and mutual respect and tolerance of different faiths and beliefs … [and calling] for the death of members of our armed forces" [99, p. 107]. Of concern here is the ambiguity surrounding the term "fundamental British values", which lacks clarity in its intended meaning. Its subjective and opaque nature can give rise to serious implementation challenges and might have adverse implications for democracy and freedom of speech [60]. This issue is evident in reports indicating that pupils in educational settings feel apprehensive about expressing their opinions on contentious and sensitive subjects due to the fear of being misconstrued by teachers, who might interpret their views as indicative of extremist inclinations [91]. Such concerns highlight the potential chilling effect on free expression and open discourse, thereby impeding the healthy exchange of ideas and perspectives. The concern lies in the potential creation of an environment wherein young individuals feel constrained from expressing themselves and articulating their grievances, thereby depriving them of a safe space to engage in peaceful democratic processes [53]. The absence of such a secure platform for free speech and open debate might compel discussions on political change to retreat to unsupervised spaces [47]. Within such spaces, there is a risk of propagating more radical messages, which could ultimately prove counterproductive, undermining the very objectives that the Prevent strategy aims to accomplish [60].

The theoretical underpinning of Prevent rests on the assumption of a linear "radicalisation process," implying a progression from radicalisation to terrorism [41]. This conceptualisation or theorisation, which has wielded considerable influence in shaping governmental policies pertaining to counter-terrorism strategies, forms a substantial basis for the Prevent initiative [87]. However, despite numerous theories offering detailed step-by-step progressions of the radicalisation process, there remains a dearth of empirical data supporting these models, particularly concerning the psychological aspects of individuals who have undergone radicalisation [54].

Furthermore, aspects of the radicalisation discourse and Prevent strategy have raised questions about the concept's unreliability, suggesting that its primary function might be to exert control over the targeted community through knowledge generation rather than relying on factual evidence [41]. Moreover, Prevent employs the notion of "vulnerability" to explicate how individuals initially considered "good Muslims" might transition into being perceived as "bad," thereby categorising its targets as both "at risk" and "risky." Thus, the failure to account for the transition from being at risk to risky suggests that the notion of radicalisation used within the context of Prevent could be deemed a performative security knowledge. This implies that the radicalisation discourse might generate the very threats it seeks to identify, serving the purpose of governance performance, rather than responding to the actual presence of these risks (Ibid.). Consequently, the lack of addressing the intermediate phase between these states blurs the boundaries between them (Ibid.).

3.5 Recommendations and Alternative Approaches

3.5.1 Building Trust and Promoting Inclusivity

As elucidated in the preceding section, the UK's ongoing implementation of its counter-terrorism strategy is likely to perpetuate a climate of unease concerning Muslims within British society, fostering and perpetuating insecurities that propagate fears and perceptions across various domains of governance [70]. In order to address this issue and achieve CONTEST's objectives effectively, it is imperative to discontinue the marginalisation of Muslim communities and alleviate their perception of being treated as suspects. Embracing them as integral members of the community, valuing their opinions [73], and affording them full freedom and equal opportunities within society [56] are essential steps in this process. Central to this endeavour is the Government's public acknowledgment of the dynamic nature of British identity, which evolves through the presence of a diverse populace residing in and constituting the UK. Such recognition underscores the right of every individual in society to contribute to shaping the nation's identity [56]. In this context, some scholars [55] have emphasised the critical role of building trust within Muslim communities for the development of effective and human rights-respecting counterterrorism strategies. Failure to establish trust might lead to an increasing sense of alienation among Muslim communities from counterterrorism laws and policies, potentially resulting in unintended negative consequences. Thus, fostering trust is crucial in ensuring inclusivity and cooperation in countering terrorism [55].

Furthermore, there is a pressing need to establish inclusive and accessible forums that facilitate diverse discussions concerning religious identity, ideology, and foreign policy. These spaces are particularly crucial for individuals who perceive themselves as disconnected from mainstream politics. Importantly, these forums should foster an environment free from judgment and should refrain from censorship that might deter individuals from expressing views that could be perceived as radical and consequently attract the attention of counter-terrorism forces or intelligence agencies [56]. In this context, the German government's "Prevention of Islamist Extremism" programme, introduced in 2016, represents an alternative approach that effectively engages Muslim populations. As part of this initiative, a Muslim working group on the prevention and intervention of Islamist extremism was established in 2019, comprising members from Muslim groups, organisations, and communities. This platform facilitates community engagement and dialogue, fostering collaboration and cooperation. An example of a project under this strategy is "Radikale Akzeptanz", which offers a secure environment for Muslim youth to discuss their issues and experiences while interacting with non-Muslim peers. Through such interactions, prejudices and stereotypes are dispelled, promoting empathy, acceptance, and understanding of diverse experiences among Muslim youth in Germany. Therefore, establishing trust and cooperation among various stakeholders, such as communities and government agencies, is essential for fostering robust partnerships and encouraging individuals and groups to report suspicious activities or potential threats [25]. These concerted efforts are essential to move away from a divisive 'them versus us' approach and, instead, work towards rebuilding the trust of Muslim communities in the UK. By minimising the marginalisation and alienation experienced by these groups, such initiatives can contribute to fostering a sense of belonging and active engagement within the broader society [33].

3.5.2 Disseminating Counter-Terrorism Narratives

An additional recommendation involves adopting innovative and alternative approaches to disseminate counter-terrorism narratives, including the use of former extremists who have successfully disengaged from extremist or terrorist activities [73]. In this context, these former extremists can play a pivotal role in constructing counter narratives by sharing their personal experiences, which can convey the harsh realities they have undergone, debunk the allure of violence, challenge ideologies that promote terrorism, and expose the contradictions inherent in the narratives and actions of terrorist groups [90]. A notable advantage of employing former extremists in this capacity is their perceived credibility and authority stemming from their firsthand experiences, which the government might lack when addressing counter-narratives (Ibid.). However, it is crucial to acknowledge that concerns surround such campaigns. For individuals actively engaged in or closely aligned with the ideological beliefs of the movement from which the former extremist has defected, these former individuals might be seen as traitors, posing a threat to their core values (Ibid.). As such, it becomes essential to consider how specific communities might perceive and react to the involvement of former extremists in counter-narrative initiatives (Ibid.). Illustrating this type of counter narrative, the independent non-profit organisation, The Institute for Strategic Dialogue (ISD), has established the Against Violent Extremism network. This network empowers former extremists to share their unique experiences, combating extremist narratives and preventing the further radicalisation of others [50]. Empirical evidence substantiates the efficacy of employing former extremists in CVE delivery models. A notable initiative in Danish schools provided concrete evidence of its effectiveness, revealing positive outcomes [72]. The program demonstrated its ability to enhance the attendees' ability to identify extremist ideas and recruitment tactics. Additionally, it bolstered their confidence in seeking assistance when exposed to extremism. Moreover, the initiative showcased effectiveness in diminishing the perceived legitimacy of resorting to political violence against public authorities (Ibid.).

Several studies provide evidence to suggest that CVE can be effective in addressing radicalisation, particularly when employing ethnographic methods to identify various contributing factors [58]. These factors include social, economic, and political influences, as well as cultural norms and values that shape attitudes towards extremism. For instance, rational choice theory, which postulates that terrorists make deliberate decisions based on weighing the outcomes of their actions, has been challenged. Alternative theories such as positivism might be more relevant in understanding the root causes of extremism [82]. Positivism emphasises that addressing underlying socioeconomic conditions can mitigate or prevent criminal activity, as environmental factors play a pivotal role in driving organised crime. It also highlights the concept of "blocked opportunities" for juveniles, where societal constraints prevent them from achieving legitimate goals, leading to resorting to criminal behaviour [26]. Furthermore, perspectives from individuals and organisations engaged in CVE initiatives outside of formal government structures play a crucial role in countering violent extremism, fostering trust, and engaging with at-risk communities [58]. A meta-analysis of 19 randomised controlled trials (RCTs) further supports the positive impact of CVE interventions on reducing extremist attitudes and behaviours [24]. The most effective interventions were those adopting a multi-component approach, targeting various risk factors for extremism, and tailoring strategies to meet the specific needs of the targeted population [24]. However, it is essential to acknowledge that the CVE strategy's reliance on a linear model of ideology development might oversimplify the complex nature of terrorism's rise [34]. Consequently, CVE campaigns might be less effective in adequately addressing the underlying causes of extremism [43]. To address this limitation, greater cooperation with nongovernmental organisations and community leaders from civil society is recommended. This collaboration will enable the promotion of social inclusion and the inclusion of perspectives from victims and former offenders alongside policymakers and law enforcement [4]. Nonetheless, the efficacy of CVE programs might be hindered by inadequate funding, potentially limiting their ability to address the magnitude of the problem and achieve significant impact [104].

Another viable approach that provides valuable insights in relation to addressing counterterrorism (CT) and CVE issues is that implemented by the Netherlands. The Dutch government's national action plan for 2007–2011 introduced a combined CT and CVE strategy, wherein larger cities in the country were tasked with developing unique and tailored programs to address the specific needs and challenges of their respective areas. This approach acknowledges that the underlying causes of radicalisation might vary across different communities and regions within the nation, prompting the need for locally focused initiatives [103]. By allowing funding to be allocated based on specific needs, this approach optimises the use of resources and enhances the effectiveness of counterterrorism efforts across different areas. In understanding the dynamics of radicalisation, Situational Action Theory [106] could also offer a valuable perspective. This theory provides a lens through which one can analyse how various components interact in the context of radicalisation. It emphasises the role of the environment in shaping extremist views and highlights the significance of addressing regional variations to effectively tackle the underlying causes of terrorism. Another alternative approach to counterterrorism that emphasises cooperation with non-governmental organisations (NGOs) and community leaders is the Aarhus model, implemented in Denmark. The Aarhus model involves regular engagement of NGOs, stakeholders, and Muslim community leaders as part of the program, which incorporates a range of social interventions, including education and employment opportunities [1]. An essential aspect of this model is the provision of a "hotline" for reporting concerns about radicalisation, staffed by trained social workers and psychologists rather than law enforcement officers. This approach ensures that individuals who come forward with concerns are not automatically treated as suspects or subjected to surveillance but are instead offered support and assistance [35].

Similarly, there are low-cost community-based CVE programs, such as the Active Change Foundation (ACF), which operates the 'Football for Unity' project, bringing together young people from diverse backgrounds to engage in football and build relationships. The project has demonstrated success in engaging with at-risk youth and fostering understanding and social cohesion between different communities [92]. Funding for such programmes is provided by private donors and government grants. However, it is important to recognise that current CVE programs might inadvertently adhere to masculine notions of strength and autonomy, primarily designed with male radicals in mind. This can lead to a lack of consideration or effective engagement with female extremists or those at risk. Furthermore, employees working on disengagement and deradicalisation programmes might exhibit gender stereotypes, potentially resulting in biases in their approach to female extremists. Such biases could lead to compromised effectiveness of these programmes, for instance, if female terrorists are perceived as less dangerous or are given preferential treatment [51]. Consequently, further research is needed to better understand how gender influences individuals' engagement with and exit from extreme movements [81]. See also Chap. 4, which assesses the role of women in terrorism.

3.5.3 Supporting Individuals at Risk of Radicalisation

A further potential recommendation that could be considered for implementation in the UK to counter radicalisation involves recognising the significant role of women, as exemplified by the Women, Peace, and Security Program (WPS) in Canada. The

WPS program acknowledges the potential of women in identifying and addressing instances of radicalisation within their families, social circles, and communities. This programme provides training and support to women, equipping them with the knowledge and skills needed to identify warning signs of radicalisation and effectively intervene [36]. In 2018, the program trained many women from diverse communities across Canada to act as "change agents" in recognising and preventing radicalisation [27]. The WPS has demonstrated some significant successes in promoting gender equality within counterterrorism efforts. Therefore, considering the implementation of such an approach might contribute to effective counter-radicalisation efforts in the UK. Furthermore, enhancing the UK's Channel and Prevent Multi-Agency Panel (PMAP) as part of the Prevent strategy could strengthen counterterrorism efforts. The Channel Vulnerability Assessment Framework [97] employs a collaborative approach and multi-agency approach to identify and support individuals susceptible to radicalisation [7]. Using indicators such as social isolation and exposure to extremist materials, the framework assesses the potential vulnerability of individuals. If considered suitable for Channel support, a tailored support plan is devised in cooperation with the individual and their family, including access to educational or employment opportunities, mentoring, and counselling (Ibid.). Empirical evidence indicates that this approach can effectively influence extremists by challenging the alignment of their acts with their "sacred values" [38]. Such influence is largely achieved through targeted messaging designed to engage the audience's rational decision-making processes, prompting extremists to contemplate a cost–benefit analysis of available choices [64].

3.5.4   Investing in Think Tanks and Research Institutes

Investing in think tanks and research institutes that have been influential in shaping counterterrorism policies is also essential. For example, the International Centre for the Study of Radicalisation (ICSR), based at King's College London, is a research institute that specialises in the study of violent extremism and radicalisation. The centre has played a crucial role in informing policy development in the UK and worldwide, providing valuable insights into the causes of radicalisation and potential countermeasures [69]. Moreover, the ICSR has employed social media analysis to gain a deeper understanding of how radical organisations utilise these channels to disseminate their messages and recruit new members [9]. Such research work is complemented by community-based CVE programs such as Shout Out UK [86], which creates diverse media content, including news articles, opinion pieces, and podcasts focusing on a range of political and social issues. The content is tailored for young audiences, offering fresh and engaging perspectives on these topics.

3.5.5 Addressing Accountability Gaps

Although there has been some level of evaluation concerning CONTEST and the Prevent duty, a robust and independent qualitative review of the strategy is still lacking. This gap in research poses challenges in providing a comprehensive analysis of the effectiveness of the UK's counter-terrorism strategy [17]. Investigations into the presence of suitable review mechanisms to ensure government accountability in countering terrorism reveal gaps and blind spots in these areas. Notably, there is a lack of periodic review processes and an absence of designated bodies responsible for surveying all counter-terrorism measures [102]. Moreover, the reviews that have been conducted thus far have remained internal, compromising their credibility, transparency, and true accountability, as the same entity being reviewed is conducting the evaluation [31]. To address these shortcomings, it is suggested that an Independent Reviewer of Terrorism Legislation be permanently appointed, provided with adequate resources, and entrusted with the commissioning of detailed evaluative reviews of UK counter-terrorism efforts [13]. These reviews should consider both quantitative and qualitative evidence, and measure the strategies against a range of established standards (Ibid.). Such an approach would facilitate the identification of effective components within the strategy and shed light on areas that require improvement, ultimately working towards a stronger, more effective, and accountable counter-terrorism strategy. The absence of an independent framework for evaluating counter-terrorism strategies might perpetuate existing failures and deepen societal mistrust in the government [17]. Consequently, expediting the development of an independent evaluation framework would enable policymakers to better discern the strengths and weaknesses of the strategy and make informed decisions towards enhancing its overall effectiveness (Ibid.).

3.6 Discussion

This chapter has underscored key problems, including the further marginalisation of Muslim communities, resulting in their classification as a 'suspect community.' Another notable criticism concerns the ambiguity of the definition of extremism used within the strategy, particularly the emphasis on 'fundamental British values'. This poses challenges for those tasked with implementing the Prevent duty and identifying individuals who might be susceptible to radicalisation. As a result, there has been a tangible impact on individuals' perceived freedom of speech, leading them to self-censor or withhold their thoughts and opinions due to fears of being misinterpreted as endorsing extremist views. This chapter has also examined a third aspect of CONTEST, which has raised concerns regarding its simplistic assumption of a linear "radicalisation process", wherein certain ideologies progress into violent extremism. Drawing from these highlighted points, the chapter proceeds to reemphasise three recommendations, discussed previously, aimed at enhancing the implementation of counter-terrorism measures in the UK. The first recommendation centres on countering the marginalisation of Muslim communities, urging the government to publicly embrace and accept diverse ideas. Creating a safe space wherein concerns can be openly expressed without fear of reprisals fosters an environment that encourages diverse communities to collaborate in the fight against radicalisation and terrorism. The second recommendation centres on exploring innovative methods to disseminate counter-narratives that can help prevent the radicalisation of individuals. In this context, the chapter has suggested employing former terrorists in campaigns, as evidence has shown their effectiveness in changing public opinions on extremism. Utilising such individuals can serve as an educational tool, enlightening individuals about the perils of engaging in terrorism. The third and arguably most significant recommendation involves the establishment of a fully independent and comprehensive review of the UK's counter-terrorism strategy. Such a review would ensure the strategy's effectiveness by leveraging the expertise of experts and academics to identify and address non-effective policies. This would lead to a more accountable and adaptive version of CONTEST.

However, it is imperative to acknowledge that these recommendations only mark the initial steps toward improving the UK's counter-terrorism strategy. There remains much research to be conducted, given the incomplete understanding of the underlying causes of violent extremism and how to effectively address them. Therefore, the chapter underscores the necessity of accepting an ongoing process of learning and policy amendments to facilitate continuous improvement in countering terrorism. Furthermore, it is important to acknowledge the challenges that the UK encounters in implementing more stringent jail sentences for terrorism-related offenders, primarily driven by capacity constraints. This concern arises despite the notable increase in terrorism-related arrests and prosecutions, as highlighted in the annual report of the [48]. Such a surge in cases places considerable strain on law enforcement agencies and the criminal justice system, underscoring the inherent limitations of the UK's justice systems.

3.7 Conclusion

Understanding the motivating factors behind individuals resorting to terrorist activities is instrumental in formulating comprehensive and targeted approaches to combatting this complex issue. The choice between hard and soft approaches to counterterrorism involves multifaceted and complex considerations of effectiveness, ethics, and public perception. While hard tactics might yield immediate results in eliminating threats, they raise moral and legal concerns. Conversely, soft interventions such as CVE programs prioritise prevention and reintegration, aiming to address the underlying drivers of extremism and fostering a more inclusive approach to countering terrorism. As such, a balanced approach that combines elements of both hard and soft strategies might be most effective in the fight against terrorism, while respecting fundamental human rights and societal values. It is also essential to engage various stakeholders, including Muslim communities, in the process of crafting counterterrorism measures in order to ensure inclusive, just, and effective policies that respect fundamental rights. In relation to the CONTEST strategy's four elements, they have demonstrated efficacy in preventing terrorist attacks in the UK. For instance, the exchange of information and intelligence, along with international cooperation and the deportation of foreign nationals involved in terrorism, have contributed to improved threat assessments.

However, these elements are not without challenges and inherent limitations. For instance, issues such as discrepancies in data privacy and human rights standards beyond the EU, the potential for stigmatisation and marginalisation of specific communities [62], and resource disparities across domains warrant attention. Drawing insights from the experiences of other countries, such as the Netherlands and the USA, can provide valuable perspectives on adapting counterterrorism policies to local contexts. Such alternative approaches might complement and enhance the UK's counterterrorism strategy. Therefore, exploring diverse strategies and best practices from different nations could yield valuable insights for enhancing the efficacy and adaptability of the UK's counterterrorism efforts. Furthermore, fostering relationships of trust and collaboration among diverse stakeholders can strengthen existing efforts and encourage individuals to report threats or suspicious activities. Similarly, formulating counterterrorism measures with due consideration for human rights, avoidance of unintended consequences, and optimal resource utilisation is crucial for effectively preventing terrorist acts and ensuring public safety. Finally, a balanced and nuanced approach is indispensable in addressing the multifaceted challenges posed by terrorism and radicalisation.

References

1. Agerschou T (2014) Preventing radicalization and discrimination in Aarhus. J Deradicalization 1:5–22
2. Aldrich RJ (2009) US–European intelligence co-operation on counter-terrorism: low politics and compulsion. Br J Polit Int Rel 11(1):122–139
3. Altheide DL (2007) The mass media and terrorism. Discourse Commun 1(3):287–308
4. Aly A, Balbi AM, Jacques C (2015) Rethinking countering violent extremism: implementing the role of civil society. J Policing Intell Counter Terrorism 10(1):3–13
5. Ambrozik C (2019) Countering violent extremism globally: a new global CVE dataset. Perspect Terrorism 13(5):102–111
6. Appleby N (2010) Labelling the innocent: how government counter-terrorism advice creates labels that contribute to the problem. Critical Stud Terrorism 3(3):421–436
7. Augestad Knudsen R (2020) Measuring radicalisation: risk assessment conceptualisations and practice in England and Wales. Behavioral Sci Terrorism Polit Aggression 12(1):37–54
8. Awan I (2012) "I am a Muslim not an extremist": how the prevent strategy has constructed a "suspect" community. Polit Policy 40(6):1158–1185. https://doi.org/10.1111/j.1747-1346.2012.00397.x
9. Baldauf J, Ebner J, Guhl J (2019) Hate speech and radicalisation online. Institute for Strategic Dialogue, Washington DC
10. Bast S (2018) Counterterrorism in an era of more limited resources. Center for Strategic & International Studies

11. Beliveau AM (2018) Hate speech laws in the United States and the Council of Europe: the fine balance between protecting individual freedom of expression rights and preventing the rise of extremism and radicalization through social media sites. Suffolk UL Rev 51:565
12. Berton B (2015) The dark side of the web: ISIL's one-stop shop? Report of the European Union Institute for Security Studies
13. Blackbourne J, De Londras F (2019) Accountability and review in the counter-terrorist state. Policy Press
14. Boer MD (2015) Counter-terrorism, security and intelligence in the EU: governance challenges for collection, exchange and analysis. Intell Nat Secur 30(2–3):402–419
15. Botha A (2008) Challenges in understanding terrorism in Africa: a human security perspectives. African Secur Rev 17(2):28–41
16. Bowman N, Lewis RJ, Tamborini R (2014) The morality of May 2, 2011: a content analysis of US headlines regarding the death of Osama bin Laden. Mass Commun Soc 17(5):639–664
17. Brady E (2016) An analysis of the UK's counter-terrorism strategy, CONTEST, and the challenges in its evaluation. Universitätsbibliothek Johann Christian Senckenberg, Frankfurt
18. Briggs R (2010) Community engagement for counterterrorism: lessons from the United Kingdom. Int Aff 86(4):971–981
19. Brooke N (2022) What are the root causes of terrorism? Contemp Terrorism Stud 157
20. Bunn ME (2007) Can deterrence be tailored? Institute for National Strategic Studies, National Defense University, Washington, DC
21. Byman D (2006) Do targeted killings work. Foreign Aff 85(2):95–111
22. Campbell A (2005) 'Taqiyya': how islamic extremists deceive the west. Nat Observer 65:11–23
23. Carson JV (2017) Assessing the effectiveness of high-profile targeted killings in the "war on terror" a quasi-experiment. Criminol Public Policy 16(1):191–220
24. Carthy N, Paget S, Rothon C (2020) A meta-analysis of the effectiveness of interventions to prevent violent radicalization and extremism. J Exp Criminol 16(4):513–537
25. Choudhury T, Fenwick H (2011) The impact of counter-terrorism measures on Muslim communities. Int Rev Law Comput Technol 25(3):151–181
26. Cloward RA, Ohlin LE (1960) Delinquency and opportunity: a theory of delinquent gangs. The Free Press, Glencoe
27. Cohn C, Kinsella H, Gibbings S (2004) Women, peace and security: resolution 1325. Int Fem J Polit 6(1):130–140
28. Corsi JL (2017) Drone deaths violate human rights: the applicability of the ICCPR to civilian deaths caused by drones. Int Human Rights Law Rev 6(2):205–241
29. Counter-Terrorism and Security Act 2015. Available at https://www.legislation.gov.uk/ukpga/2015/6/contents/enacted. Accessed 06 Sept 2023
30. Counter Terrorism Policing (2023) Counter terrorism policing urge vigilance ahead of coronation. Available at https://www.counterterrorism.police.uk/counter-terrorism-policing-urge-vigilance-ahead-of-coronation/. Accessed 06 Sept 2023
31. De Londras F (2018) CONTEST: 'world-leading' oversight, scrutiny and transparency? University of Birmingham. Available at https://blog.bham.ac.uk/counterterrorismreview/2018/06/29/228/. Accessed 27 Sept 2023
32. Doosje B, Moghaddam FM, Kruglanski AW, De Wolf A, Mann L, Feddes AR (2016) Terrorism, radicalization and de-radicalization. Curr Opin Psychol 11:79–84
33. Edwards J, Gomis B (2011) Islamic terrorism in the UK since 9/11: reassessing the 'soft' response. Available at https://www.chathamhouse.org/sites/default/files/pp_edwardsgomis.pdf. Accessed 27 Sept 2023
34. Gielen AJ (2019) Countering violent extremism: a realist review for assessing what works, for whom, in what circumstances, and how? Terrorism Polit Violence 31(6):1149–1167
35. Gielen AJ (2015) Supporting families of foreign fighters. A realistic approach for measuring the effectiveness. J Deradicalization 2:21–48

36. Global Affairs Canada (2023) Women, peace and security. Government of Canada. Available at https://www.international.gc.ca/world-monde/issues_development-enjeux_developpement/gender_equality-egalite_des_genres/women_peace_security-femmes_paix_securite.aspx?lang=eng. Accessed 27 Sept 2023

37. Gunaratna R (2017) Strategic counter-terrorism: a game changer in fighting terrorism? Counter Terrorist Trends Anal 9(6):1–5

38. Hamid N (2020) The ecology of extremists' communications: messaging effectiveness, social environments and individual attributes. RUSI J 165(1):54–63

39. Hardy K (2017) Preventive justice principles for countering violent extremism. In: Regulating preventive justice. Routledge, pp 117–135

40. Hathaway OA, Crootof R, Levitz P, Nix H, Nowlan A, Perdue W, Spiegel J (2012) The law of cyber-attack. California Law Rev: 817–885

41. Heath-Kelly C (2013) Counter-terrorism and the counterfactual: producing the 'radicalisation' discourse and the UK PREVENT strategy. Br J Polit Int Relations 15(3):394–415

42. Hedges P (2017) What is (wrong with) radicalisation? A response to Manchester bombing

43. Hemmingsen AS, Castro Møller KI (2017) Why counter-narratives are not the best responses to terrorist propaganda. In: Challenges, risk and propaganda (DIIS report 2017: 1). DIIS, Copenhagen

44. Hernández FO (2023) Independent/hybrid P/CVE youth work in the UK: grassroots work beyond prevent. Perspectives on countering extremism: diversion and disengagement, 163

45. House of Commons (2009) Project CONTEST: the government's counter-terrorism strategy (HC 212). https://publications.parliament.uk/pa/cm200809/cmselect/cmhaff/212/212.pdf

46. Howie E (2018) Protecting the human right to freedom of expression in international law. Int J Speech Lang Pathol 20(1):12–15

47. Independent (2015) PREVENT will have a chilling effect on open debate, free speech and political dissent. Available at https://www.independent.co.uk/voices/letters/prevent-will-have-a-chilling-effect-on-open-debate-free-speech-and-political-dissent-10381491.html. Accessed 27 Sept 2023

48. Independent Reviewer of Terrorism Legislation (2018) Terrorism acts in 2018. Available at https://terrorismlegislationreviewer.independent.gov.uk/terrorism-acts-in-2018. Accessed 27 Sept 2023

49. Innes M (2006) Policing uncertainty: countering terror through community intelligence and democratic policing. Ann Am Acad Pol Soc Sci 605:1–20

50. Institute for Strategic Dialogue (2022) Against violent extremism (AVE) network. Available at https://www.isdglobal.org/against-violent-extremism-ave/. Accessed 27 Sept 2023

51. Jackson SM, Ratcliff K, Gruenewald J (2021) Gender and criminal justice responses to terrorism in the United States. Crime Delinquency 69(5)

52. Kearns EM, Betus AE, Lemieux AF (2019) Why do some terrorist attacks receive more media attention than others? Justice Q 36(6):985–1022

53. Khaleeli H (2015) 'You worry they could take your kids': is the prevent strategy demonising Muslim schoolchildren? The Guardian. Available at https://www.theguardian.com/uk-news/2015/sep/23/prevent-counter-terrorism-strategy-schools-demonising-muslim-children. Accessed 27 Sept 2023

54. King M, Taylor DM (2011) The radicalisation of homegrown Jihadists: a review of theoretical models and social psychology evidence. Terrorism Polit Violence 23(4):602–622

55. Klausen J (2009) British counter-terrorism after 7/7: adapting community policing to the fight against domestic terrorism. J Ethn Migr Stud 35(3):403–420

56. Kundnani A (2015) A decade lost: rethinking radicalisation and extremism. Claystone. Available at https://mabonline.net/wp-content/uploads/2015/01/Claystone-rethinking-radicalisation.pdf. Accessed 27 Sept 2023

57. Kundnani A, Hayes B (2018) The globalisation of countering violent extremism policies: undermining human rights, instrumentalising civil society. Transnational Institute, Amsterdam

58. Lee A (2019) Informal CVE actors: assessing the role of civil society and the private sector in countering violent extremism. Stud Conflict Terrorism 42(7):638–657
59. Lewis JA (2002) Assessing the risks of cyber terrorism, cyber war and other cyber threats. Center for Strategic & International Studies. Washington, DC
60. Lowe D (2017) Prevent strategies: the problems associated in defining extremism: the case of the United Kingdom. Stud Conflict Terrorism 40(11):917–933
61. Maclean K, Cuthill M, Ross H (2014) Six attributes of social resilience. J Environ Planning Manage 57(1):144–156
62. Mason PL, Matella A (2014) Stigmatization and racial selection after September 11, 2001: self-identity among Arab and Islamic Americans. IZA J Migration 3(1):1–21
63. McGoldrick D (2013) The limits of freedom of expression on Facebook and social networking sites: a UK perspective. Hum Rights Law Rev 13(1):125–151
64. Mishan EJ, Quah E (2020) Cost-benefit analysis. Routledge
65. Moffat A, Gerard FJ (2020) Securitising education: an exploration of teachers' attitudes and experiences regarding the implementation of the Prevent duty in sixth form colleges. Critical Stud Terrorism 13(2):197–217
66. Moggridge E, Montasari R (2022) A Critical analysis of the dark web challenges to digital policing. In: Montasari R (ed) Artificial intelligence and national security. Springer, Cham, pp 157–167
67. Murphy A (2021) Political rhetoric and hate speech in the case of Shamima Begum. Religions 12(10):834
68. Mythen G, Walklate S, Khan F (2009) "I'm a Muslim, but I'm not a terrorist": victimization, risky identities and the performance of safety. Br J Criminol 49(6):736–754
69. Neumann PR (2014) The New Jihadism: a global snapshot. International Centre for the Study of Radicalisation and Political Violence
70. O'Toole T, Meer N, DeHanas DN, Jones SH, Modood T (2016) Governing through prevent? Regulation and contested practice in state-Muslim engagement. Sociology 50(1):160–177. https://doi.org/10.1177/0038038514564437
71. Pantazis C, Pemberton S (2009) From the 'old' to the 'new' suspect community: examining the impacts of recent UK counter-terrorist legislation. Br J Criminol 49(5):646
72. Parker D, Lindekilde L (2020) Preventing extremism with extremists: a double-edged swords? An analysis of the impact of using former extremists in Danish schools. Educ Sci 10(4):111
73. Powell L (2016) Counter-productive counter-terrorism. How is the dysfunctional discourse of prevent failing to restrain radicalisation? J Deradicalisation 8:46–99
74. Price M (2013) National security and local police. Brennan Center for Justice, New York
75. Poynting S, Mason V (2007) The resistible rise of islamophobia: anti-Muslim racism in the UK and Australia before 11 September 2001. J Sociol 43(1):61–86
76. Poynting S, Mason V (2006) "Tolerance, Freedom, Justice and Peace"? Britain, Australia and anti-Muslim Racism since 11 September 2001. J Intercult Stud 27(4):365–391
77. Robinson M, Jones K, Janicke H (2015) Cyber warfare: issues and challenges. Comput Secur 49:70–94
78. Sabir R (2017) Blurred lines and false dichotomies: integrating counterinsurgency into the UK's domestic 'war on terror.' Crit Soc Policy 37(2):202–224
79. Salasznyk PP, Lee EE, List GF, Wallace WA (2006) A systems view of data integration for emergency response. Int J Emergency Manage 3(4):313–331
80. Schmid AP (2012) The revised academic consensus definition of terrorism. Perspect Terrorism 6(2):158–159
81. Schmidt R (2022) Duped: examining gender stereotypes in disengagement and deradicalization practices. Stud Conflict Terrorism 45(11):953–976
82. Scott J (2000) Rational choice theory. In: Browning G, Halcli A, Webster F (eds) Understanding contemporary society: theories of the present, 129, 126–138
83. Security Service MI5 (2022) Threat levels. Available at https://www.mi5.gov.uk/threat-levels. Accessed: 27 Sept 2023

84. Sheerin P (2023) The coronation and terrorism: know the risk. Club Insure Ltd. Available at https://www.club-insure.co.uk/the-coronation-and-terrorism/. Accessed 27 Sept 2023

85. Shehabat A, Mitew T, Alzoubi Y (2017) Encrypted jihad: investigating the role of Telegram App in lone wolf attacks in the west. J Strat Secur 10(3):27–53

86. Shout Out UK (2023) Shout Out UK—the home of political literacy and youth voice. Available at https://www.shoutoutuk.org/. Accessed 27 Sept 2023

87. Silva DMD (2018) 'Radicalisation: the journey of a concept', revisited. Inst Race Relations 59(4):34–53

88. Snaith J, Stephenson K (2016) The prevent duty: a step too far. Educ LJ: 56

89. Stern J (2015) Obama and terrorism: like it or not, the war goes on. Foreign Affairs. Available at https://www.foreignaffairs.com/articles/obama-and-terrorism. Accessed 27 Sept 2023

90. Tapley M, Clubb G (2019) The role of formers in countering violent extremism. International Centre for Counter-Terrorism

91. The Muslim Council of Britain (2016) The impact of prevent on Muslim communities. Available at http://archive.mcb.org.uk/wp-content/uploads/2016/12/MCB-CT-Briefing2.pdf

92. UEFA Foundation (2023) Football for unity. UEFA Foundation. Available at https://uefafoundation.org/action/football-for-unity/#Context. Accessed 27 Sept 2023

93. UK Government (2023) CONTEST: the United Kingdom's strategy for countering terrorism 2023 (CP 903). Available at https://www.gov.uk/government/publications/counter-terrorism-strategy-contest-2023. Accessed 06 Sept 2023

94. UK Government (2021) Revised prevent duty guidance: for England and wales. Available at https://www.gov.uk/government/publications/prevent-duty-guidance/revised-prevent-duty-guidance-for-england-and-wales. Accessed 06 Sept 2023

95. UK Government (2020) Channel duty guidance: protecting people vulnerable to being drawn into terrorism. Available at https://assets.publishing.service.gov.uk/government/uploads/system/uploads/attachment_data/file/964567/6.6271_HO_HMG_Channel_Duty_Guidance_v14_Web.pdf. Accessed 06 Sept 2023

96. UK Government (2018) CONTEST: the United Kingdom's strategy for countering terrorism (Cm 9608). Available at https://www.gov.uk/government/publications/counter-terrorism-strategy-contest-2018. Accessed 06 Sept 2023

97. UK Government (2012) Channel vulnerability assessment. Available at https://www.gov.uk/government/publications/channel-vulnerability-assessment. Accessed 07 Sept 2023

98. UK Government (2011a) CONTEST: the United Kingdom's strategy for countering terrorism (Cm 8123). Available at https://www.gov.uk/government/publications/counter-terrorism-strategy-contest. Accessed 06 Sept 2023

99. UK Government (2011b) Prevent strategy (Cm 8092). Available at https://www.gov.uk/government/publications/prevent-strategy-2011. Accessed 06 Sept 2023

100. UK Government (2009) The United Kingdom's strategy for countering international terrorism (Cm 7547). Available at https://assets.publishing.service.gov.uk/government/uploads/system/uploads/attachment_data/file/228644/7547.pdf. Accessed 06 Sept 2023

101. UK Government (2006) Countering international terrorism: the United Kingdom's strategy (Cm 6888). Available at https://assets.publishing.service.gov.uk/government/uploads/system/uploads/attachment_data/file/272320/6888.pdf. Accessed 06 Sept 2023

102. University of Birmingham (2022) Counter-terrorism: how do we ensure the state is accountable? https://www.birmingham.ac.uk/research/quest/towards-a-better-society/counter-terrorism-state.asp

103. Vermeulen F, Visser K (2021) Preventing violent extremism in the Netherlands: overview of its broad approach. Revista CIDOB d'Afers Internacionals 128:131–151

104. Vidino L, Hughes S (2015) Countering violent extremism in America. Center for Cyber and Homeland Security: The George Washington University

105. White J (2004) Should religious education be a compulsory school subject? Br J Relig Educ 26(2):151–164

106. Wikström POH, Bouhana N (2016) Analyzing radicalization and terrorism: a situational action theory. In: The handbook of the criminology of terrorism. Wiley, Chichester, UK, pp 175–186
107. Zammit A (2017) New developments in the Islamic state's external operations: the 2017 Sydney plane plot. Combating Terrorism Center 10(9)

Chapter 4
Understanding and Assessing the Role of Women in Terrorism

Abstract This chapter undertakes a comprehensive analysis of the multifaceted role of women in the realm of terrorism, asserting that a departure from entrenched dualistic paradigms is essential to capture the intricacies inherent to their engagement. It contends that a more holistic understanding of women's involvement in terrorism can be achieved by transcending rigid categorisations. To this end, the chapter delves into both facets of the discourse, beginning with a critical examination of the prevalent notion that women predominantly assume victim roles within terrorism, occupying subservient positions such as sympathisers, spouses, and maternal figures. Subsequently, the chapter examines the burgeoning corpus of scholarship that repositions women not only as perpetrators of violence but also as active participants within the intricate web of terrorist organisations. Central to this analysis is the nuanced concept of agency—an individual's capacity to exercise personal choices and actions—while also acknowledging the constraints that might curtail such agency. Throughout the discourse, emphasis is placed upon cultivating a discerning understanding that incorporates agency in tandem with the intricacies of choice, thus enabling a more accurate portrayal of the multifarious roles that women assume in the context of terrorism. Ultimately, the chapter calls into question the proposition that tends to homogenise women into a facile dichotomy, alternating solely between the poles of vulnerable victims and empowered agents.

Keywords Terrorism · Radicalisation · Violent extremism · Social media platforms · Propaganda · Recruitment · Female terrorists · Women in terrorism · Gender dynamics · Dualistic paradigms · Victim roles · Women · Perpetrators

4.1 Introduction

Terrorism constitutes a global phenomenon that has exerted its influence across virtually all geographical regions across the world. However, the establishment of a universally accepted definition for terrorism remains an ongoing endeavour as discussed throughout this book. As opposed to achieving consensus, the terminology continues

R. Montasari, *Cyberspace, Cyberterrorism and the International Security in the Fourth Industrial Revolution*, Advanced Sciences and Technologies for Security Applications, https://doi.org/10.1007/978-3-031-50454-9_4

to be a topic of fervent discourse and contention within scholarly circles. In the context of this chapter, terrorism can be broadly construed as the use or threat of violent actions with the primary purpose of instilling fear, primarily aimed at achieving political goals [6]. The inclusion of the term 'violence' within the conceptual framework of terrorism assumes a pivotal role by contributing to the nuanced analysis of gendered interpretations surrounding this phenomenon. Embedded within society is the prevalent presumption that men inherently harbour proclivities towards violence, in contrast to the prevailing belief in the intrinsic pacifism of women. These ingrained societal expectations significantly influence the construction of gender-specific roles and influence the extent of female engagement in acts of violence. Consequently, historical portrayals have predominantly cast females in the light of passive victims of terrorism, subjected to manipulation and coercion orchestrated by their male counterparts. In an effort to rectify this gender-biased narrative, scholarly discourse has progressively embraced the recognition of women's agency and their active involvement within the ranks of terrorist organisations [12]. This trajectory of scholarship has consequently engendered a dialectical interplay between the roles of victim and agent, persistently perpetuating a discursive dichotomy.

This chapter aims to perform a critical analysis of the role of women in terrorism. The chapter argues that the trajectory of scholarly work must transcend the confines of the prevailing dualistic discourse so as to capture the intricate nuances inherent to the participation of women in terrorism. In order to examine the false dichotomy characterising the roles of women, this study is poised to navigate through the contours of both dimensions constituting this discourse. To this end, the chapter begins by meticulously analysing the entrenched notion that women predominantly assume the mantle of victimhood within the purview of terrorism, thus relegating them to subservient positions as sympathisers, spouses, and maternal figures. Subsequently, the discourse pivots to a comprehensive evaluation of the burgeoning collection of scholarship that repositions women as not only perpetrators of violence but also as proactive agents deeply implicated within the fabric of terrorist organisations. Throughout the analytical trajectory of this study, emphasis will be accorded to the intricate construct of agency—an individual's inherent ability to exercise volition in their decision-making and ensuing actions, or the attendant absence thereof. An insightful understanding which duly recognises agency in tandem with the constraints that circumscribe choice will provide a more accurate representation of the diverse roles that women assume in the realm of terrorism.

The remainder of this chapter is structured as follows. Section 4.2 examines women's role as passive victims in the Islamic State. Section 4.3 analyses women's role as active agents in the Tamil Tigers. Section 4.4 investigates gender-biased assumptions in relation to why and how women participate in terrorism. Finally, the chapter is concluded in Sect. 4.5.

4.2 Women's Role as Passive Victims in the Islamic State

In order to deconstruct the aforementioned dichotomous categorisation, it becomes imperative for this study to engage in a comprehensive critical assessment of both facets constituting the debate. Commencing with the initial premise that women's engagement within the domain of terrorism is often relegated to the position of passive and incapacitated victims, it is essential to embark upon an analysis that acknowledges the encompassing cultural and societal milieu within which both men and women function. The narrative portraying women as victims is not confined solely to the discourse surrounding terrorism; instead, it is deeply entrenched within pervasive gender-based stereotypes that construe women as inherently lacking the strength to safeguard themselves. Consequently, women are conventionally depicted as proponents of nonviolence, aligned with the role of peacekeepers [9]. Gendered narratives exhibit extensive prevalence and often become deeply rooted in academic literature, policy documents, and mainstream media. A notable example can be found in the United Nations Security Council Resolution 1325 [16], which offers a limited interpretation of the relationship between gender and violence. This resolution assumes that increased participation of women in conflict resolution processes can effectively promote peace, while also concentrating exclusively on portraying women as inherent terrorism victims necessitating protection. Consequently, when examining the participation of women in acts of violence, it is crucial to acknowledge the influence of gender roles and stereotypes that can shape one's perspective.

Owing to divergent ideologies and operational approaches, the tasks undertaken by women diverge across terrorist organisations. Thus, across historical contexts, women have assumed a multitude of roles within terrorism, including combatants, mothers, recruiters, wives, sympathisers, and even subjected to roles such as sex slaves. In jihadist militant groups such as the Islamic State of Iraq and Syria (ISIS), women have often been portrayed as "jihadi brides" who are victimised [10]. This portrayal relegates women to a subservient position, portraying them as merely manipulated into domestic roles within terrorist activities. The persistence of gendered assumptions continues to impact the contemporary analysis of women's involvement in the sphere of terrorism. Echoing this perspective, Besenyő [4] emphasises that within ISIS, women are frequently subjected to slavery and compelled into matrimonial unions. He also underscores the pivotal role played by propaganda in influencing Muslim women's decision to join ISIS, positioning them as supportive mothers and wives for the group. While Besenyő's [4] study makes a significant contribution by examining sexual exploitation in the context of terrorism, it predominantly centres on the portrayal of women as coerced victims of manipulation and mistreatment. This perspective, however, might be considered by some as a simplified and reductionist viewpoint that potentially overlooks the agency of women and their potential political or ideological motivations for aligning with ISIS. In order to comprehensively evaluate the involvement of women in terrorism, it is crucial to take into account the motivating factors behind their decision to join. Martini [10], a feminist researcher, criticises the problematic use of the label docile and obedient jihadi brides. She argues that depicting women as proactive and willing individuals capable of violence is often disregarded, as this characterisation undermines deeply ingrained gender norms. Rather than being seen as active participants, female members of ISIS are often presented as symbols of victimisation. Consequently, a one-dimensional assumption persists, suggesting that women participating in terrorism are merely coerced into doing so (Ibid.). Martini [10] concludes that the conventional analysis of women's roles in ISIS demonstrates a gender bias that predominantly emphasises the victim status of women.

The roles of both men and women in terrorism frequently mirror the prevailing gender norms of the specific region. The ideology embraced by the Islamic State is deeply entrenched in a patriarchal framework that enforces a rigid segregation between genders. In this context, men are assigned the role of combatants, while women are tasked with fulfilling their responsibilities as mothers and wives. As a result, women affiliated with ISIS tend to assume supportive and domestic positions, and only a small minority have been involved in acts of violent jihad. Nonetheless, Lahoud [9] argues against the notion that Islamic law prohibits women from taking on frontline combat roles. In reality, the doctrine stipulates that defensive jihad is an obligation for all Muslims, regardless of gender. According to Lahoud [9], the restriction on women's participation in jihad stems from societal perceptions favouring gender segregation. The ideologies of the Islamic State promote the idea that women should adopt conventional, feminine, and non-violent positions. Consequently, entrenched stereotypes concerning femininity and masculinity play a pivotal role in shaping the distinct dynamics of terrorism.

Typically, the literature has disproportionately emphasised the discourse of victimisation. However, it is important not to entirely dismiss the narrative of victimhood, as not all women involved in terrorism do so willingly. This is evidenced by the gender-based atrocities perpetrated by ISIS, particularly the abduction and widespread sexual violence against numerous women and girls. In a study conducted by Ibrahim et al. [7], 416 Yazidi girls and women who survived ISIS enslavement in Iraq were interviewed. The findings revealed that around 90% of the interviewed survivors experienced post-traumatic stress disorder, an exceptionally high rate. The participants disclosed that sexual violence was a central element of ISIS's strategies (Ibid.). This assertion finds support in Amnesty International's report [3], which documents that ISIS has carried out killings, rapes, and abductions of thousands of men and women. The report indicates that the atrocities committed by ISIS have impacted approximately half a million individuals. Accounts reveal that numerous women were coerced with threats of forced marriage or sale to men [3]. This underscores the unprecedented magnitude and extent of sexual violence within terrorist organisations. Evidently, it is crucial to recognise that the victimisation of women in terrorism is a deeply pervasive problem. Nevertheless, discussions and literature become problematic when the label of victimhood is solely ascribed to women.

In spite of the documented instances of atrocities perpetrated by ISIS affiliates against women, these incidents have not been successful in dissuading a substantial portion of individuals. The aspiration of ISIS to establish a self-proclaimed 'caliphate', characterised by a genuine Islamic State committed to religious principles, has proven to be an appealing enticement for women, children, and families to engage with their domain. Within this context, women continue to play a pivotal role in the sustenance of the terrorist organisation. Their responsibilities encompass tending to their husbands, nurturing the forthcoming generation of jihadi combatants, and engaging in online recruitment activities [14]. However, the fundamental tenets of the group's ideology preclude women from participating in frontline conflicts. Consequently, while women associated with ISIS predominantly assume roles 'behind the scenes', they play an integral role in providing support and perpetuating the caliphate. Contemporary scholarship underscores that the roles of women within ISIS extend beyond the archetype of victimised 'jihadi brides'. Spencer's [14] investigation, which focused on women residing in territories controlled by ISIS, revealed that 55% were engaged in online recruitment activities and 48% fulfilled domestic responsibilities. Typically, female recruits operate within the domestic sphere, facilitating the indoctrination of others in accordance with the ideology propagated by ISIS. This demonstrates that women actively partake in significant capacities as recruiters and propagandists.

Traditionally, gender stereotypes have exerted influence over theoretical frameworks pertaining to women's roles and motivations for joining terrorist organisations. Prevailing narratives often contend that Muslim women are compelled into providing support for the Islamic State, consequently presuming that women exist under male dominance and lack their own autonomous extremist perspectives [12]. However, female terrorists do not form a monolithic and uniform entity. Their roles and motivations exhibit diversity contingent on factors such as age, educational background, nationality, political grievances, and other circumstances unique to each individual. Manifestly, the engagement of women encompasses a significant array of distinctions, underscoring the necessity for analytical perspectives to transcend the confines of a victim-agent dichotomy. Speckhard and Ellenberg [13] conducted an extensive study commissioned by the International Center for the Study of Violent Extremism (ICSVE), which diverges from a simplistic binary perspective that portrays women solely as victims. The research comprised of interviews with 220 individuals affiliated with ISIS, including 38 women. The participants were selected using convenience sampling, a method that has limitations in terms of demographic representation and potential sampling bias. Therefore, the findings cannot be extrapolated to the broader population of ISIS associates. Nevertheless, the study offers substantial insights into the experiences, roles, and motivations of both male and female affiliates within the context of ISIS. The interviews illustrated that the majority of women chose to join ISIS voluntarily, enticed by the prospects of an Islamic Caliphate. Speckhard and Ellenberg [13] revealed that female recruits were primarily influenced by their relationships with partners and family members. However, it is crucial to acknowledge that even though many female participants were encouraged by others to join ISIS, this does not negate their agency. The bulk of the interviewed women were politically active agents who knowingly made the decision to travel to the Islamic State and reside under the regime of ISIS.

In conclusion, the prevailing portrayal of women, particularly Muslim women, frequently adheres to a homogenised narrative of victimhood. This societal construct, depicting women as apolitical entities devoid of agency, has historically dominated literature and informed the theoretical underpinnings of terrorism studies. While it is evident that there are regrettably numerous victims of terrorism, the emphasis on female victimisation tends to be magnified [3, 7]. It is imperative to acknowledge that construing women solely within the framework of victimhood overlooks their individual agency. Although women within ISIS primarily occupy roles of support and domesticity, the importance and willingness to engage in these roles should not be underestimated. Similar to men, women possess the capacity to engage in intricate decision-making processes when affiliating with terrorist organisations [12]. Nonetheless, such decisions are not made in isolation; women's involvement often contends with deeply ingrained gendered assumptions and constraints. Thus, to offer a comprehensive analysis, the roles of women in terrorism should be examined through a "gendered lens", acknowledging the interplay between terrorist roles, personal agency, and structural limitations (Ibid.).

4.3   Women's Role as Active Agents in the Tamil Tigers

As mentioned earlier, scholarly investigation into women's engagement with terrorism often delineates two primary perspectives: the 'victim' and the 'agent' roles. To further deconstruct this binary, a critical examination of women's function as proactive agents within terrorism is imperative. The historical examination of female participation in terrorism has led to substantial gaps in understanding, specifically pertaining to the assertive and militant roles women assume within these contexts. Conventional discourse has largely been dominated by narratives of victimisation. More recently, feminist scholarship has undertaken the task of shifting the analytical focus to illuminate women's dynamic participation within terrorism. However, it is essential to note that the mere presence of women in active combat positions does not automatically equate to agency and emancipation.

In general, the portrayal of women within the Liberation Tigers of Tamil Eelam (LTTE), an ethno-nationalist terrorist group, has tended to depict them as proactive and empowered agents. The LTTE sought to establish an independent Tamil state in response to the marginalisation and disenfranchisement of Tamil civilians in Sri Lanka [17]. In the initial stages, Tamil women were primarily assigned to auxiliary roles such as propagandists and recruiters, mirroring the patterns observed in the regime of ISIS. However, during the 1980s, the organisation experienced a notable reduction in male fighters, prompting the recruitment of women to assume more active combatant roles. As a result, a new unit was founded, termed the Women's Front of the Liberation Tigers, often referred to as the 'Birds of Freedom'. Estimates indicate that women constituted approximately one-third of the total frontline fighters. This 'Women's Front' division was committed to achieving an independent Tamil Eelam state, eradicating cultural subjugation, and attaining social, economic, and political parity for women [17, p. 102]. Despite these developments, the incorporation of women combatants within the LTTE gave rise to a substantial breach in traditional Tamil values and gender roles. The cultural norms in Sri Lankan Tamil society predominantly adhere to traditional gender norms that confine women to domestic spheres. This adherence to traditional norms enforces stringent segregation between men and women, akin to the dynamics observed within the Islamic State. However, in contrast to ISIS, the Tamil Tigers permitted and mobilised female cadres for active deployment.

Terrorism scholar Schalk [11] examines the LTTE's propaganda and gender ideology. Schalk asserts that the involvement of women was actively encouraged through the construction of the 'new modern Tamil woman' construct. Aligned with their male counterparts, these 'new women' were portrayed as resolute, valiant, disciplined combatants, and martyrs. This elucidates that the LTTE's discourse, which championed the agency and empowerment of women, challenged the entrenched gender norms within Tamil culture. Consequently, the pursuit of female emancipation was employed as a recruitment mechanism, motivating numerous women to enlist in the movement. However, the professed ideology of a terrorist organisation often diverges from the actual experiences of its members. Echoing this, Stack-O'Connor [15] argues that the LTTE employed women as operatives not primarily to advance women's liberation, but due to their constrained circumstances, marked by a shortage of male fighters. The inclusion of women in more prominent and combat-oriented roles emerged during this phase. However, despite such developments, the LTTE maintained stringent gender-based policies that preserved high levels of segregation and disempowerment between men and women. According to Stack-O'Connor [15], the Tamil Tigers mould women to appear and behave similar to men while denying them the privileges afforded to men. Therefore, it is essential to exercise caution in overstating the correlation between women's active engagement in terrorism and the presence of genuine agency or transformative societal change.

The intricate interplay between prominent roles within terrorist organisations and women's empowerment is substantiated by several scholars, among them [8]. Their research entails a comprehensive qualitative investigation involving four individuals in Sri Lanka who were impacted by the LTTE regime. The findings of their study unveiled that despite the progressive ideologies professed by the LTTE, patriarchal norms and subjugation persisted. One interviewee corroborated that female training camps were subjected to stricter regulations in comparison to those for male recruits. For instance, women were prohibited from engaging in premarital relations, stringent dress codes were enforced, and there was a notable emphasis on women marrying fellow Tamil recruits. In summation, Jordan and Denov [8] assert that while female fighters appeared to experience some emancipation within the LTTE, such liberation was ultimately circumscribed and overseen by men. This study acknowledges methodological constraints due to its limited examination of a small sample size. However, it is crucial to recognise that when evaluating the roles of women in terrorism, the acquisition of firsthand information is often challenging due to restricted participant access. Nevertheless, the study conducted by Jordan and Denov

[8] furnishes a valuable and illuminating perspective on the real-life encounters of Sri Lankan civilians affected by the consequences of war.

The employment of suicide bombing as a tactic within terrorism came to epitomise the strategies of the LTTE. The members of this group directed their efforts towards critical infrastructure, high-ranking officials, and civilians in Sri Lanka, culminating in the loss of numerous lives and the injury of many. As a consequence, the roles of women within the Tamil Tigers underwent a discernible shift towards involvement in suicide attacks. Among these acts, the most notorious was orchestrated by Dhanu, a female operative of the LTTE, who assassinated the former Prime Minister of India, Rajiv Gandhi. It is approximated that approximately 30% of suicide attacks were executed by the 'Birds of Freedom' unit [15]. Tamil women assumed prominent roles as agents of violence. However, several scholars question whether women's active engagement in terrorism genuinely leads to their emancipation and the realisation of equal rights. To illustrate this, Alison [2] highlights that the LTTE recognised the tactical and strategic advantages of employing female suicide bombers. Women aroused less suspicion due to several factors. For instance, traditional attire made it easier to conceal explosive devices, and cultural sensitivities often restrained security forces from thoroughly searching females. Additionally, research indicates that women often lack the perceived agency to be regarded as significant security threats (Ibid.). This study implies that the LTTE manipulated prevailing gender norms, consequently allowing females to infiltrate targets with greater ease and evade detection. Thus, while assuming roles as frontline operatives might evoke feelings of empowerment and gender equality, women were ultimately exploited for tactical gains.

In a similar vein, Bloom [5] undertook an examination of the phenomenon concerning female suicide operatives. Within her analytical framework, Bloom delineates five distinct motivational drivers that propel women toward engagement in terrorism, encompassing themes of Redemption, Revenge, Respect, Relationships, and Rape [5]. Of noteworthy significance, Bloom [5] accentuates the role of sexual abuse as a pivotal determinant compelling women's involvement in the ranks of suicide bombers. Within the traditional context of Tamil culture, the act of rape is perceived as a stain upon the honour of women, an ethos deeply entrenched within the framework of patriarchal oppression. Consequently, instances of rape are construed as inflicting ignominy upon the afflicted woman's family. Bloom [5] posits that the LTTE capitalised upon the vulnerabilities of these women through the invocation of martyrdom rhetoric, expounding that the conduit of suicide terrorism offers avenues for both redemption and emancipation. However, a scholarly rejoinder offered by Sjoberg and Gentry [12] levies critical evaluation upon Bloom's analysis of the LTTE, underscoring a concentration upon the personalised motivational underpinnings that impel women's participation. Sjoberg and Gentry [12] enumerate motivations spanning shame, humiliation, revenge, and redemption, thereby concurring that while sexual abuse and other individualistic catalysts warrant meticulous examination, an interpretative framework which primarily construes female Tamil Tigers as victims of coercion inadvertently diminishes their agency within the overarching narrative.

Therefore, while Bloom's [5] scholarship laudably addresses the multifarious dimensions of women's involvement within the realm of terrorism, it inadvertently tends to amplify a paradigm in which these women predominantly inhabit the role of 'victim', an ontological binary that potentially obscures the intricacies of their agency and active participation [5].

In summary, the extant body of literature often displays a proclivity to present victimhood and agency within women's participation as discrete and incompatible categories. This dichotomous portrayal, however, tends to overlook the intricate tapestry characterising women's involvement in terrorism. The emergence of female terrorists, though not an unprecedented occurrence, has been accompanied by a tendency to oversimplify their roles into the dualistic labels of 'victims' or 'agents'. This simplification, unfortunately, fails to adequately capture the multifaceted dimensions inherent to women's engagement in acts of terrorism. In the context of the LTTE, women assumed active roles, functioning as combatants, recruiters, and even suicide bombers, thereby constituting indispensable assets to the organisation. Nevertheless, their capacity for agency was frequently delimited and constrained by prevailing Tamil cultural norms and stereotypes. The pervasive influence of gendered norms and values fundamentally shapes the contours of terrorist practices and ideologies. Consequently, a critical question has arisen among scholarly circles, analysing the extent to which women's participation in acts of terrorism indeed advances the cause of female empowerment and gender equality.

4.4 Gender-Biased Assumptions

Throughout historical contexts, women's engagement in acts of terrorism has been a recurrent phenomenon; however, persisting gender-biased assumptions continue to underpin investigations into the motives and modalities of their participation. Within the academic literature, a discernible tendency exists wherein the motivations and roles of women are prone to undue simplification. This tendency manifests in a dualistic portrayal, one that primarily casts women as passive victims [4, 5], while, simultaneously, there exists a proclivity to excessively emphasise their dynamic agency [17]. Nonetheless, it is imperative to recognise that women's multifarious roles within the spectrum of terrorism elude facile reduction to a singular, dichotomous construct. The descriptors of 'active agent' or 'passive victim' lack rigid boundaries and are not mutually exclusive, frequently, they overlap and coexist in complex ways. As succinctly articulated by Alison [1], the discourse surrounding the dichotomy of agency versus victimhood, liberation versus subjugation, and emancipation versus oppression concerning women affiliated with the LTTE appears to be an unnecessary and simplistic binary. In its place, Alison [1] advocates for the adoption of the concept of 'ambivalent empowerment', a conceptual paradigm which recognises that women's positionality within the sphere of terrorism transcends binary categorisations of victimisation and empowerment. Within this nuanced framework, women's agency is not monolithic, and their experiences are not confined to unilateral victimhood. Rather, the paradigm of ambivalent empowerment underscores the intricate interplay of diverse factors, ultimately shaping their roles and trajectories within the realm of terrorism.

In light of the foregoing discussion, a critical imperative emerges for an increased acknowledgment of women as simultaneous agents and victims. This recognition should be underpinned by a discerning awareness of the sociocultural and environmental constraints that circumscribe women's choices within the realm of terrorism. Such a holistic perspective stands to catalyse a more nuanced discourse. This perspective transcends the confines of the conventional agent-victim dichotomy, thereby enriching understanding of the intricacies characterising women's engagement in acts of terrorism.

4.5 Conclusion

In conclusion, the involvement of women in contemporary terrorism constitutes an enduring and recurrent phenomenon rather than a novel occurrence. Historical antecedents reveal that women have historically assumed prominent and impactful roles within terrorist organisations, influenced by the underpinnings of extremist ideologies. Therefore, it is essential to acknowledge that both genders are driven to engage in acts of political violence by a multifaceted array of motivations. In the analysis of women's presence within the sphere of terrorism, the imperative arises to transcend deeply ingrained gender stereotypes that pervade the discourse. Conventionally, prevailing gender norms have constructed a dichotomy where men are ascribed traits of rationality, logical thinking, and proclivity towards violence, while women are portrayed as passive, apolitical, emotionally inclined, and oriented towards peaceful pursuits. It is evident that women's representation within the predominantly male-dominated landscape of terrorism remains a numerical minority; however, the impact of this minority contingent is far-reaching and profound. This subset of women not only perpetrates acts of violence, resulting in casualties and injuries on a substantial scale but also fulfils pivotal roles in the operational machinery of terrorist organisations. These roles encompass functions such as recruitment, dissemination of propaganda, frontline combat, and domestic responsibilities.

The discourse advanced in this chapter has brought into focus the variegated roles that women undertake within the domain of terrorism, accentuating the intricate interplay between agency and victimhood. Moreover, this study has undertaken an exploration of the historical framing of women as victims within this context, counterbalanced by an alternative narrative advocating for the recognition of women's agency and underlying motivations. While this analysis has delineated the contours of the agent-victim binary, it has concurrently grappled with the nuanced constraints and ambivalent dimensions of women's empowerment within the sphere of terrorism. Furthermore, this study underscores the imperative for a departure from the confines of the 'agent' versus 'victim' dualism. Embracing a more gender-sensitive and nuanced approach in the assessment of women's roles within terrorism stands as a critical scholarly trajectory. This trajectory acknowledges the intricate web of agency, victimisation, and empowerment, ultimately enriching the discourse and engendering a more comprehensive understanding of the multifaceted dimensions consisting of the participation of women in the realm of terrorism.

References

1. Alison M (2003) Cogs in the wheel? Women in the liberation tigers of Tamil Eelam. Civil Wars 6(4):37–54
2. Alison M (2004) Women as agents of political violence: gendering security. Secur Dialog 5(4):447–463
3. Amnesty International (2014) Escape from hell: torture and sexual slavery in Islamic State captivity in Iraq. https://www.amnesty.org/en/documents/MDE14/021/2014/en/. Accessed 27 Sept 2023
4. Besenyő J (2016) The Islamic State and its human trafficking practice. Strat Impact 60(3):15–21
5. Bloom M (2011) Bombshell: women and terrorism. University of Pennsylvania Press
6. Hoffman B (2017) Inside terrorism, 3rd edn. Columbia University Press
7. Ibrahim H, Ertl V, Catani C, Ismail AA, Neuner F (2018) Trauma and perceived social rejection among Yazidi women and girls who survived enslavement and genocide. BMC Med 16(1):154
8. Jordan K, Denov M (2007) Birds of freedom? Perspectives on female emancipation and Sri Lanka's Liberation Tigers of Tamil Eelam. J Int Women's Stud 9(1):42–62
9. Lahoud N (2018) Empowerment or subjugation: an analysis of ISIL's gender messaging. UN Women
10. Martini A (2018) Making women terrorists into "Jihadi brides": an analysis of media narratives on women joining ISIS. Crit Stud Terror 11(3):458–477
11. Schalk P (1994) Women fighters of the liberation tigers in Tamil Ilam: the martial feminism of Atel Palacinkam. South Asia Res 14(2):163–183
12. Sjoberg L, Gentry CE (eds) (2011) Women, gender, and terrorism. The University of Georgia Press
13. Speckhard A, Ellenberg MD (2020) ISIS in their own words: recruitment history, motivations for joining, travel, experiences in ISIS, and disillusionment over time analysis of 220 In-depth interviews of ISIS returnees, defectors and prisoners. J Strat Secur 13(1):82–127
14. Spencer AN (2016) The hidden face of terrorism: an analysis of the women in Islamic State. J Strat Secur 9(3):74–98
15. Stack-O'Connor A (2007) Lions, tigers, and freedom birds: how and why the liberation tigers of Tamil Eelam employs women. Terror Polit Viol 19(1):43–63
16. United Nations Security Council (2000) Resolution 1325 (2000). S/RES/1325. https://documents-dds-ny.un.org/doc/UNDOC/GEN/N00/720/18/PDF/N0072018.pdf?OpenElement. Accessed 07 Sept 2023
17. Wang P (2011) Women in the LTTE: birds of freedom or cogs in the wheel? J Polit Law 4(1):100–108

Part II
Cyberterrorism Landscape

Chapter 5
Exploring the Current Landscape of Cyberterrorism: Insights, Strategies, and the Impact of COVID-19

Abstract Governments worldwide are confronted with the formidable challenge of safeguarding their populations and critical infrastructures against ever-evolving cyber threats posed by malicious actors, including terrorist groups and their sympathisers. In response to this challenge, this chapter aims to illuminate the intricate relationship between terrorists and the Internet, providing valuable insights to assist policymakers, researchers, and security professionals in devising proactive strategies to combat cyberterrorism. To achieve this goal, the chapter begins by offering a comprehensive examination of the current cyberterrorism landscape, with a specific focus on the impact of the COVID-19 pandemic on online terrorism. It highlights how terrorist groups leverage the Internet's interactivity and global reach for various nefarious activities, such as propaganda dissemination, recruitment, training, planning, funding, and executing attacks. Furthermore, the chapter emphasizes the paramount importance of national security in addressing challenges associated with cyberterrorism, highlighting the arduous task governments worldwide face in safeguarding their populations and critical infrastructures from evolving cyber threats posed by terrorist actors. Moreover, it delves into the multifaceted nature of cyberterrorism, elucidating the characteristics of cyberattacks, their most common forms, and objectives. Understanding these tactics is crucial for developing effective countermeasures and enhancing cybersecurity at both the national and international levels. This chapter contributes to the existing body of knowledge on cyberterrorism by providing valuable insights into its current state, along with a specific examination of the impact of the COVID-19 pandemic on its proliferation.

Keywords National security · Cyberterrorism · Extremism · Internet · Propaganda · Cyber-attacks · Machine learning · Artificial intelligence · The Dark Web

© The Author(s), under exclusive license to Springer Nature Switzerland AG 2024   65
R. Montasari, *Cyberspace, Cyberterrorism and the International Security in the Fourth Industrial Revolution*, Advanced Sciences and Technologies for Security Applications, https://doi.org/10.1007/978-3-031-50454-9_5

5.1 Introduction

In today's interconnected world, governments are confronted with the formidable challenge of safeguarding both their populations and their critical infrastructures from ever-evolving cyber threats posed by malicious actors, including terrorist groups and their sympathisers. The Internet's interactivity and global reach have become potent tools exploited by terrorist entities for various nefarious purposes, such as propaganda dissemination, recruitment, training, planning, funding, and execution of attacks [73]. As the digital landscape continues to evolve, the menace of cyberterrorism looms large, prompting the need for a comprehensive understanding and proactive strategies to counter it effectively. This chapter aims to explore the current landscape of cyberterrorism, with a particular emphasis on its evolution and impact in the wake of the COVID-19 pandemic. The pandemic has intensified the already complex interplay between terrorist organisations and the Internet, heightening concerns about national security. In the face of this pressing challenge, governments worldwide strive to protect their populations and critical infrastructures from cyber threats. This necessitates a deeper understanding of the multifaceted nature of cyberterrorism. Therefore, to provide a comprehensive overview, the chapter delves into the various ways in which terrorist groups exploit the interactive and global nature of the Internet to advance their agendas. Understanding the tactics employed by these groups in the cyberspace is crucial for devising effective countermeasures and bolstering cybersecurity on both national and international levels. Additionally, the study examines the characteristics of cyberattacks perpetrated by terrorist entities, highlighting the most common forms and objectives. By illuminating the intricate interplay between terrorists and the Internet, this chapter seeks to assist policymakers, researchers, and security professionals in devising proactive strategies to combat cyberterrorism more effectively.

The subsequent sections of this chapter are organised as follows. In Sect. 5.2, a contextual background is provided to the subject matter. Section 5.3 examines the emergence of the Internet and the influence of widely adopted social media platforms (SMPs) on the landscape of terrorism. To this end, the section investigates how terrorist groups effectively leverage the Internet and SMPs for various criminal activities, including propaganda dissemination, financing, training, planning, and execution. Following that, Sect. 5.4 delves into the concept of national security, while Sect. 5.5 explores the characteristics of cyber-attacks. In Sect. 5.6, existing literature on cyberterrorism is reviewed, providing insights into its current landscape. Section 5.7 investigates the impact of the COVID-19 pandemic on cyberterrorism and its operational dynamics. Moving forward, Sect. 5.8 discusses the findings, and, finally, the chapter concludes in Sect. 5.9.

5.2 Background

Ontologies are valuable tools as they provide a comprehensive and shared framework for a particular subject, utilising both internal and external factors to construct models [70]. Research suggests that ontologies are crucial for defining the scope and content of a field, as well as establishing a coherent structure, which is the objective of this chapter (Ibid.). To explore the current landscape of cyberterrorism effectively, this section aims to establish context by defining key terms, highlighting relevant concepts, and elucidating relationships (Ibid.). This contextual foundation is essential for a deeper understanding of the subject matter. Legal definitions of terrorism are widely problematic and have faced criticism due to differing interpretations of politically motivated violence [59]. Consequently, definitions of cyberterrorism are even more complex and misunderstood. It is suggested that "the continuing popularity of cyberterrorism as a concept – and fear – has been underpinned by established economic and political interest" [30, p. 658]. Jarvis and Macdonald [30] identify two primary areas of definitional challenges: referential problems and relational problems. Referential problems relate to what the term cyberterrorism should encompass, while relational problems pertain to how cyberterrorism differs or relates to other behaviours or acts of violence (Ibid.). Within the academic community, a contentious discourse prevails in relation to the categorisation of cyberterrorism, wherein the central issue pertains to whether it should be construed as a subset of conventional terrorism or, rather, as a discrete entity characterised by unique attributes, thereby embodying an autonomous and distinct phenomenon (Ibid.). The combination of the terms "terrorism" and "cyber" engenders a transformative impact on the landscape of inquiry, rendering the establishment of precise definitions considerably more challenging.

Denning [17] defines cyberterrorism as the merging of terrorism and cyberspace, encompassing unlawful attacks and threats targeting computers, networks, and the data they contain when carried out with the intention of intimidating or coercing a government to advance political or social objectives [30]. Extremism is often considered to fall under the same umbrella as terrorism, with definitions encompassing the glorification of violence and the incitement of political violence [58]. In conflict settings, extremism often manifests as a form of engagement in the conflict. However, as Coleman and Bartoli [13] suggest, defining what is considered normal and what constitutes extremism will always be subjective. They propose that the focus should shift towards understanding that acts of extremism can change as circumstances evolve, and also emphasise the importance of power dynamics (Ibid.). Groups with little power are more likely to resort to episodic and direct forms of violence, such as suicide bombings. In contrast, groups with more power adopt more structured or institutionalised approaches, which might involve the cyber domain (Ibid.). Multiple perspectives within this field attempt to explain the formation and origins of terrorism and extremism. One viewpoint asserts that extremism emerges as a result of adverse conditions and unmet human needs, leading to acts of extremism. Other perspectives propose that extremism is constructed, suggesting that political leaders or authorities capitalise on these unfavourable conditions to incentivise terrorism, promote their beliefs, and gain power (Ibid.). Alternative perspectives on the construction of extremism explore how dominant groups can portray the actions of marginalised groups as "extremist" and create a self-fulfilling prophecy that elicits increasingly extreme actions from these marginalised groups. This suggests that maintaining power within society is achieved through such actions (Ibid., p. 3).

Additionally, the term "radicalisation" will be frequently used in this chapter and throughout the book, considering that law enforcement agencies (LEAs) and governments employ artificial intelligence (AI) to detect "radical behavior" or "radical" content online. The European Commission [20] defines "violent radicalisation" as "the phenomenon of people embracing opinions, views and ideas which could lead to acts of terrorism as defined in Article 1 of the Framework Decision on Combating Terrorism" (p. 2). However, this definition has faced criticism as the terms "opinions," "views," and "ideas" can be vague, and the word "could" introduces broad possibilities. Moreover, it fails to fully address the complexity of the issue, as radicalisation can lead to forms of conflict beyond terrorism and other types of violence [64]. In contrast, the Netherlands General Intelligence and Security Service (AIVD) defines radicalisation as:

> The (active) pursuit of and/or support to far-reaching changes in society which may constitute a danger to (the continued existence of) the democratic legal order (aim), which may involve the use of undemocratic methods (means) that may harm the functioning of the democratic legal order (effect). (Avid 2004, as cited in The Ministry of the Interior and Kingdom Relations (BZK)[1] [67, p. 13])

This definition places substantial emphasis on radicalisation as a process encompassing actions that significantly deviate from societal norms and have the potential to subvert the established legal order [64]. Furthermore, radicalisation intertwines with extremism and, specifically, violent extremism. Throughout history, extremism has been affiliated with ideological convictions such as fascism, designating individuals categorised as extremists as political agents who exhibit a disposition to disregard the rule of law, owing to their alignment with far-right, ultranationalist, and authoritarian ideologies (Ibid.). In contrast, radicals might or might not espouse democratic values, and their proclivity for resorting to violence remains uncertain. Conversely, extremists disavow democracy and display an unwavering intolerance for diversity (Ibid.), opting to employ force as a means to preserve and consolidate political power. An important distinction exists, as radicals are deemed "redeemable" (Ibid., p. 10), for they might be more amenable to compromise without compromising their core beliefs and strategic objectives. This understanding of the motivations and beliefs underlying radicalism establishes a foundational framework for the subsequent sections of this chapter and indeed the entire book. Such knowledge proves indispensable for governmental entities and stakeholders tasked with detecting and pre-empting cyberterrorism effectively.

---

[1] The Ministry of the Interior and Kingdom Relations (BZK) constitutes one of the eleven ministries within the Dutch central government.

Prior to concluding this Section, it is also imperative to undertake a brief examination of the impact of the COVID-19 pandemic on cyber security. As the virus disseminated worldwide, diverse facets of society experienced profound disruptions, including economic activities and interpersonal interactions. Pertaining to the impact on terrorism, it is crucial to acknowledge the remarkable adaptability of terrorist organisations to their surroundings, a characteristic that has historically contributed to their success [1]. While the pandemic did affect certain operational aspects within their organisations, it did not deter them from capitalising on vulnerabilities in the public domain for their benefit (Ibid.). Notably, limited research on the intersection of the pandemic and terrorism reveals that most governments globally are navigating uncharted territory. The COVID-19 pandemic represents the most significant global health crisis since the influenza pandemic of 1918–1919, which occurred over a century ago (Ibid.). However, during that time, technological advancements were not as prevalent, and the Internet did not exist, rendering contemporary society confronted with an entirely novel and technologically advanced set of challenges. The COVID-19 pandemic has introduced increased "friction" into terrorist operations (Ibid.). However, despite the potential risk of members contracting the virus, their unwavering ideological commitment and motivations for perpetrating their terrorist acts are unlikely to be overridden during the final stages of their actions. In essence, the presence of the virus does not deter their intent or convictions (Ibid.). Nonetheless, it is worth noting that the training and recruitment phases of their operational procedures might experience notable ramifications due to the pandemic's influence (Ibid.). For a comprehensive analysis of the relationship between COVID-19 and cyberterrorism, it is imperative to delve further into the attractiveness of the Internet to these groups, their exploitation of digital platforms, and the nature of cyber threats they pose (Ibid.). Consequently, a more extensive discussion of the pandemic's implications will be presented in the forthcoming "Related Work in the Cyberterrorism Landscape" Section of this chapter.

5.3 Transition to the Internet

The proliferation of the Internet has exhibited a rapid trajectory since the 1990s, witnessing a considerable surge in users, from approximately 25 million in 1994 to over 3.5 billion by 2016 [33]. The advent of the Internet and the widespread adoption of SMPs have had a significant impact on the landscape of terrorism, fundamentally altering the modalities through which terrorist groups communicate and disseminate their radical views and ideologies (Ibid.) [31]. Moreover, the Internet's pervasive reach enables it to circumvent barriers imposed by governments and security organisations, thus affording diverse opportunities for terrorist entities [18]. Prominent SMPs such as Twitter, Facebook, and YouTube have provided extremist groups with direct channels to manipulate and groom vulnerable individuals, exploiting their insecurities and grievances [4]. In response, the UK government has undertaken concerted efforts to eliminate "jihadist propaganda" materials as part of their counter-terrorism initiatives. Nevertheless, the resourcefulness of terrorists and their sympathisers remains evident as they persistently devise novel approaches to circumvent security mechanisms, presenting formidable challenges to counter-terrorism organisations (Ibid.).

SMPs offer a compelling capability to reach an extensive and diverse audience, virtually unrestricted by temporal or geographical boundaries, thereby rendering them an attractive tool for terrorist organisations [18]. Notably, research indicates that social media offer a cost advantage, facilitating swift and efficient messaging in comparison to traditional mass media channels such as radio or television (Ibid.). Terrorists adeptly leverage social media to orchestrate a process of cultivating online hate [4]. The escalating engagement in online discussions is increasingly recognised as a new manifestation of political activism, representing a shift from passive to active discourse concerning political violence (Ibid.) [50]. In response to these challenges, the UK government is contemplating the introduction of a new Communications Data Bill aimed at empowering authorities to filter extremist content, regulate its dissemination, and collaborate closely with internet service providers (ISPs) to expunge material that manipulates or incites individuals towards violent extremism or terrorism [32]. Moreover, the establishment of the TERFOR task force by the government serves to explore strategies for restricting Internet access to specific materials [4]. While numerous internet-based companies maintain the prerogative to remove any content endorsing or promoting violence or terrorist acts [75], reporting suspected terrorist activities on SMPs remains a complex endeavour for internet users, partly due to a lack of clarity and the perception that such actions are not necessarily illegal or an immediate threat [3].

The influence of online preachers and internet propaganda on a substantial number of individuals has captured the attention of both scholars and governments, prompting scrutiny of the perceived role of the internet in what is commonly referred to as "radicalisation towards violent extremism" [3, p. 1]. Consequently, researchers endeavour to comprehend the motivations behind terrorist groups and organisations' adoption of the internet as a medium for conducting their activities, seeking avenues to prevent its exploitative deployment. Weimann [73] identifies five primary reasons underpinning the appeal of the internet for terrorists. Firstly, the internet allows for remote engagement, eliminating concerns related to time zones and facilitating extensive communication outreach. Secondly, the internet's reach is of considerable magnitude, enabling these groups to connect with individuals from any region worldwide who share similar interests and ideologies [3, 74]. Moreover, the Internet's capacity for instantaneous information dissemination across "vast geographical regions" renders it a crucial facilitator in the process of radicalisation and preparations for terrorist acts [31, p. 375]. The United Nations Office on Drugs and Crime (UNODC) further acknowledges that the Internet's potential reach will augment the size of the affected audience (2012). Additional factors contributing to the appeal of the internet to terrorists include its relative anonymity compared to traditional methods, the seemingly limitless range of potential targets, and, significantly, the cost-effectiveness it affords when compared to other means [73]. This is further evidenced by recent developments where terrorist groups have increasingly utilised the internet for recruitment, training, fundraising, and operational purposes, employing it to "seduce, radicalise, and recruit" individuals [3, p. 50].

In order to understand how terrorists exploit the internet for recruitment and training purposes, Freiburger and Crane [21] apply the social learning theory, which suggests that individuals learn deviant behavior through association, imitation, and differential reinforcement (Ibid.). Extremist groups adeptly harness this theory as a means to facilitate recruitment and operational attacks, employing videos or images depicting violent acts or confrontations [4]. Importantly, the dynamics of cybert-errorism and violent extremism in the online realm evolve alongside changing behaviours and social practices, necessitating an ongoing refinement of our under-standing [21]. Furthermore, while grooming individuals, these groups rationalise violence as retribution for perceived oppression and invasions against Muslims worldwide [72]. In this context, Bandura [5] contends that the social cognitive theory assumes significance in elucidating motivating factors, asserting that the theory furnishes "an agentic conceptual framework" for investigating the "determinants and mechanisms" underlying "such effects" [5, p. 265]. Bandura's model postulates human behaviour as either shaped and controlled by environmental influences or internal dispositions, depicting a unidirectional causation [5]. Furthermore, terrorist groups adeptly leverage the internet in numerous ways to their advantage, encom-passing a wide spectrum of activities that span from the dissemination of propa-ganda to the comprehensive training and execution of their operations [68]. This multifaceted utilisation of the online sphere by extremist organisations underscores the need to delve deeper into specific aspects of their online activities. Therefore, in the following subsections, some of the key dimensions of these activities will be explored, illuminating their methods, strategies, and the evolving landscape of cyberterrorism and online extremism.

5.3.1 Propaganda

Terrorist organisations have leveraged the internet to their significant advantage, exploiting its expansive reach for the dissemination of their propaganda [57]. In this context, propaganda is defined by the UNODC as a multifaceted medium encom-passing multimedia communications that offer ideological or practical instructions, explanations, justifications, or promotion of terrorist activities [68]. Such mate-rials might encompass a variety of forms, including textual messages, audio and video files, and even video games, originating from terrorist organisations or their supporters (Ibid.). The intricacy of this issue emerges from the inherent tension between the need to combat terrorist propaganda and the imperative to uphold funda-mental human rights, particularly freedom of speech and expression. Navigating this complex landscape hinges on the challenging task of distinguishing between the exercise of one's right to express opinions and the endorsement of terrorist propa-ganda, a determination that ultimately rests on subjective judgment (Ibid.). At its core, terrorism-related violence, in its simplest form, serves to promote and encourage acts of violence. Its potency lies in its adaptability, allowing content to be finely tailored to suit specific target groups or individuals, with the overarching aim of recruitment, radicalisation, or incitement by instilling a sense of pride and motivation toward extreme objectives. Furthermore, this propaganda can be meticulously tailored to accommodate varying demographic factors, such as age, gender, and socio-economic circumstances, with the overarching goal of instigating heightened levels of fear and panic. Importantly, this psychological manipulation can inflict lasting harm on its victims, leaving deep-seated scars in its wake (Ibid.).

This adaptability and effectiveness of propaganda as a tool for recruitment and radicalisation becomes particularly apparent when considering how terrorist organisations strategically employ the internet to establish relationships with those most receptive to their targeted messages (Ibid.). Utilising restricted access chat rooms, these groups engage in recruitment processes away from public scrutiny, providing a conducive environment to train and indoctrinate potential recruits (Ibid.). The recruitment messages aim not only to entice individuals by highlighting the benefits of joining such organisations but also to encourage their active involvement, including providing material support [63]. Consequently, implementing barriers to restrict entry to these recruitment chats poses challenges for intelligence and LEAs attempting to monitor and analyse terrorism-related activities, which largely remain obscured from their view [68]. Scanlon and Gerber suggest the development of an automated system to identify and understand the attractiveness and incitement mechanisms operating within these clandestine online communities (2014). Extensive research endeavours have been devoted to understanding the process of radicalisation online and the motivational factors prompting individuals to embrace extremist behaviours [62]. The efficacy of propaganda in radicalisation varies depending on the intended audience, necessitating nuanced analyses [68]. The clandestine nature of online radicalisation has grown more pronounced, with the identification of warning signs becoming increasingly crucial. These signs include detecting extreme language, sharing extremist material, and observing deviations in behavior [62]. To address this growing concern, the burgeoning field of research seeks to enhance automatic and predictive tools for analysing radicalisation on SMPs. Nonetheless, questions remain concerning the effectiveness and integration of these tools in conjunction with police investigations and existing policy practices [60]. Crucially, it is imperative to distinguish between general propaganda and content specifically intended to incite acts of terrorism, as deterring the incitement into terrorism is crucial for safeguarding national security [68].

In summary, within the contemporary context, where the internet's ubiquity has made it an indispensable tool for communication and information dissemination, the challenge of addressing terrorist propaganda remains an ongoing concern for governments, international organisations, and civil society alike. As technology continues to evolve, so too must strategies for countering the pernicious influence of online terrorist propaganda.

5.3.2 Financing, Training, Planning and Execution

Cyberterrorists adeptly exploit the internet for a range of financial pursuits, including fundraising to sustain their operations and manipulating critical infrastructures for illicit monetary gains [26]. This strategic use of the digital realm to target financial and commercial systems, government databases, medical records, and air traffic control systems underscores their potential to wield considerable control over a nation (Ibid.). It raises fundamental questions about the security and resilience of our critical infrastructure in the face of evolving cyber threats. Terrorist organisations employ a diverse array of tactics to exploit the internet for financial gains. One disturbing example involves the misuse of government-supported charities that receive financial aid, diverting these funds for nefarious purposes [68]. Additionally, certain groups impersonate corporations or organisations, ostensibly supporting environmental or humanitarian goals while channelling received funds into acts of terrorism (Ibid.). These activities highlight vulnerabilities in financial systems and call for more robust mechanisms to detect and prevent such misuse. Direct solicitation through chat rooms or targeted communications is another method through which terrorists solicit financial aid from their supporters (Ibid.). They further leverage online payment tools, resorting to various forms of fraud such as wire or auction fraud to facilitate fund transfers between parties. Notably, platforms such as PayPal and Skype are exploited for their ease and anonymity (Ibid.). This underscores the pressing need for improved regulation and security measures in the online financial sphere.

Furthermore, as previously discussed, the rise of social media has significantly facilitated the recruitment and training of individuals into extremist groups, due to technological advancements that allow terrorists to reach and communicate with potential recruits globally. The online instructional content disseminated through these platforms includes training in not only physical acts of terrorism but also encryption methods aimed at ensuring anonymity during hacking activities and securing communication within the group (Ibid.). This blurring of the lines between online and offline extremism raises important questions about monitoring and countering extremist ideologies in the digital age. Additionally, the interactive nature of social media fosters a sense of community among individuals from diverse backgrounds, united by the shared objective of embracing violent extremism [71]. This phenomenon necessitates a critical examination of the role SMPs play in disseminating extremist content, and the responsibilities of tech companies in curbing the spread of such dangerous ideologies.

Therefore, the ease with which individuals can be trained and recruited into extremist groups has become a pressing concern. As previously discussed, this accessibility is facilitated by technological advancements that enable terrorists to reach a vast pool of potential recruits. The dissemination of instructional content includes tools and techniques such as encryption methods, intended to ensure anonymity during hacking activities and enhance communication security among group members (Ibid.). This highlights the need for enhanced cybersecurity measures to protect against these threats. It is evident that the internet has become a battleground for both physical acts of terrorism and the spread of extremist ideologies, raising fundamental questions about law enforcement and intelligence agency capabilities. Moreover, the internet presents logistical advantages to terrorists, minimising the risk of detection and facilitating the concealment of their identities [68]. Cyberattacks, in some instances, constitute the primary objective for these groups. They seek to exploit computer networks by disrupting their proper functioning through tactics such as hacking, deploying viruses, employing Phlooding attacks, or deploying malware (Ibid.). These actions can cause significant disruptions to critical infrastructure and have far-reaching implications for national security, demanding robust cybersecurity strategies and international cooperation to counter cyberterrorism effectively.

5.4 National Security

In a broader context, national security pertains to safeguarding the integrity of a state and implementing measures to defend its institutions against external threats [8]. Technology has emerged as a pivotal instrument in this endeavour, prompting governments to enhance security measures for their computer systems and networks to prevent exploitation (Ibid.). Debates surrounding national security have attracted numerous political actors, but discussions concerning cyberterrorism have often become sensationalised and exaggerated in the media (discussed in further detail in Chap. 6), rather than offering informative and precise definitions [73]. Nonetheless, it remains evident that terrorist attacks constitute a primary cause of national instability [29]. Traditionally, national security has been defined as the "capacity to deter or to resist the invasion of one's territorial borders by foreign military, naval, or air forces" [25, p. 68]. However, this definition has evolved to encompass a broader range of factors, including threats to public health, education, welfare, and the national economy. In this evolving landscape, the recent global COVID-19 pandemic, for example, has not only disrupted the global economy but has also underscored the importance of addressing cyber threats. As computer networks heavily rely on secure telecommunication systems, the introduction of novel communication technologies becomes imperative to enhance network survivability [35]. Technologically advanced countries have developed robust communication systems, maintaining different modes of communication, which enhances their resilience to various cyber threats (Ibid.). In contrast, less technologically developed nations are more susceptible to cyber threats on their infrastructure [73]. These security threats, faced by national infrastructures, consist of a wide spectrum of social, economic, and physical dimensions [76]. For instance, digital security threats can manifest as terrorists launch hacking attacks and exploit vulnerabilities in national networks, potentially leading to severe economic consequences. Political threats involve the use of social bots to manipulate public opinion, while social threats exploit SMPs to divulge online privacy breaches (Ibid.).

The rapid proliferation of digital data has introduced numerous challenges to national security, necessitating the adoption of more advanced analytic tools to effectively address and mitigate these threats (Ibid.). In response to the multifaceted challenges related to national security, artificial intelligence (AI) has emerged as a potential asset. AI not only supports international security initiatives but also enhances national competitiveness across various domains, including the military, information, and the economy [2]. In this context, machine learning (ML), a branch of AI, stands as a key discussion point, warranting a comprehensive analysis of its role in addressing national security challenges within subsequent chapters of this book. This is especially pertinent given the rapid advancement of technology, which, coupled with heightened military dependence and the escalating use of the internet for extremist activities, has placed significant strain on national security.

One particularly prominent concern in this context is Offensive Information Warfare (OIW). OIW is a form of cyber warfare that consists of a collection of activities and tactics intended to utilise information and communication technologies to achieve "specific political and strategic objectives" [69, p. 17], often in the context of military or national security operations. OIW constitutes a wide range of cyberattacks, such as network intrusions, malware distribution, and hacking, to gain unauthorised access to critical systems, disrupt operations, or steal sensitive information. It can be characterised by its focus on compromising "the integrity, availability, and confidentiality" of collected, stored, and transferred data [69, p. 17], information systems; and communication networks. These activities are driven by specific political or strategic objectives, which can encompass disrupting an adversary's military capabilities and influencing public opinion or decision-making. OIW often involves altering or manipulating data to deceive or misinform the targeted organisation or individuals, including spreading false information or altering critical data to create confusion or undermine trust. It typically targets information systems across various critical sectors, such as government networks, military infrastructure, and financial institutions. OIW activities are carried out covertly, making attribution challenging, and malicious actors might utilise various techniques to hide their identity or the source of the attacks [14].[2] To this end, national security agencies and organisations often employ a wide array of countermeasures and cybersecurity practices to identify and defend against OIW tactics (Ibid.). In the context of the potential exploitation of OIW by terrorist organisations, there are debates about whether terrorists might use OIW to directly undermine national security [37, 69]. This concern is particularly significant due to the growing reliability and accuracy of computer sensors, which have increased their capacity to exploit and maintain systems.

In addition to OIW, intelligence warfare, often referred to as information warfare or cyber intelligence warfare, is another critical aspect of national security. It involves the direct integration of intelligence into operations, rather than solely utilising it as an input for overall system control [37]. Intelligence warfare distinguishes

---

[2] For additional reading, please refer to Cordesman's [14] work, in which he also discusses deterrence models, including the retaliatory doctrine.

itself from traditional military operations by focusing on the strategic use of intelligence and information-related activities to achieve specific political, military, or strategic goals [44]. In this context, intelligence is not merely collected but actively integrated into military and strategic operations, influencing decision-makers and shaping events (Ibid.). This integration involves utilising technology for information exploitation, which includes activities such as hacking, cyber espionage, and data manipulation [12, 38, 65]. These tactics allow for the gathering, analysis, and dissemination of information, often involving breaches of critical systems. Deception plays a significant role in intelligence warfare, with strategies that include modifying data and manipulating communication channels [10]. The ultimate aim is to confuse adversaries and influence public opinion (Ibid.).

Targeted attacks such as advanced persistent threats (APTs) [41] are another central component of intelligence warfare (discussed in more detail in Sect. 5.5). They encompass actions such as hacking, social engineering, spear phishing, and watering hole attacks, malware distribution, network intrusions, and denial-of-service (DoS) attacks on critical infrastructure and government networks [7, 16]. These operations are driven by political and strategic objectives, ranging from disrupting military capabilities to influencing crucial decision-making processes. One of the challenges in intelligence warfare concerns the intricacy of attributing actions to their source [36, 65]. These operations are conducted covertly, rendering it arduous to determine the responsible party [12]. In response to these threats, national security entities implement countermeasures that involve strengthening cybersecurity, sharing information, and developing strategies to combat misinformation. As modern conflicts continue to evolve, intelligence warfare underscores the interconnectedness of military, political, and information operations [65]. This interconnectedness accentuates the paramount importance of safeguarding information infrastructure. Therefore, in an era where nations rely heavily on information and communication technologies, intelligence warfare has become increasingly vital in contemporary conflicts and national security considerations.

The impact of cyberterrorism has been amplified by the global immediacy of the internet and social media, which have transformed how terrorists utilise technology while preserving certain aspects of their traditional methods [69]. However, it is important to acknowledge that while the internet has undoubtedly provided terrorist groups with enhanced communication and propaganda instruments to facilitate their activities, some argue that this transformation cannot be classified as revolutionary. In light of these considerations, addressing the challenges of countering cyberattacks and disruptive activities necessitates governments to devise more advanced policies that might, in some cases, compromise the flexibility and accessibility of the internet but are essential to maintaining a positive public outlook [69]. One approach employed by governments to combat cyber threats is wiretapping and data interception, however, these efforts may inadvertently weaken the security of the devices [8]. critical national information infrastructure (CNII) holds vital significance for any nation and its government, as its destruction would have far-reaching consequences, impacting security, public health and safety, the nation's image, and economic strength [77]. Consequently, disruption of CNII represents a high-impact target for terrorists seeking to inflict significant damage on a country (Ibid.). In conclusion, as substantiated by the foregoing discussions, intelligence warfare emphasizes the need for nations to protect their critical information infrastructure, invest in cybersecurity, and develop strategies to defend against information-based threats. It serves as a reminder of how military, political, and information operations are intricately linked in modern conflicts.

5.5 Characteristics of Cyber Attacks

The complexity of cyber-attacks continues to escalate, fuelled by cyber warfare and the rapid technological advancements [22]. It has been projected that cyber-attacks cost approximately $3 trillion annually in 2015, with estimates indicating an increase to $6 trillion annually by 2021 (Ibid.). A comprehensive understanding of the scale of cyberterrorism demands an understanding of the functioning of the cyber world and how cyber-attacks exploit security and infrastructures [56]. Scholarly research has categorised cyberterrorism into three primary clusters: destructive, disruptive, and enabling [9]. Destructive cyberterrorism, also known as pure cyberterrorism, involves physical acts of terrorism that focuses on corrupting and manipulating systems to inflict damage on virtual or physical assets of national infrastructure [24]. On the other hand, disruptive cyberterrorism, commonly referred to as jihadist hacking, centres on disrupting normal cyber functions to expose and deface critical infrastructures [9]. This disruptive approach is prevalent and pursued by the majority of terrorist organisations, as it enables them to assert control and power. Disrupting national infrastructure conveys a potent threat of power and demonstrates the groups' ability to breach any network (Ibid.). Lastly, enabling terrorism, also termed Enabling Cyber Militancy (ECM), focuses on activities such as recruitment, training, radicalisation, and incitement for cyberterrorism.

Perpetrators of such acts may not inherently necessitate religious motivation; instead, political aspirations play a pivotal role within this context (Ibid.). This distinction is vital as the absence of political motivation would preclude the classification of a cyberattack as a terrorist act, relegating it to conventional criminal activity [42]. According to the World Economic Forum, "cyberthreats are potential online events that might cause detrimental outcomes to individuals, organisations and countries" [19, p. 206]. Extremist groups employing the Internet for disruptive or destructive operations in the digital domain aim to instil fear through violence or the threat thereof, driven by a militant belief system [3, p. 12]. In certain cases, terrorists have engaged in the illicit acquisition of credit card information and financial data to support their campaigns [35]. Existing research indicates that, as of the present time, no critical infrastructures have experienced devastating cyber-attacks that resulted in shutdowns. The observed cyber activities have mostly revolved around intelligence collection, digital graffiti, and defacement or disruption of other groups' websites (Ibid.).

Furthermore, in line with certain research perspectives, cyber threats can be categorised into two main types: non-kinetic and kinetic. Non-kinetic threats do not involve physical violence but can compromise data confidentiality and availability [19]. On the other hand, kinetic threats result in actual violence, potentially leading to severe consequences, including loss of life, by targeting critical infrastructures as well as exploiting data. The burgeoning financial resources at the disposal of organisations such as Al-Qaeda or the Islamic State in Iraq and Syria (ISIS) have significantly enhanced their capacity to pose cyber threats and exploit vulnerabilities in national security (Ibid.). Marsilli, in particular, examines the distinctions between cyberattacks carried out by non-state entities, such as rebel groups or terrorist organisations, and those sponsored by governments (2019). He contends that non-state cyberterrorism should be treated as a form of ordinary crime, applying the tools and measures already available. On the other hand, state-sponsored cyberattacks could warrant the application of the NATO doctrine, which treats the cyber domain akin to the three traditional domains of warfare (Ibid.). An essential concern lies in the potential of a cyber-attack to complement a physical attack. Although it might potentially evade detection amidst the customary tumult of everyday existence, the ramifications can be profound. Therefore, the identification and mitigation of such threats assume paramount significance, particularly in light of the heightened threats posed by terrorists who employ this strategy [35]. To effectively identify such threats, it becomes crucial to comprehend the attack methods utilised and how they exploit networks.

There exist five fundamental stages involved in cyber-attacks [11]. The first step entails reconnaissance to gather information about the target system, including details about its software, network, or personal data related to the victim. Subsequently, the attack focuses on penetrating any remaining defences in the system, and various attack methods might be employed during this stage (Ibid.). Once the initial breach of security is achieved, the attacker proceeds to modify settings that accompany security measures, facilitating convenient and continuous access to the system in the future. After the system is entirely compromised, the attacker utilises this initial breach as a foundation and reference point for all subsequent attacks (Ibid.). Consistency is maintained throughout the attack process, employing the same tools, techniques, and methods in all stages. Finally, the attacker faces a pivotal decision: whether to inflict damage and paralyse the network by deleting critical system files or installing software that renders the computer inoperable (Ibid.). These systematic stages form a strategic process for cyber-attacks, presenting significant challenges in ensuring the security and resilience of targeted systems.

APT is a prevalent attack method employed by cybercriminals to acquire highly sensitive information, with a particular focus on a specific target [55]. APTs have gained popularity among terrorist organisations due to their ability to infiltrate vulnerable systems, operate covertly, and leave no discernible traces, making them an attractive option for conducting cyber and intelligence espionage (Ibid.). Contrary to popular belief, the security industry contends that APT attacks are not necessarily more technologically advanced than other cyberattacks, despite their persistence

[66]. The GhostNet case serves as an illustrative example of APT in practical application. It emanated from China and entailed a malevolent email chain, housing an attachment that, upon activation, executed malicious software on the victim's system (Ibid.). This malware facilitated the download and installation of the Ghost Remote Administration Toolkit, thereby granting the perpetrators remote control over the compromised system and unauthorised access to personal data (Ibid.). APTs often employ sophisticated techniques to exploit network vulnerabilities, posing challenges for detection methods and rendering it difficult to prevent such attacks [22]. In an effort to counter APTs, Intrusion Detection Systems (IDS) and Prevention Systems (PS) have received significant investment from governments and intelligence organisations, as they are perceived as effective detection tools (Ibid.). It is worth noting that although many attacks are categorised as APTs, some researchers, such as Sood and Enbody [66], argue that APTs should be regarded as a subset within the broader category of targeted attacks. Sood and Enbody [66] introduce a targeted attack model comprising three phases: intelligence gathering, threat modelling, and attacking and exploiting targets. These phases encapsulate the strategic methodology employed in APTs to effectively compromise specific systems and gain unauthorised access to sensitive information.

The process of intelligence gathering in cyberattacks involves the collection of extensive data to develop targeted attack vectors against the victim. Open-Source Intelligence (OSINT) is often the preferred approach for gathering information, as it involves extracting data from publicly available resources (Ibid.). Data acquisition can be achieved passively, semi-passively, or actively, with the passive methods designed to go unnoticed by the victim, while the active approach involves interactions with the victim to access further resources (Ibid.). Once the intelligence is gathered, attackers proceed to construct a threat model, where they map the victim's environment using the acquired information. The threat model allows them to assess the risks involved, determine the most effective approach, and identify weaknesses in the victim's network for exploitation (Ibid.). Subsequently, the actual attack is launched, with the primary objective being to deploy malware on the victim's system to extract additional information (Ibid.). APTs pose significant risks to organisations, companies, and governments even though they are seldom utilised within the realm of cyberterrorism. Nonetheless, it is crucial to maintain awareness of APTs, as their use might evolve over time [22]. Vigilance and understanding of APTs can aid in implementing effective countermeasures to protect against potential future threats. Distributed Denial of Service (DDoS) is another frequently employed attack method, which, although technically less complex than other techniques, should not be underestimated due to its potential political motivations and capacity to lead to more severe subsequent attacks [55]. In a successful DDoS attack, certain services on the target machine are blocked, but the internal functioning of the system remains unaffected. Companies and organisations that adhere to proper guidelines and security practices can mitigate the impact of DDoS attacks (Ibid.). This attack type is characterised by an attempt to flood the network by obstructing the normal flow of traffic. Consequently, it disrupts connections, hinders certain individuals' access to particular services, and ultimately culminates in a complete system disruption [34]. The repercussions of a DDoS attack can extend beyond merely uncovering companies' trading networks and their financial information, it has the potential to compromise the integrity of the entire network infrastructure, exposing valuable data to the public domain [55].

5.6 Related Work in the Cyberterrorism Landscape

This section primarily centers on the substantial contributions of two key scholars, Gill et al. [23] and Weimann [73, 74], within the realm of cyberterrorism and its contemporary landscape. These scholars have played pivotal roles in advancing the understanding of this subject matter through their valuable research and insights. The deliberate and purposeful focus on Gill et al. [23] and Weimann [73, 74] is motivated by their authoritative standing in this domain, marked by their comprehensive insights and rigorous analyses. By concentrating on these works, this section aims to conduct an in-depth examination of fundamental perspectives and concepts in cyberterrorism, facilitating a more nuanced understanding of the subject matter. This targeted approach enables a thorough exploration of their ideas and methodologies while offering a framework upon which subsequent discussions and analyses can be built.

Gill et al. [23] emphasise the shifting focus from viewing the internet as a potential causal factor to understanding its use by individuals based on their needs, expectations, and motivations (Ibid.). The internet is regarded as a sophisticated tool that operates in a two-way process, accommodating diverse usage patterns in various environments by different individuals. This perspective challenges the notion of a simple dichotomy between online and offline radicalisation and suggests that such a dichotomy may be a false construct [33]. The research reveals significant differences in online behaviour influenced by ideology, for instance, right-wing offenders are more inclined to acquire behaviours online compared to Jihadist-inspired perpetrators. This divergence is attributed to the limited physical environments available for individuals identifying with right-wing groups to gather and learn (Ibid.). The internet is perceived as a facilitative tool that provides ample opportunities for violent radicalisation and attack planning, even though radicalisation and planning are not exclusively dependent on the internet. Notably, the internet's continuous development since its conceptualisation in a 1936 paper demonstrates its capacity to process, create, and share information, as envisioned in that early work (Ibid.).

Gill et al.'s [23] research provides a critical insight that underscores the imperative need to examine the internet's role in terrorist attacks. Their work illuminates the evolving landscape of modern terrorism, where online platforms have become pivotal tools for recruitment, radicalisation, and communication among extremist groups. By emphasising the internet's significance in this context, Gill et al. [23] highlight the urgency of addressing the digital dimension of counterterrorism strategies and the critical role of online surveillance and regulation. Effectively countering cyberterrorism might require researchers and intelligence organisations to delve into the complexities of terrorist activities within the metaphorical "cyber rabbit hole" [33, p. 97].

Akin to Gill et al.'s [23] research, Weimann's [73, 74] works also constitutes a notable contribution to the contemporary understanding of the cyberterrorism landscape. His research illuminates the internet's multifaceted role as a tool employed by terrorists, encompassing activities such as psychological warfare, publicity, propaganda dissemination, fundraising, recruitment, networking, information sharing, and strategic planning. As previously discussed, Gill et al. [23] underscore the significant threat posed by cyberterrorism, an issue widely acknowledged by security experts and scholars due to the potential risks associated with cyberterrorists infiltrating private computer systems and exploiting various facets of national security (Ibid.). Weimann also theorises that the cyber realm offers modern-day terrorists an attractive option, as it has the potential to cause substantial destruction and wield a significant psychological impact (2006). The anonymity afforded by the cyber space appeals to terrorists, as it enables them to evade security measures while planning and executing their acts, subsequently disappearing from scrutiny. This anonymity becomes especially crucial due to the extremist beliefs and unique values held by terrorists, allowing them to operate covertly in certain social environments that might not share their particular ideology [74].

Weimann's extensive research has delved into various strategies employed by terrorists in their utilisation of the Internet (Ibid.). These strategies encompass launching campaigns, directing and recruiting volunteers, fundraising for planning and training, and facilitating networking and attack coordination (Ibid.). As previously discussed in this chapter, terrorists exploit interactive technology, engaging in chat rooms and online 'cafes' as a means of recruitment [4]. This recruitment tactic is further substantiated by Magouirk et al., who emphasise the crucial role played by such interactive exchanges in chat rooms, fostering ideological relationships and contributing to the radicalisation of young individuals (2008). Moreover, contemporary terrorist organisations maintain dedicated websites for propaganda dissemination and communication with potential members [74]. These online platforms are closely monitored by government entities through conventional means, prompting terrorists to seek evasion from detection and surveillance by resorting to the Dark Web as a clandestine platform [40, 43, 51–53, 74]. It is asserted that establishing a robust cyber presence has become paramount for the long-term success of terrorist organisations, potentially surpassing the significance of physical threats [74]. This shift underscores the evolving dynamics within the realm of terrorism in the digital age, where effective utilisation of the internet has become integral to their operations and strategies.

Weimann's research involves monitoring websites operated by terrorist organisations, downloading their content, translating messages, and archiving them. This comprehensive approach allows for diverse content analyses, focusing on attributes such as content characteristics, intended recipients, and the technologies employed by these groups [74]. Communication facilitated through computer networks is particularly advantageous for terrorists, as it evades control or restriction, remains uncensored, and enables unrestricted access for anyone interested [35, p. 9]. Consequently, terrorists are inclined to exploit the internet to gather data on potential targets, and intelligence services might use covert means to penetrate computer networks and acquire information not publicly accessible to others [35, 74]. Furthermore, Weimann [74] highlights that irrespective of terrorists' motivations or geographical locations, these groups predominantly employ the internet for two primary purposes: to advocate freedom of expression and to rally support for individuals they consider political prisoners [74]. Their online campaigns are geared towards reinforcing the ideology of unrestricted speech and communication, often justifying their actions as a response to perceived oppression against Muslims worldwide [72]. This notion of political prisoners aims to elicit sympathy from their target audience, exploiting emotions such as shame and guilt to make them more susceptible to their messaging and recruitment efforts [74].

5.7   COVID-19 and Terrorism

The impact of the COVID-19 pandemic on cyberterrorism and its operational dynamics is highly significant in contemporary society. With the increased reliance on internet-based communication during the pandemic, vulnerabilities in both society and critical infrastructures have become more pronounced, rendering them susceptible to exploitation [39]. Cybercriminals have leveraged this vulnerability to perpetrate scams and victimise individuals online (Ibid.). Additionally, terrorist and violent extremist groups have sought to disseminate their propaganda online, but countermeasures, such as de-platforming certain individuals, have led these groups to migrate to smaller or obscure platforms, including the Dark Web, to propagate their ideologies [15]. As platforms endeavour to curb misinformation online and combat the surge in conspiracy theories, they are compelled to remove content deemed extremist, thereby raising concerns about the potential impact on freedom of speech and expression online (Ibid.). The pandemic has also witnessed a rise in fraudulent activities, with terrorist and criminal organisations exploiting COVID-19 support programs to launder illegally obtained funds without detection by authorities (Ibid.). However, it is noteworthy that certain measures implemented during the pandemic, such as border control restrictions, have temporarily hindered the rise of terrorism by limiting their ability to travel and carry out attacks abroad (Ibid.). Nonetheless, as commercial aviation has resumed and adopted contactless technology, airports and border control systems have become more vulnerable to exploitation and hacking. Furthermore, it has been suggested that terrorist groups might have shifted to utilising cargo and parcel services for their illicit activities during the period of reduced air travel (Ibid.).

The ramifications of the pandemic extend beyond terrorist organisations and violent extremist groups, impacting counter-terrorism responses as well. The CTED under the United Nations has highlighted issues concerning budgeting and resource allocation during the COVID-19 pandemic, necessitating significant reallocation of resources [15]. Many states that allocate budgets for counter-terrorism initiatives have had to reevaluate their priorities, while those reliant on external resource support have encountered challenges in maintaining their operations (Ibid.). However, this evolving landscape has prompted institutions and organisations involved in counter-terrorism to reassess their objectives and strategies. Furthermore, there is a renewed focus on prioritising counter-terrorism efforts and enhancing international cooperation to effectively address both cyber and physical threats that continue to evolve amidst the pandemic (Ibid.). In their notable study, Ackerman and Peterson [1] present a comprehensive examination of the impact of the pandemic on terrorist organisations. The researchers classify these organisations into ten distinct groups, which will be briefly outlined and discussed as follows:

The first group involves terrorists engaging in pro-social activities. Some larger terrorist organisations seized the pandemic as an opportunity to recruit and fund their activities through propaganda. However, some aimed to gain positive attention by providing assistance to pandemic-affected communities. This pro-social engagement raises questions about solely categorising these groups as terrorists, given potential alignments with far-right ideologies or other motivations (Ibid.). The second group focuses on increased susceptibility to radicalisation due to the pandemic's psychological impact. Job losses, bereavements, and mental health issues left individuals vulnerable, exploited by terrorist groups. Notably, organisations such as ISIS used COVID-19-related hashtags to redirect individuals to extremist propaganda online, facilitating recruitment and fundraising. Terrorists capitalised on anxieties and vulnerabilities, leveraging reduced social contact to obscure radical behavior (Ibid.). The third group discusses a rise in anti-government attitudes, fuelled by conspiracy theories and dissatisfaction with governmental responses. Some governments' perceived incompetence in handling the pandemic led individuals, regardless of their ideologies, to develop critical attitudes. Extremist groups found this situation opportune for further blame attribution and frustration, fostering an environment conducive to radicalisation (Ibid.). The fourth group examines how the pandemic inspires apocalyptic-millenarian extremists who believe their actions are necessary to initiate apocalyptic events for salvation. The pandemic motivated some to act on these beliefs due to the virus's destructive impact (Ibid.).

The fifth group observes terrorists working from home, engaging in cybercrime during lockdowns. More free time enabled planning, training, and online skill development. However, this exposed them to countermeasures, including detection and removal from online platforms. The pandemic also saw a surge in cyber-attacks on critical infrastructures, exacerbating the challenges faced by governments (Ibid.). The sixth group considers the potential for bioterrorism as the pandemic emphasises the catastrophic potential of biological agents. Even in developed nations, containing the spread of a virus presents challenges. Terrorist acts involving biological agents can lead to societal disruption, governance vulnerabilities, loss of life, and psychological trauma (Ibid.). The seventh group explores the weaponisation of COVID-19 by certain extremists, particularly far-right groups inciting followers to deliberately contract and spread the virus to specific targets (Ibid.). The eighth group presents a paradox: the pandemic offers opportunities and challenges for terrorist attacks. While some vulnerabilities emerge, the pandemic's severity and media coverage may reduce the impact of major attacks. Traditional targets such as airports and train stations have become less attractive, while medical facilities might become potential targets (Ibid.). The ninth group concerns less-secure facilities during lockdowns, with reduced personnel presence potentially rendering them attractive targets for terrorists. High-risk facilities, including those housing weapons and prisons, require careful attention (Ibid.). The tenth group discusses distractions faced by counter-terrorism officials, as the pandemic disrupts their operations. Resource diversion, remote work, and ongoing virus spread challenge counter-terrorism efforts, potentially eroding their effectiveness (Ibid.).

In view of the foregoing discussions, a critical imperative emerges for the government to comprehensively assess the ramifications of the COVID-19 pandemic on counter-terrorism endeavours. In this regard, a nuanced understanding necessitates the differentiation of these impacts into short-term, medium-term, and long-term categories. Much like the persistent vigilance required in countering cyberterrorism, it becomes evident that the repercussions stemming from COVID-19 are poised to exert a protracted influence, particularly with regard to the terrorism threat landscape. Foreseeing the years subsequent to the pandemic, it becomes apparent that the efforts made by extremist organisations to recruit, radicalise, and train individuals might yield palpable results. This assertion gains credence due to the widespread unemployment and substantial psychological trauma experienced by segments of the population during the pandemic era. Such socio-economic and psychological vulnerabilities could potentially render certain individuals susceptible to extremist ideologies and recruitment efforts. It is noteworthy that despite facing resource and funding constraints, the collaborative efforts of researchers and counter-terrorism agencies have already yielded valuable insights. These insights enable a degree of predictive analysis and have led to the prioritisation of preparedness measures aimed at identifying potential cyber threats. Consequently, the recognition of the enduring impact of the COVID-19 pandemic on counter-terrorism strategies underscores the importance of vigilance and strategic foresight in mitigating emerging threats in the post-pandemic landscape.

5.8 Discussion

Each sovereign government bears a set of core responsibilities, which many experts contend should command future work and policy endeavours [28]. Primarily, it is advocated that governments allocate resources for expanded research into automated and authentic technologies, notably AI, to enhance the capability of agencies in detecting cyberterrorist threats [45-47, 53, 54]. Concurrently, emphasis is placed on fostering collaboration and synergy across different jurisdictional domains to combat the transnational nature of cyberterrorism [6, 28, 51]. Recognising that terrorism transcends national boundaries, international collaboration becomes pivotal in fortifying counterterrorism efforts. Therefore, by fostering collaboration and enhancing knowledge-sharing, the global community can work together to tackle this menace and secure cyberspace for the benefit of all. Another imperative for governments is the cultivation of robust relationships between LEAs and the general public. This entails public education and training programs aimed at equipping individuals to navigate cyber threats and safeguard their privacy. Equally significant is the establishment of legal frameworks to underpin the prosecution of cybercriminals engaged in hacking and propagating violent extremism [28]. Demonstrating innovation is also incumbent upon governments, achieved through the formation of multidisciplinary teams entrusted with diverse responsibilities [8].

A pivotal facet in enhancing governmental awareness of cyberterrorism pertains to profiling and predictive policing, particularly focusing on the pre-terrorist profile [48, 49]. This proactive approach serves to bolster knowledge, enhance preparedness, and refine prediction capabilities. Rae [61] underscores that terrorist profiling "would maximise the efficiency of prophylactic resource allocation, increasing the likelihood of the interception of a terrorist attack" (p. 64). Therefore, to identify the warning behaviours delineated in the pre-terrorist profile, meticulous scrutiny of evolving individual behaviours becomes imperative. Similarly, Johansson et al. accentuate the salience of "perseveration and negative characterisation", identifying them as critical indicators (2016). However, profiling efforts encounter inherent challenges, as inaccuracies can lead to false flags or unjust prosecutions of innocent individuals [48, 49]. Therefore, balancing the advantages and ethical concerns associated with profiling is a delicate enterprise, demanding judicious calibration of policies and practices (Ibid.). By navigating these complexities, governments can strive to bolster counterterrorism strategies while safeguarding individual rights and upholding justice.

Consistent with the discourse presented in Chap. 3, an additional proposition emerges pertaining to the sustained application of the Prevent strategy in the context of counter-radicalisation endeavours [27]. This strategy, notable for its emphasis on pre-crime interventions, rehabilitation initiatives, and group risk assessments, bears relevance to the discourse on combating cyberterrorism. Although not readily apparent, offering support plays a role in addressing cyberterrorism, particularly when an individual might be ensnared in radicalisation, seeking answers through misguided channels. Strategies such as Prevent recognise that a punitive approach might not always be optimal, instead, prioritising support and rehabilitation can yield more effective outcomes for both the individual and the overarching governmental objective. Notably, the Prevent strategy embodies a commitment to community resilience, accentuating the significance of fostering cohesion and mutual support within communities (Ibid.). Sustaining the salience of this strategy, remains pivotal for bolstering counterterrorism efforts and warrants heightened prioritisation. In light of its pre-crime interventions, support-oriented initiatives, and community resilience components, the Prevent strategy stands as a multifaceted instrument in addressing the complex and evolving landscape of cyberterrorism. Hence, its perpetuation and evolution merit concerted attention and strategic implementation to fortify the counterterrorism paradigm.

5.9   Conclusion

In the rapidly evolving and interconnected world, governments face the formidable task of safeguarding their populations and critical infrastructures from ever-evolving cyber threats posed by malicious actors, including terrorist groups and their sympathisers. The internet's interactivity and global reach have become powerful tools exploited by these entities for various nefarious purposes, such as propaganda dissemination, recruitment, training, planning, funding, and execution of attacks [73]. Within this context, the emergence of cyberterrorism poses a significant challenge, necessitating comprehensive understanding and proactive strategies to effectively counter it.

To provide a comprehensive overview, this chapter delved into the contemporary landscape of cyberterrorism, elucidating the impact of the COVID-19 pandemic on online terrorism dynamics. It clarified the rationale behind extremist groups' migration to digital platforms, a trend largely attributed to the perceived security and anonymity the internet affords in facilitating global outreach instantaneously. This level of interactivity and global accessibility underscores its potent capacity, rendering it an inherently hazardous and influential medium for such groups [18]. Within this context, an emphasis was placed on the significance of national security in addressing this issue, encapsulating the essence of terrorists' exploitation objectives and governments' protective aspirations. Subsequently, the narrative extended to a review of Weimann's works (2004, 2006), elucidating the diverse ways through which terrorists exploit the internet within their operational paradigm. This multifaceted role encompasses activities spanning from propaganda dissemination, recruitment, training, and planning, to securing funding and executing attacks. This breadth of utility underscores the internet's versatile and integral role in facilitating the modus operandi of such groups (Ibid.). Subsequent sections examined the characteristics of cyber-attacks, meticulously analysing prevalent and infrequent attack vectors, along with their targeted vulnerabilities. By scrutinising the underlying mechanisms and objectives of various attack strategies, the chapter also illuminated the multifaceted nature of cyber-terrorism operations, offering a comprehensive exploration of the digital battleground.

Finally, as the threat of cyberterrorism continues to evolve, staying abreast of the latest developments is critical. By illuminating the intricate interplay between terrorists and the internet, this chapter can assist policymakers, researchers, and security professionals in devising proactive strategies to combat cyberterrorism more effectively.

References

1. Ackerman G, Peterson H (2020) Terrorism and COVID-19: actual and potential impacts. Perspect Terrorism 14(3):59–73
2. Allen G, Chan T (2017) Artificial intelligence and national security. Belfer Center for Science and International Affairs, Harvard Kennedy School
3. Aly A, Macdonald S, Jarvis L, Chen T (eds) (2016) Violent extremism online new perspectives on terrorism and the internet. Routledge
4. Awan I (2017) Cyber-extremism: ISIS and the power of social media. Society 54(2):138–149
5. Bandura A (2001) Social cognitive theory of mass communication. Media Psychol 3(3):256–299
6. Baraz A, Montasari R (2023) Law enforcement and the policing of cyberspace. In: Montasari R, Carpenter V, Masys AJ (eds) Digital transformation in policing: the promise, perils and solutions. Springer, Cham, pp 59–83
7. Bencsáth B, Pék G, Buttyán L, Félegyházi M (2012) The cousins of Stuxnet: Duqu, flame, and Gauss. Future Internet 4:971–1003. https://doi.org/10.3390/fi4040971
8. Battaglini M (2020) How the main legal and ethical issues in machine learning arose and evolved. Technology and Society. Available at: https://www.transparentinternet.com/techno logy-and-society/machine-learning-issues/. Accessed: 27 Sept 2023
9. Brickey J (2012) Defining cyberterrorism: capturing a broad range of activities in cyberspace. Combating Terrorism Center West Point 5(8):4–7
10. Caddell JW (2004) Deception 101—primer on deception. Strategic Studies Institute, US Army War College. Available at: http://www.jstor.org/stable/resrep11327. Accessed: 08 Sept 2023
11. Ciampa M (2009) Security+ guide to network security fundamentals, 3rd edn. Cengage Learning
12. Clark DD, Landau S (2010) Untangling attribution. In: Proceedings of a workshop on deterring cyberattacks: informing strategies and developing options for U.S. policy, pp 25–40
13. Coleman PT, Bartoli A (2003) Addressing extremism. The International Center for Cooperation and Conflict Resolution (ICCCR). White paper. Available at: https://fpamed.com/wp-content/ uploads/2015/12/WhitePaper-on-Extremism.pdf. Accessed: 27 Sept 2023
14. Cordesman AH (2001) Cyber-threats, information warfare, and critical infrastructure protection: defending the U.S. Homeland. Bloomsbury Publishing, USA
15. CTED (2021) Update on the impact of the COVID-19 pandemic on terrorism, counter-terrorism and countering violent extremism. United Nations Security Council, Counter-Terrorism Committee Executive Directorate (CTED). Available at: https://www.un.org/securitycouncil/ ctc/sites/www.un.org.securitycouncil.ctc/files/files/documents/2021/Jun/cted_covid_paper_1 5june2021_1.pdf. Accessed: 27 Sept 2023
16. Cybersecurity and Infrastructure Security Agency (CISA) (2022) Russian state-sponsored and criminal cyber threats to critical infrastructure. America's Cuber Defense Agency. Available at: https://www.cisa.gov/news-events/cybersecurity-advisories/aa22-110a. Accessed: 08 Sept 2023
17. Denning D (2000) Cyberterrorism: testimony before the special oversight panel on terrorism. Committee on Armed Services, U.S. House of Representatives
18. Denning D (2001) Activism, hacktivism, and cyberterrorism: the internet as a tool for influencing foreign policy. In: Arquilla J, Ronfeldt D (eds) Networks and netwars: the future of terror, crime and militancy. RAND Corporation, pp 239–288
19. Dillon L (2019) Cyberterrorism: using the internet as a weapon of destruction. In: National security: breakthroughs in research and practice. IGI Global, pp 206–230
20. European Commission (2005) Terrorist recruitment: addressing the factors contributing to violent radicalisation, COM/2005/0313 final. Available at: https://eur-lex.europa.eu/legal-con tent/EN/TXT/PDF/?uri=CELEX:52005DC0313. Accessed: 07 Sept 2023
21. Freiburger T, Crane JS (2008) A systematic examination of terrorist use of the internet. Int J Cyber Criminol 2(1):309–319

22. Ghafir I, Hammoudeh M, Prenosil V, Han L, Hegarty R, Rabie K, Aparicio-Navarro FJ (2018) Detection of advanced persistent threat using machine-learning correlation analysis. Futur Gener Comput Syst 89:349–359
23. Gill P, Corner E, Conway M, Thornton A, Bloom M, Horgan J (2017) Terrorist use of the internet by the numbers. Criminol Public Policy 16(1):99–117
24. Gordon S, Ford R (2002) Cyberterrorism? Comput Secur 21(7):636–647
25. Grabosky P (2015) Organized cybercrime and national security. In: Cybercrime risks and responses. Palgrave Macmillan, London, pp 67–80
26. Hansen J, Lowry PB, Meservy R, McDonald D (2007) Genetic programming for prevention of cyberterrorism through dynamic and evolving intrusion detection. Decis Support Syst 43(4):1362–1374
27. Heath-Kelly C (2017) The geography of pre-criminal space: epidemiological imaginations of radicalisation risk in the UK prevent strategy, 2007–2017. Crit Stud Terrorism 10(2):297–319
28. Hua J, Bapna S (2012) How can we deter cyber terrorism? Inf Secur J Glob Perspect 21(2):102–114
29. Huamani EL, Alicia AM, Roman-Gonzalez A (2020) Machine learning techniques to visualize and predict terrorist attacks worldwide using the global terrorism database. Int J Adv Comput Sci Appl 11(4):526–570
30. Jarvis L, Macdonald S (2015) What is cyberterrorism? Findings from a survey of researchers. Terrorism Polit Violence 27(4):657–678
31. Johansson F, Kaati L, Sahlgren M (2016) Detecting linguistic markers of violent extremism in online environments. In: Artificial intelligence: concepts, methodologies, tools, and applications. IGI Global, pp 2847–2863
32. Kohlmann EF (2006) The real online terrorist threat. Foreign Aff 85(5):115–124
33. LaFree G (2017) Terrorism and the internet. Criminol Public Policy 16(1):93–98
34. Lau F, Rubin SH, Smith MH, Trajkovic L (2000) Distributed denial of service attacks. In: IEEE international conference on systems, man and cybernetics, Nashville, TN, USA, pp 2275–2280
35. Lewis JA (2002) A assessing the risks of cyber terrorism, cyber war and other cyber threats. Center for Strategic & International Studies, Washington, D.C., pp 1–12
36. Levite AE, Lee J (2022) Attribution and characterization of cyber attacks. Carnegie Endowment for International Peace. Available at: https://carnegieendowment.org/2022/03/28/attribution-and-characterization-of-cyber-attacks-pub-86698. Accessed: 08 Sept 2023
37. Libicki MC (1995) What is information warfare? Center for Advanced Concepts and Technology, Institute for National Strategic Studies, National Defense University
38. Lubin A (2018) Cyber law and espionage law as communicating vessels. In: 10th International conference on cyber conflict (CyCon), Tallinn, Estonia, pp 203–226
39. Luknar I (2020) Cyber terrorism threat and the pandemic. In: International scientific conference the Euro-Atlantic values in the Balkan countries, pp 29–38
40. Magouirk J, Atran S, Sageman M (2008) Connecting terrorist networks. Stud Conflict Terrorism 31(1):1–16
41. Maras MH (2016) Cybercriminology. Oxford University Press
42. Marsili M (2019) The war on cyberterrorism. Democr Secur 15(2):172–199
43. Moggridge E, Montasari R (2023) A critical analysis of the dark web challenges to digital policing. In: Montasari R (eds) Artificial intelligence and national security. Springer, Cham, pp 157–167
44. Molander RC, Riddile AS, Wilson PA (1996) Strategic information warfare: a new face of war. RAND Corporation, Santa Monica, California, USA
45. Montasari R (2023a) Countering cyberterrorism: the confluence of artificial intelligence, cyber forensics and digital policing in US and UK national cybersecurity. Springer Nature
46. Montasari R (2023b) National artificial intelligence strategies: a comparison of the UK, EU and US approaches with those adopted by state adversaries. In: Countering cyberterrorism: the confluence of artificial intelligence, cyber forensics and digital policing in US and UK national cybersecurity. Springer, Cham, pp 139–164

47. Montasari R, Carpenter V, Masys AJ (eds) (2023c) Digital transformation in policing: the promise, perils and solutions. Springer Nature
48. Montasari R (2023d) The potential impacts of the national security uses of big data predictive analytics on human rights. In: Countering cyberterrorism: the confluence of artificial intelligence, cyber forensics and digital policing in US and UK national cybersecurity. Springer, Cham, pp 115–137
49. Montasari R (2023e) The application of big data predictive analytics and surveillance technologies in the field of policing. In: Countering cyberterrorism: the confluence of artificial intelligence, cyber forensics and digital policing in US and UK national cybersecurity. Springer, Cham, pp 81–137
50. Montasari R (2023f) Cyber threats and the security risks they pose to national security: an assessment of cybersecurity policy in the United Kingdom. In: Countering cyberterrorism: the confluence of artificial intelligence, cyber forensics and digital policing in US and UK national cybersecurity. Springer, Cham, pp 7–25
51. Montasari R, Boon A (2023) An analysis of the dark web challenges to digital policing. In: Jahankhani H (eds) Cybersecurity in the age of smart societies. Advanced sciences and technologies for security applications. Springer, Cham, pp 371–383
52. McIntyre MW, Montasari R (2023) The dark web and digital policing. In: Montasari R (eds) Artificial intelligence and national security. Springer, Cham, pp 193–203
53. Montasari R (ed) (2022) Artificial intelligence and national security. Springer Nature
54. Montasari R, Jahankhani H (eds) (2021) Artificial intelligence in cyber security: impact and implications: security challenges, technical and ethical issues, forensic investigative challenges. Springer Nature
55. Mukaram A (2014) Cyber threat landscape: basic overview and attack methods. Recorded Future. Available at: https://www.recordedfuture.com/cyber-threat-landscape-basics. Accessed: 27 Sept 2023
56. Neely P, Allen M (2018) Policing cyber terrorism. J Cybersecur Res 3(1):13–19
57. Nouh M, Nurse JRC, Goldsmith M (2019) Understanding the radical mind: identifying signals to detect extremist content on Twitter. In: IEEE international conference on intelligence and security informatics (ISI), pp 98–103
58. Onursal R, Kirkpatrick D (2021) Is extremism the 'New' terrorism? The convergence of 'Extremism' and 'Terrorism' in British parliamentary discourse. Terrorism Polit Violence 33(5):1094–1116
59. Palmer D, Whelan C (2006) Counter-terrorism across the policing continuum. Police Pract Res 7(5):449–465
60. Pelzer R (2018) Policing of terrorism using data from social media. Eur J Secur Res 3:163–179
61. Rae JA (2012) Will it ever be possible to profile the terrorist? J Terrorism Res 3(2):64–74
62. Rowe M, Saif H (2016) Mining pro-ISIS radicalisation signals from social media users. In: Proceedings of the international AAAI conference on web and social media, vol 10(1), pp 329–338
63. Scanlon JR, Gerber MS (2014) Automatic detection of cyber-recruitment by violent extremists. Secur Inf 3(1):1–10
64. Schmid AP (2013) Radicalisation, de-radicalisation, counter-radicalisation: a conceptual discussion and literature review. International Centre for Counter-Terrorism. ICCT research paper, vol 97(1), p 22
65. Schmitt MN (ed) (2013) Tallinn manual on the international law applicable to cyber warfare. Cambridge University Press
66. Sood AK, Enbody RJ (2013) Targeted cyberattacks: a superset of advanced persistent threats. IEEE Secur Priv 11(1):54–61
67. The Ministry of the Interior and Kingdom Relations (BZK) (2004) From Dawa to Jihad: the various threats from radical Islam to the democratic legal order. Available at: https://irp.fas.org/world/netherlands/dawa.pdf. Accessed: 07 Sept 2023
68. United Nations Office on Drugs and Crime (UNODC) (2012) The use of the internet for terrorist purposes. United Nations. Available at: https://www.unodc.org/documents/frontpage/Use_of_Internet_for_Terrorist_Purposes.pdf. Accessed: 27 Sept 2023

69. Valeri L, Knights M (2000) Affecting trust: terrorism, internet and offensive information warfare. Terrorism Polit Violence 12(1):15–36
70. Veerasamy N, Grobler M, Von Solms B (2012) Building an ontology for cyberterrorism. In: Proceedings of the 11th European conference on information warfare and security, Laval, France, pp 286–295
71. Verhelst HM, Stannat AW, Mecacci G (2020) Machine learning against terrorism: how big data collection and analysis influences the privacy-security dilemma. Sci Eng Ethics 26:2975–2984
72. Verton D (2003) Black ice: the invisible threat of cyber-terrorism. McGraw-Hill Education
73. Weimann G (2004) Cyberterrorism: how real is the threat? United States Institute of Peace. Special report 119. Available at: https://www.usip.org/sites/default/files/sr119.pdf. Accessed: 27 Sept 2023
74. Weimann G (2006) Terror on the internet: the new arena, the new challenges. United States Institute of Peace Press, Washington D.C.
75. Weimann G (2015) Terrorism in cyberspace: the next generation. Columbia University Press
76. Yu S, Carroll F (2021) Implication of AI in national security: understanding the security issues and ethical challenges. In: Montasari R, Jahankhani H (eds) Artificial intelligence in cyber security: impact and implications. Springer, Cham, pp 157–175
77. Yunos B (2009) Putting cyber terrorism into context. Available at: https://www.cybersecurity.my/data/content_files/13/526.pdf. Accessed: 27 Sept 2023

Chapter 6
Exploring the Imminence of Cyberterrorism Threat to National Security

Abstract The rapid advancement of technology has ushered in significant progress across various domains of human life but has also introduced a spectrum of complex challenges. While technological innovation is widely regarded as a positive stride towards the future, safeguarding cyberspace against formidable adversaries such as cyberterrorists and hackers remains an intricate and yet-to-be-achieved undertaking. This chapter examines cyberterrorism and its potential implications as a new form of terrorism, considering the evolving landscape of technology. To this end, the chapter aims to assess the realistic nature of the cyberterrorism threat to national security, clarifying whether the prevailing perception of imminent danger might be inflated due to limited understanding. Through a comprehensive analysis of existing literature and expert insights, this research illuminates the complex relationship between technological progress, vulnerabilities, and the perceived cyberterrorism threat, providing a more informed perspective on its implications for national security.

Keywords Cyberterrorism · Cyber-attack · Critical infrastructure · Technology

6.1 Introduction

Cyberterrorism constitutes a recent phenomenon that amalgamates two fundamental components: the tactical strategies of terrorism and their execution through cyber-attacks. In 1990, the National Academy of Sciences articulated the notion that forthcoming terrorists might possess the capability to inflict greater harm through the manipulation of digital interfaces than through conventional explosive devices [1]. Given the progression into the digital age over the three decades since the aforementioned assertion, the phenomenon of cyberterrorism has emerged as a disconcerting actuality warranting proactive readiness. Nonetheless, deliberations have arisen regarding the immediacy of the cyberterrorist threat and the extent to which it embodies the succeeding tier of terrorist methods, juxtaposed with the possibility of its magnification due to fearmongering tactics and its enigmatic characteristics. This research endeavours to elucidate these concerns and construct a well-rounded

R. Montasari, *Cyberspace, Cyberterrorism and the International Security in the Fourth Industrial Revolution*, Advanced Sciences and Technologies for Security Applications, https://doi.org/10.1007/978-3-031-50454-9_6

discourse that underscores the significance of prudent preparedness for potential cyberterrorist attacks, while concurrently avoiding an undue amplification of the matter that could evoke fear within the general populace—an outcome that could inadvertently align with the strategic objectives of terrorist entities.

To this end, this study will delve into the grounds necessitating vigilance and apprehension regarding the potentiality of a cyberterrorist attack. As discussed in the previous chapters, terrorists exhibit adeptness in leveraging contemporary technologies, particularly through the use of social media platforms (SMPs), to facilitate their recruitment campaigns and the widespread dissemination of propagandistic content. Nevertheless, a pertinent question arises as to whether this technical adeptness extend to the orchestration and execution of complex cyber operations with the potential to inflict catastrophic consequences. Furthermore, it is noteworthy that terrorist organisations have issued explicit threats regarding the prospect of executing cyberterrorist attacks [14]. Similarly, in light of the ongoing digital evolution that characterises our contemporary existence—where technology profoundly permeates our daily lives—terrorist entities would endeavour to remain abreast of technological advancements and potentially exploit them for malicious purposes, thereby maximising the scope of destruction they can effectuate. To support this perspective, Conway [2] postulates that cyberterrorism serves to merge two prevailing contemporary fears: the trepidation associated with technological advancements and the apprehension stemming from terrorist activities. This fusion of concerns underscores the possibility that fear could engender an exaggerated perception of the severity of a given issue, surpassing its actual implications. Furthermore, the media often exhibit a propensity to inflate the significance of the cyberterrorism issue. They occasionally resort to sensationalist strategies to invoke fear among their audience, ultimately serving the purpose of boosting viewership and engagement.

This inclination towards amplification may, in part, arise from the definitional ambiguities surrounding the term 'cyberterrorism'. Numerous interpretations extend beyond a specific, tightly focused delineation, as advanced by Weimann [14], a subject that will be further explored in subsequent sections of this study. Furthermore, it becomes imperative to assess the operational capabilities of terrorist organisations. While the basic requirements for engaging in cyberterrorist activities are limited to computer access and an internet connection, acquiring advanced technical skills remains a formidable challenge. This holds particularly true when considering the intricacies involved in targeting complex operating systems. It is essential to emphasise that acquiring such proficiency is not devoid of considerable financial costs, a factor that may pose a substantial impediment to many organisations seeking to secure the necessary resources. Moreover, it is essential to consider the inherent tendency of terrorists to adhere to well-established methods, often gravitating toward tried and proven techniques. Therefore, if a novel approach introduces a significant degree of uncertainty, it is conceivable that they might choose conventional methods, ensuring a more assured means of causing maximum human casualties. However, a notable observation emerges from the absence of any documented instance of cyberterrorism hitherto. Therefore, this observation warrants careful analysis, and this study aims to identify and elucidate the diverse factors contributing to this absence. Nonetheless, it is crucial to acknowledge that the absence of historical incidents in the realm of cyberterrorism should not lead to complacency or a disregard for preparatory measures. The increasing prevalence of hacking-related cybercrimes highlights the potential inherent in this evolving digital landscape, emphasising the need to carefully anticipate and strategise for the possibility of cyberterrorism.

The subsequent sections of this chapter are structured as follows. Section 6.2 offers essential background information. Section 6.3 delves into the reasons for concern regarding cyberterrorism. In this regard, it examines the current applications and potentialities of technology employed by cyberterrorists and elucidate why cyberterrorism might hold greater appeal for these actors compared to conventional terrorism. Conversely, Sect. 6.4 analyses the rationales behind not being unduly concerned about cyberterrorism. Section 6.5 discusses the findings, and finally, the chapter is concluded in Sect. 6.6.

6.2 Background

In the wake of the September 11, 2001 attacks, Osama Bin Laden notably issued a threat pertaining to the potentiality of a cyberterrorist attack, asserting support from a contingent of "hundreds of Muslim scientists" [14, p. 11]. The disclosure of such intentions evoked a heightened sense of apprehension, prompting questions into the conceivable misuse of the relatively nascent phenomenon of the internet for malevolent purposes, including the prospect of orchestrating mass casualties. The realisation of such widespread devastation would be consistent with Al-Qaeda's overarching objectives of inflicting maximum fatalities and fomenting widespread fear among populations. It is pertinent to note that these efforts could even yield the inadvertent consequence of dissuading individuals from embracing technological advancements, potentially delaying societal progress within the digital realm, a sphere where rapid acceleration is increasingly indispensable in contemporary society. Weimann posits that a somewhat paradoxical outcome of achieving success in the broader 'war on terror' might entail an increased inclination of terrorists towards the adoption of unconventional instruments, such as cyberterrorism, as a strategic recourse (2006). This perspective underscores the adaptability and determination of terrorist organisations to sustain their mission in the face of formidable challenges. Nonetheless, the trajectory of progress since that time has witnessed an exponential surge in technological advancements, along with an escalation in instances of cyber-attacks perpetrated by highly proficient criminal hackers. In light of this evolving landscape, many have raised concerns about an imminent cyberterrorist attack. However, despite Osama Bin Laden's aforementioned pronouncement, there has been no documented instance of a cyberterrorist attack. This circumstance raises questions about the credibility of the perceived threat and leads to reflections on its immediate plausibility, in contrast to the prevailing perception that has been constructed.

6.3 Reasons for Concern

6.3.1 Current Trends and Future Prospects

A primary consideration revolves around the contemporary tactics adopted by terrorist organisations in leveraging the internet as a strategic tool. ISIS, in particular, stands out for its comprehensive use of digital platforms. Within this paradigm, SMPs assume a central role, facilitating a range of activities including the dissemination of propaganda, financial transactions, training exercises, operational planning, execution of actions, and even orchestrating cyberattacks [11]. These multifaceted applications collectively contribute to the overarching objectives of recruitment, incitement, and radicalisation, domains in which ISIS particularly displays adeptness. The organisation has also extended its influence to encompass virtual entrepreneurs, who provide guidance to individuals interested in participating in terrorist activities, often using a professional and conventional script style [7]. In this evolving digital landscape, terrorist entities exploit the advantages of internet interconnectivity to their benefit, effectively adapting to the rapid pace of the cyber domain. For instance, the English-language 'Dabiq' magazine disseminated by ISIS through online channels serves the purpose of appealing to individuals who have undergone radicalisation or manifest susceptibility to such processes [5]. This instance underscores the absence of geographical confines facilitated by the internet, allowing global access to ISIS' ideology and guidance for those inclined. Additionally, in 2016, a faction of ISIS-affiliated hackers known as the CCA issued a compendium of targets comprising names, email addresses, and physical locations of 4000 individuals sourced from SMPs, accompanied by directives for pro-IS adherents to promptly execute lethal actions against them [6].

The aforementioned instances demonstrate how ISIS and similar terrorist entities exploit the capabilities of the internet, along with its associated advantages, to further their respective objectives. Nevertheless, it is important to note that these activities do not necessarily constitute manifestations of cyberterrorism. Instead, they exemplify a proficient utilisation of technology, effectively contributing to organisational progress in a manner that could be characterised as 'professional'. While the evolution of the internet and technological advancements might appear familiar due to their enduring presence, it is crucial to acknowledge the substantial strides made within a relatively short period. This situation prompts reflection about potential future trajectories in human endeavours in the coming decades. Consequently, questions arise regarding what future prospects may materialise.

In light of these considerations, the assertion that the internet already constitutes an eminent facet of contemporary society is undeniable. It has permeated nearly every aspect of daily life, from communication and commerce to education and entertainment. However, what is firmly established is the expectation that the internet will continue to evolve and exert an even more profound influence on the future. The corollary inference from this is that an ever-increasing proportion of societal activities will inevitably become contingent upon the internet, potentially surpassing the already extensive reach it has today. This ongoing integration of the digital realm into daily life not only amplifies its importance but also raises concerns about its vulnerabilities. The scenario of an increasingly interconnected world, where critical functions rely on digital networks, creates an opportune focal point for malicious exploitation by terrorist elements. Thus, as the dependence on the internet deepens, so does the potential impact of cyber threats. Terrorist organisations, as discussed earlier, have displayed adaptability and proficiency in exploiting digital platforms. In this context, it becomes crucial to address the evolving landscape of cyber threats and develop comprehensive strategies to protect against them. Ensuring the security and resilience of digital infrastructure is not only a matter of technological advancement but also a fundamental aspect of national and global security. The future trajectory of society is inextricably linked with the internet, and safeguarding it from malicious exploitation is a paramount challenge that must be confronted.

Furthermore, the contemplation of this emerging landscape instils a disconcerting verity: the next generation of terrorists will have matured within an exclusively digital milieu, wherein technology permeates diverse facets of their existence. It is theorised that a direct relationship exists between the degree of technological advancement a nation attains and its susceptibility to cyberattacks targeting its critical infrastructure [14]. As a consequence, nations such as the United States (US), the United Kingdom (UK), and various European countries, despite being pioneers in technological innovation, paradoxically emerge as prime targets for cyberterrorism. This juxtaposition stems from the reality that the very technological strides these nations foster also provide potential vulnerabilities. Similarly, the continuous emergence of novel tools, while contributing to technological progress, concurrently provide criminal elements, including those predisposed to cyberterrorism, with avenues to harness such tools for illicit ends. In response, an imperative exists to emphasise judicious innovation alongside the implementation of robust measures to curtail these risks, thereby forestalling the malevolent appropriation of such tools. Therefore, it is vital to navigate a delicate equilibrium between innovation and responsible usage, given the confluence of factors encompassing transparency concerns and the latent potential for constructive deployment. This balance is further complicated by the intricate interplay between the propensity for misuse and the ethical use of these tools for beneficial purposes. In an era where the impact of technology on society deepens with each advancement, the imperative for individuals, organisations, and policymakers is clear: to safeguard against the risks while harnessing the potential for a future where innovation serves as a force for positive change.

6.3.2 The Growing Appeal of Cyberterrorism

As the global trajectory rapidly advances into a digitally interconnected world, terrorists are finding a range of advantages within this burgeoning cyber landscape, both realised and anticipated. A noticeable advantage pertains to the elimination of conventional physical barriers that aspiring terrorists might have previously encountered as impediments. This dissolution eliminates the need for risky physical journeys, making it easier for individuals to become involved in terrorist activities. However, it also raises concerns about the potential for more frequent and sophisticated attacks. For example, joining a jihadist training facility in Syria, which was once perilous, is now facilitated by this shift in dynamics. This poses a significant security challenge that necessitates a comprehensive response. Equally disconcerting is the reduced mortality risk associated with cyberterrorism. In comparison to its traditional counterpart, cyberterrorism involves less physical proximity to the target, reducing the risk of casualties. This contrast between cyberterrorism and its conventional variant could conceivably render the former more appealing to individuals undergoing the radicalisation process. This appeal might particularly resonate with those disinclined towards self-sacrifice yet harbour an inclination to foment disorder and devastation. This complicates counterterrorism efforts, thereby disrupting traditional strategies that rely on identifying and neutralising physical threats.

In addition to the aforementioned considerations, the absence of a physical presence at the targeted site reduces the immediate risk of apprehension for cyberterrorists, affording them valuable time to effectuate escape. This attribute, in conjunction with the propensity for cyberterrorism to engender a diminished personal risk of fatality, underscores the modus operandi's potential to yield a significantly broader impact on the general populace. This capacity arises from the ability to carry out cyberattacks that are not territorially limited, thereby allowing for widespread devastation. This prospect is further amplified when coordinated attacks, combining physical and cyber methods, are orchestrated. To illustrate, a conventional terrorist act, reminiscent of the Manchester Arena bombings, could coincide with a simultaneous cyberterrorist effort aimed at disrupting telecommunications, emergency services, or healthcare systems such as the National Health Service (NHS). In such a joint attack, the incapacitation of vital emergency response mechanisms could result in increased loss of life due to delayed medical assistance. Furthermore, the complex technical composition of prospective targets increases the likelihood of exploitable vulnerabilities. Operating systems are inherently complex, rendering comprehensive defences a formidable challenge [14] as discussed in Chap. 5. These vulnerabilities can be exploited for destructive purposes, continuously challenging LEAs. Consequently, the ever-evolving nature of technology makes it difficult to stay ahead of potential threats, requiring ongoing investment in research, development, and training to effectively defend against cyberterrorism. Therefore, as the digital world becomes increasingly interconnected, it provides terrorists with distinct advantages that demand a comprehensive response from governments, LEAs, and organisations. Addressing the dissolution of physical barriers, the reduced mortality risk, diminished susceptibility to immediate apprehension, potential for widespread impact, and exploitable vulnerabilities requires a holistic and adaptable approach.

6.3.3 Potential and Possibilities

While instances of explicit cyberterrorism remain absent from the records, it is imperative to acknowledge that daily occurrences of cyber-attacks persist as a pressing concern. This poses a significant threat to societal well-being, especially in cases of inadequate or compromised cyber security measures. Among the prevalent choices in these malicious efforts are ransomware attacks, which hold the potential for substantial disruption. A prominent case in point is the WannaCry ransomware incident. In this instance, an individual employed ransomware to orchestrate a global network attack, encrypting essential files and coercing payment for the decryption key and file restoration. The ramifications, however, extended beyond the intended scope, profoundly impacting multiple computer networks within the NHS in the UK. Emergency medical departments across the nation were incapacitated due to this attack, an outcome that was not the primary objective but underscores the pervasive chaos that can ensue when confronted with compromised cyber infrastructure [4]. While the NHS was not deliberately singled out as the primary target, this incident serves as a poignant illustration of the potential for widespread turmoil, particularly in the event of a recurrence. The prospect of intentionally causing harm and destruction becomes even more disconcerting when considering the feasible realisation of simultaneous attacks, combining cyber-attacks with physical acts of terrorism as discussed in the previous section.

Despite terrorists effectively exploiting social media and various other platforms to propagate their ideologies, there currently exists no substantiated evidence indicating a notable degree of sophistication in their hacking capabilities, particularly with respect to orchestrating large-scale destructive cyber-attacks. However, the absence of overt proficiency does not discount the potential threat, as terrorists could conceivably circumvent their own limitations by enlisting the services of adept hackers-for-hire. In this context, the collaboration with cybercriminal entities becomes a tangible avenue, warranting consideration. It is pertinent to acknowledge that cybercriminals, owing to their criminal predisposition, can be motivated by financial incentives. Consequently, the potential for leveraging their skills and expertise for ulterior motives, including terrorist objectives, must not be dismissed. An illustrative precedent underscores this possibility. An experiment conducted by the National Security Agency (NSA) involved authorised ethical hackers who successfully penetrated several critical computer systems within the Pentagon, attesting to the vulnerability of high-security networks [12]. This example accentuates the plausibility of external actors, including cybercriminals, impacting critical infrastructures, underscoring the significance of pre-emptive measures against such collaborations. However, it is important to note that the scope of the aforementioned experiment was confined to the use of software and hacking tools available for public download from the internet, highlighting their accessibility to a broad spectrum of individuals. This constraint reflects the experiment's controlled nature, resulting in no actual destructive consequences. Nonetheless, this exercise vividly illustrates the potential consequences that might ensue if adept criminal hackers, acting on behalf of a terrorist organisation, manage to infiltrate crucial systems within the defence departments, thereby causing widespread havoc. This scenario raises compelling questions regarding the efficacy of cybersecurity measures in the private sector. If even a security institution of the stature of the Pentagon can exhibit vulnerabilities despite its heightened security stature, it is reasonable to deduce that similar vulnerabilities could exist within the private sector. An NSA official contended that the experiment not only highlighted flaws in the Pentagon's computer systems but also underscored the plausible capacity to inflict "strategic damage to the US 'money-supply'" [12]. This assertion poignantly illustrates the potential and far-reaching consequences associated with malicious hacking intent. In the absence of comprehensive cybersecurity measures to mitigate these vulnerabilities, the potential fallout could indeed be of a devastating nature.

In scenarios where targeting complex operating systems poses a considerable challenge even for proficient hackers, terrorist organisations might resort to alternative strategies. This could involve the recruitment of dissatisfied personnel from the targeted organisation, or potentially coercing employees through threats of dire consequences for non-compliance, including the spectre of death. Such an approach could pivot towards an 'inside' job, capitalising on the insider knowledge and access possessed by individuals already affiliated with the target institution. Furthermore, another avenue for exploitation involves hacktivists, a distinct group characterised by their engagement in cyber activities for political motives. It is important to recognise that, despite their shared underpinning of political motivations, terrorism and hacktivism are inherently distinct, with the latter not intrinsically driven by the intent to cause loss of life. Nevertheless, the potential for unscrupulous actors devoid of "moral restraint" to repurpose tactics developed by hackers to sow chaos remains a concerning prospect [14, p. 4]. This underscores the versatility of cyber methods, which individuals with malicious intentions could potentially employ for disruptive purposes beyond their original scope.

6.4 Reasons Not to Be Concerned

6.4.1 Lack of Capacity

It is crucial to underscore the fact that there exists no documented occurrence of a cyberterrorist attack executed by a conventional terrorist organisation [14]. However, it should be duly recognised that instances of cyber-attacks perpetrated by these same organisations have indeed materialised [6]. Furthermore, in the landscape of contemporary cyber activities, while the prevalence of cyber-attacks is palpable, their underlying objectives seldom align with the level of profound devastation intrinsic to the concept of terrorism. This progression of thought leads this study to a pivotal juncture: an investigation into the essential question of whether terrorist organisations possess

the requisite capability to orchestrate complex and potentially catastrophic cyberterrorist attacks. This pivotal question delves into the dynamic interplay between technical capacity and the intent intrinsic to acts of terrorism, illuminating the inherent
complexities within this domain of study. According to an article on CNet.com,
it was determined that executing a successful cyberterrorist attack would necessitate the involvement of a well-resourced syndicate, requiring a substantial financial
commitment of approximately $200 million, access to "country-level intelligence"
capabilities, and an extensive preparation period spanning 5 years [14, p. 10]. This
assessment highlights the formidable requisites, encompassing financial backing and
meticulous planning, which such an attempt demands. Additionally, the effort would
involve a profound understanding of complex operating systems, a domain known
to be mastered by a select few.

Furthermore, it is prudent to consider whether a terrorist organisation would find it
strategically advantageous to allocate such considerable resources and time towards
a solitary attack, particularly considering the inherent risks associated with increased
visibility and the potential for apprehension. Pertinently, an investigation conducted
by US troops in Afghanistan yielded insights from seized Al-Qaeda laptops, leading
to the conclusion that no concrete evidence of a cyberterrorist plot was present within
the group. Rather, it was observed that the internet was predominantly employed
by Al-Qaeda for communication and coordination of physical assaults [14, p. 9].
While this evidence suggests a lack of imminent intent, the temporal context of
this discovery, nearly two decades ago, should be duly acknowledged. Even in the
hypothetical scenario where a terrorist organisation were to engage a hacker for the
purpose of infiltrating intricate operating systems, the compatibility of motivations
between hackers and terrorists remains uncertain. The driving forces underlying
hackers' activities and the motives driving terrorists fundamentally diverge. Hackers
are motivated by a wide array of factors, encompassing showcasing their prowess,
data acquisition, or orchestrating attacks for financial gain. These motivations are
inherently disparate from the core objectives of terrorism, which revolve around
instilling fear and chaos through the disruption of critical infrastructures and the mass
targeting of civilians [8]. The leap from hacking for monetary rewards or personal
status to committing acts of terrorism involving the potential loss of life and profound
societal impact is a considerable one. It is plausible that moral constraints would
intervene, particularly for individuals primarily motivated by financial incentives or
the allure of notoriety. Additionally, as noted by Conway, while hackers might be
inclined towards elaborate mischief and disruptive actions, this does not necessarily
translate to a willingness to endanger lives or perpetrate violence in the name of
a political cause (2014). This crucial distinction between the objectives and moral
boundaries of hackers and terrorists serves to temper the alignment of their interests.

Furthermore, the execution of a cyberterrorist attack, inherently involving the
substantial "disruption of information infrastructure", contradicts the "self-interest"
typically associated with hackers [14, p. 10]. This disagreement underscores a
misalignment of objectives, as hackers, driven by their own motives, would likely
be disinclined to engage in actions resulting in far-reaching chaos and detriment
to the very digital landscapes they navigate. Additionally, the majority of hackers

lack the requisite skills and expertise needed to orchestrate cyberterrorism [14]. Even if prospective cyberterrorists were to succeed in enlisting hackers willing to contribute their expertise, the essential technical acumen necessary for such a complex undertaking might still be elusive. This accentuates the multifaceted challenge of assembling a cooperative unit capable of effectively executing a cyberterrorist attack. Consequently, given the inherent disparities in motivations, objectives, and technical proficiencies, the collaborative synergy between hackers and terrorists appears unlikely in the context of mounting a complex and time-intensive cyberterrorist attack. Moreover, an additional facet worth considering pertains to employees within critical infrastructure domains, including those responsible for power grids. These professionals are accustomed to managing challenges arising from natural disasters and possess contingency plans tailored to mitigate the consequences of substantial disruptions. Insights garnered from addressing natural disaster scenarios could potentially be adapted and applied to counteract the impact of cyberterrorist attacks, enhancing preparedness and response mechanisms [14, p. 10]. Furthermore, it is noteworthy that a substantial portion of critical infrastructure, such as air traffic control systems and agencies such as the FBI, are maintained as air-gapped environments. This implies that these systems remain disconnected from the internet, thereby precluding external hacking attempts [14, p. 10]. This precautionary measure substantially heightens the challenge posed to potential cyberterrorists seeking to infiltrate such systems. As a result, the necessity for an insider element to disrupt operations becomes a pertinent consideration, aligning with previous discussions that highlighted the practical obstacles associated with this approach. While the prospect of insider involvement should not be dismissed, it remains a scenario of heightened difficulty and limited probability, requiring a meticulous evaluation of potential vulnerabilities.

6.4.2 Definitional Ambiguity

In the process of delving into the subject of cyberterrorism, one encounters a plethora of definitional uncertainties. These ambiguities engender challenges in understanding the precise focus of policy makers and intelligence agencies' defensive strategies. Moreover, they give rise to concerns within the general public, as the human tendency to fear the unfamiliar is exacerbated when confronted with concepts that remain largely uncharted, a sentiment particularly applicable to both the domains of terrorism and technology [10]. This quandary extends to the realm of defining cyberterrorism itself. Despite dedicated efforts to formulate a comprehensive and universally agreeable definition, the prevailing lack of consensus remains a pervasive challenge. This lack of accord not only complicates the task of securing unanimous acceptance within a single nation but also extends its ramifications across the complex global landscape. A parallel discord is evident in the field of terrorism, where an absence of a universally embraced definition persists even among United Nations (UN) member states [9]. This disarray in conceptual clarity poses significant challenges for establishing effective policies and countermeasures, further exacerbating the complexities inherent to addressing the evolving threat landscape.

The historical lineage of traditional terrorism far precedes that of cyberterrorism, contributing to the complexities inherent in achieving a comprehensive and widely accepted definition for either phenomenon. Given the complex nature of these concepts and the multifaceted perspectives from which they are studied, the prospects for establishing universally embraced definitions in the foreseeable future appear challenging. Compounding this challenge is the interchangeable use of terms such as 'terrorist', 'violent extremism', 'extremists', and others, resulting in linguistic ambiguity that further complicates the efforts of arriving at precise definitions. Notably, the term 'terrorist' frequently assumes varying connotations depending on the context and perspective, leading to divergent interpretations. Denning sought to encapsulate cyberterrorism by describing it as the intersection of cyberspace and terrorism [3]. However, this attempt to define cyberterrorism remains constrained by the absence of an unequivocal definition of terrorism itself [13]. This circularity underscores the intricacies inherent in attempting to delineate these evolving and multifaceted concepts, further accentuating the challenges associated with conceptual clarity in this domain.

Indeed, the existence of divergent perspectives comprising narrow and broad definitions of cyberterrorism precipitates a lack of consensus regarding the requisite criteria for classifying an occurrence as an instance of cyberterrorism. This definitional ambiguity engenders a climate where the boundaries of cyberterrorism become blurred, resulting in the misclassification of certain cyber-attacks, including those attributed to hacktivists. Hacktivists, while employing cyber-attacks as a means to promote or protest particular political ideologies, fundamentally lack the violent and lethally motivated disposition characteristic of cyberterrorists. This distinction underscores a clear contrast between the intentions of hacktivists and those of cyberterrorists. The frequent incidence of cyber-attacks often leads to a context in which the broader significance of cyberterrorism might be unintentionally inflated. This conflation of distinct phenomena, such as hacktivism, with cyberterrorism can potentially cultivate an undue sense of importance and apprehension surrounding the latter. Consequently, the complexities of defining and distinguishing cyberterrorism, combined with the regularity of cyber incidents, can paradoxically magnify its perceived significance and evoke undue concern within society.

6.4.3 Media Sensationalism

The role of the media and its interaction with definitional ambiguities also warrants consideration, particularly with regard to the impact on information sources. The media exerts substantial influence over public perception. When it disseminates accurate and factually sound information, it serves as a valuable informative tool.

However, there is a notable risk associated with the media's inclination toward sensationalism, which can lead to misinformed narratives. In their pursuit of maximising reader engagement and impact, the media might inadvertently mislabel certain cyber incidents as instances of cyberterrorism. This misclassification can magnify public fears and concerns about events that do not actually fall under the purview of cyberterrorism. This underscores the need for a responsible and judicious media narrative when addressing these nuanced subjects.

As emphasized by Weimann [14], there is a significant distinction between the potential damage cyberterrorists could hypothetically inflict and the actual realisation of such harm. Mere possibility, as Weimann argues, does not necessarily translate into actual occurrence, particularly given the complexities associated with executing cyberterrorist attacks. However, the media often neglects this nuanced perspective and fails to account for these inherent limitations. Instead, media outlets might inadvertently contribute to unnecessary fear and apprehension within the public sphere, succumbing to the temptation of sensationalism. This pattern of media behavior poses multiple problems. For instance, it inadvertently aligns with the aims of terrorist organisations that seek to sow fear and disorder. Additionally, the psychological impact of cyberterrorism, devoid of immediate and direct violence, can be as potent as the threat posed by traditional terrorist attacks [14]. This highlights the intricate interplay between fear, perception, and the potential to exacerbate anxiety, even in the absence of immediate physical harm.

Consequently, the media emerges as a pivotal tool employed by terrorist organisations to propagate their narratives. This necessitates a careful consideration by media outlets when reporting on such issues, as their dissemination of information inadvertently contributes to a form of propaganda that aligns with the goals of these organisations. The media's role extends beyond mere reporting; it functions as a channel through which these narratives are amplified. Moreover, the portrayal of cyberterrorism in various forms of media, including novels and movies such as Bond's 'Goldeneye', illustrates how once far-fetched scenarios are progressively becoming plausible due to technological advancements [14]. As technology evolves, the transformation of these once otherworldly notions into feasible threats has the potential to evoke a sense of panic, driven by the human propensity to fear the unknown (Pollitt, 2018). Adding to the complexity, Weimann [14] contends that policy makers often place undue reliance on the media, potentially crafting policies based on misinformation rather than seeking insights from academic experts [2]. This tenuous relationship between policy making and media influences exacerbates the challenge of discerning credible sources of information and fostering trust. The resulting mixed messages can inadvertently amplify unwarranted anxiety and apprehension surrounding the issue of cyberterrorism.

6.5 Discussion

The assertion of whether terrorists could inflict "more damage with a keyboard than a bomb" remains a matter yet to be definitively ascertained [1, p. 117]. Nonetheless, the rapid acceleration of the technological revolution precludes the dismissal of the potentiality of cyberterrorism. Considering this uncertainty, it is imperative that policy makers, alongside potential targets such as critical infrastructure, undertake proactive measures to prepare for and address the potential repercussions of a potentially devastating cyberterrorist attack. In the contemporary landscape, terrorists adeptly harness the capabilities of the internet, leveraging SMPs and capitalising on the inherent interconnectivity it affords. This digital realm enables them to disseminate their ideologies, coordinate activities, and exploit vulnerabilities, underscoring the multifaceted dimensions through which modern terrorist actors operate. Consequently, fostering comprehensive preparedness strategies becomes paramount to navigating this evolving threat landscape and minimising the potential fallout of a cyberterrorist event.

Instances of cyberattacks have demonstrated how malicious actors can exploit publicly available tools for nefarious purposes. While some of these attacks might have limited immediate impact, it remains conceivable that with the cultivation of hacking skills and knowledge, these incidents could escalate significantly. This emerging generation of potential terrorists is poised to mature within an exclusively digital environment, in stark contrast to previous generations that embraced technology through a process of adoption and learning. This paradigm shift presents a palpable prospect: that these upcoming digital natives might inherently possess the skill set and technical acumen required to perpetrate acts of cyberterrorism. This evolving scenario reflects a dual-edged reality: the democratisation of technical expertise empowers individuals to harness technological tools but concurrently raises concerns about the misuse of these capabilities for destructive ends. This unique context underscores the urgency of cultivating a comprehensive understanding of the digital landscape from an ethical and security standpoint. Proactive efforts to instil responsible digital literacy, coupled with measures to curtail the malicious exploitation of technology, are pivotal in averting the realisation of this potential and safeguarding against the emergence of technologically enabled threats.

The question of whether terrorist organisations possess the capacity to execute cyberterrorism remains enveloped in uncertainty, as previously examined. Current evidence predominantly suggests that such organisations often lack the requisite funding, technical proficiency, and sustained presence necessary to orchestrate cyberterrorist acts. However, this assessment does not preclude the potential for their capabilities to evolve over time. While existing limitations might restrain some terrorist organisations from immediately engaging in cyberterrorism, the appeal of its benefits and potential to generate considerable havoc are not overlooked by them.

The intrinsic appeal of cyberterrorism as an innovative avenue for terrorist activities necessitates strategic adaptation to acquire the necessary resources and technical expertise. This could entail the exploration of novel methods to augment their capabilities.

Presently, the threat posed by cyberterrorist attacks might not be deemed overwhelmingly significant, despite potential disparities in how media outlets portray the situation. Indeed, a substantial factor contributing to the present state of cyberterrorism discourse is the lack of a universally accepted understanding of its scope and implications. Thus, collaborative efforts between policy makers and academics to formulate a comprehensive and comprehensible definition could serve as a pivotal step toward rectifying this challenge. A clearly delineated definition could significantly reduce the likelihood of misclassifying cyber-attacks or hacktivist activities as cyberterrorism, which often leads to unnecessary fear fuelled by misinformation. The aftermath of the 9/11 tragedy marked a period of heightened emotions and reactions, culminating in the "war on terror". This fervour might have contributed to a climate of unwarranted anxiety. However, two decades have passed since those events, prompting a reflection on the absence of significant cyberterrorist incidents. Osama Bin Laden's threats in the aftermath of 9/11 underscore the potential for such attacks, raising questions about the reasons for their non-occurrence. Capacity limitations, alongside dedicated cybersecurity measures employed by entities responsible for safeguarding critical infrastructure, are pivotal factors that could be inhibiting cyberterrorist endeavours. This dynamic interplay between evolving capacities, protective measures, and the evolving threat landscape necessitates ongoing assessment and adaptation to address the multifaceted challenges posed by cyberterrorism more effectively.

6.6 Conclusion

The rapid advancement of technology has ushered in significant progress across various domains of human life. However, this transformative wave has also presented a spectrum of complex challenges. While technological innovation is widely regarded as a positive stride towards the future, safeguarding the realm of cyberspace against formidable adversaries such as cyberterrorists and hackers remains an intricate and yet-to-be-achieved undertaking. This research sought to navigate the intricate landscape of cyberterrorism, considering its multifaceted dimensions, potential threats, and perceptual nuances. The chapter delved into the challenges posed by the evolving nature of technology and the potential for adversaries, including cyberterrorists and hackers, to exploit these advancements. The chapter underscored the importance of a nuanced understanding of the cyberterrorism threat, arguing that despite the absence of a confirmed cyberterrorism incident, it was imperative to assess the realistic nature of this threat rather than succumb to sensationalism. Furthermore, throughout this chapter, a comprehensive analysis of existing literature, trends, and expert insights were conducted to illuminate the intricate interplay between technological progress, vulnerabilities, and the perceived threat of cyberterrorism. This, in turn, highlighted that by adopting a nuanced perspective, resources could be better allocated, and effective strategies could be developed to counter potential cyberterrorism activities more effectively.

As societies move forward into an increasingly technology-driven future, it is clear that the challenges in cyberspace will continue to evolve. Armed with a deeper understanding of these challenges and a commitment to proactive measures, nations can better prepare themselves to protect their interests and populations from potential cyber threats, ensuring a more secure and resilient future in the digital age. In light of this, it is imperative for national security entities to acknowledge the genuine risks posed by cyberterrorism and the severe ramifications that could unfold if such an attack were to occur. However, simultaneously, a cautious approach is essential to ensure that undue fear is not generated based on potential scenarios rather than concrete realities. Overreacting to hypothetical threats inadvertently aligns with the strategic goals of terrorist organisations, which seek to sow fear within society. Furthermore, as part of a comprehensive preparedness strategy, the existence of well-defined emergency plans capable of mitigating the aftermath of a potentially catastrophic cyberterrorist incident is crucial. Similarly, intelligence agencies' efforts to gather pertinent information regarding the capabilities and intentions of terrorist organisations are pivotal in effective risk assessment and response planning. Additionally, in navigating the dynamic landscape of cyberterrorism, the equilibrium between vigilance and pragmatism remains vital. By adhering to a measured approach, bolstered by informed decision-making, national security agencies can position themselves to effectively address the multifaceted challenges associated with the ever-evolving realm of cyberthreats.

References

1. Cavelty MD (2013) From cyber-bombs to political fallout: threat representations with an impact in the cyber-security discourse. Int Stud Rev 15(1):105–122
2. Conway M (2014) Reality check: assessing the (un)likelihood of cyberterrorism. In: Chen T, Jarvis L, Macdonald S (eds) Cyberterrorism. Springer, New York, USA, pp 103–121
3. Denning D (2000) Cyberterrorism: testimony before the special oversight panel on terrorism. Committee on Armed Services, U.S. House of Representatives
4. Department of Health (2018) Investigation: wannacry cyber attack and the NHS (HC 414). National Audit Office
5. Gambhir H (2014) Dabiq: the strategic messaging of the Islamic State. Inst Study War 15(4):1–12
6. Giantas D, Stergiou D (2018) From terrorism to cyber-terrorism: the case of ISIS. SSRN Electr J
7. Meleagrou-Hitchens A, Hughes S (2017) The threat to the United States from the Islamic State's virtual entrepreneurs. Comb Terror Center West Point 10(3):1–35
8. McAfee (2011) 7 types of hacker motivations. McAfee. https://www.mcafee.com/blogs/family-safety/7-types-of-hacker-motivations/. Accessed 27 Sept 2023

9. Meserole C, Byman D (2019) Terrorist definitions and designations lists: what technology companies need to know. Royal United Services Institute for Defence and Security Studies, Global Research Network on Terrorism and Technology: Paper No. 7
10. Pollitt M (1998) Cyberterrorism—fact or fancy? Comput Fraud Secur 1998(2):8–10
11. United Nations Office on Drugs and Crime (UNODC) (2012) The use of the internet for terrorist purposes. United Nations. https://www.unodc.org/documents/frontpage/Use_of_Internet_for_Terrorist_Purposes.pdf. Accessed 11 Sept 1980
12. Verton D (2003) Black ice: the invisible threat of cyber-terrorism. McGraw-Hill Education
13. Weimann G (2004) How modern terrorism uses the internet, Special Report. Washington: United States Institute of Peace.
14. Weimann G (2006) Terror on the internet. The new arena, the new challenges. Int J Public Opinion Res 19(3):391–393

Part III
Countering Cyberterrorism with Technology

Chapter 7
The Impact of Technology on Radicalisation to Violent Extremism and Terrorism in the Contemporary Security Landscape

Abstract The widespread integration of digital technologies including the internet, Internet of Things (IoT), and social media platforms (SMPs) has ushered in an era of unparalleled connectivity and information dissemination. While these advancements have transformed various aspects of modern life, they have simultaneously given rise to evolving cyber threats that now extend to the realm of terrorism and extremism. As a result, these technological developments have inadvertently fostered an environment conducive to the spread, recruitment, and orchestration of terrorist activities. This convergence of technology and violent ideologies presents novel challenges and opportunities for security experts and broader society alike. Understanding the complex interplay between technology, terrorism and extremism is therefore crucial in formulating effective countermeasures and policies. To this end, this chapter aims to investigate the multidimensional impact of the internet, IoT, and SMPs on terrorism and extremism, illuminating their role in driving cyber threats. By analysing their dynamic relationship and examining real-world instances, the chapter underscores the necessity for comprehensive strategies to mitigate the negative repercussions. Through its comprehensive analysis, the chapter contributes to shaping a more secure global landscape.

Keywords Digital technologies · Internet · Internet of Things · Social media · Terrorism · Extremism · Cyber threats · Global security · IoT · Policies · Countermeasures

7.1 Introduction

The widespread adoption of digital technologies such as the internet, the IoT and SMPs has heralded an era of unparalleled interconnectedness and information dissemination. While technological breakthroughs have transformed many elements of modern life, they have concurrently given rise to the emergence of ever-evolving cyber threats that now reach into the domain of terrorism and extremism. This confluence of technology and violent ideologies has created new challenges and

© The Author(s), under exclusive license to Springer Nature Switzerland AG 2024   109
R. Montasari, *Cyberspace, Cyberterrorism and the International Security in the Fourth Industrial Revolution*, Advanced Sciences and Technologies for Security Applications, https://doi.org/10.1007/978-3-031-50454-9_7

opportunities for both security experts and the broader society alike. The internet's exponential growth has produced a worldwide digital ecosystem that defies geographical boundaries, facilitating rapid connection and instant dissemination of information. Simultaneously, the IoT has connected devices and systems in unprecedented manners, promising efficiency and convenience whilst also introducing new channels for malevolent exploitation. Similarly, SMPs have transformed communication, enabling individuals and organisations to engage, share ideas, and organise with extraordinary ease. However, although technological advancements have unquestionably facilitated beneficial societal transformation, they have also inadvertently created a fertile environment for the dissemination, recruitment, and orchestration of terrorist and extremist activities.

The emergence of the internet and associated technology has had a transformative and profoundly impact on the domains of extremism and terrorism [31]. In response to this technological paradigm shift, both terrorist and extremist entities have displayed a capacity for adaptability [7]. They have skilfully recalibrated their tactics over extended periods of time [68] to effectively harness the copious resources and functionalities afforded by the digital domain. This has enabled them to advance their objectives in in manners hitherto unattainable [31]. Studies have illuminated the prevalence of problematic internet usage across various platforms. This issue consists of a spectrum of extremist ideologies, ranging from overtly political orientations [42] to affiliations with religious dogmas, and even extending to more enigmatic classifications, such as deeply entrenched misogynistic worldviews [55]. Organisational imperatives such as recruitment, financial resourcing, and public visibility can now be readily fulfilled through the use of the internet and associated technology. This digital realm serves as a cost-effective conduit for both individuals and groups not only to enlist new adherents but also to amass funds, coordinate operational initiatives, and propagate their ideological tenets [68].

Notwithstanding the presented discussion, the role of digital technologies is perceived as a catalyst for radicalisation as opposed to a primary causal factor [33]. This underscores the significance of exploring ways in which technologies support and accelerate the radicalisation process. Furthermore, in the ever-evolving digital landscape, it is vital to understand the complex interplay between technology and terrorism. This understanding will facilitate the development of effective countermeasures, policies, and interventions that leverage the potential of digital platforms whilst protecting societies against their exploitation by malicious entities. Within this context, this chapter aims to investigate the multidimensional impact of the internet, the IoT, and SMPs on terrorism and extremism. By analysing the dynamic relationship between these digital platforms within the larger digital ecosystem and extremist ideologies, the chapter illuminates the intricate mechanisms driving the evolution of cyber threats. In doing so, it contributes to our understanding of the ever-shifting landscape of contemporary security challenges. Specifically, the chapter undertakes a critical analysis of how the advent of digital technologies has brought about transformative changes in extremism and terrorism. It elucidates the ramifications of these changes on the domains of propaganda dissemination, recruitment strategies, operational planning, and financial resourcing within the realm of terrorist and extremist collectives. Furthermore, this chapter will examine case studies and real-world instances that illustrate the complex methods in which terrorists and extremists have harnessed these technologies, emphasising the necessity for comprehensive strategies to mitigate their negative effects. By scrutinising the confluence of ever-evolving cyber threats, terrorism, extremism, and technology, this chapter contributes to a more informed and secure global landscape.

The subsequent sections of this chapter follow a structured progression. Section 7.2 delves into essential definitions related to crucial concepts within the field of terrorism studies. Section 7.3 then investigates the intricate interplay between offline and online manifestations of extremism. Moving forward, Sect. 7.4 conducts an in-depth exploration of filter bubbles and echo chambers. The analysis proceeds in Sect. 7.5, where a critical examination takes place regarding the impact of the internet and associated technologies on the process of radicalisation to violent extremism. Building upon the foundations laid in Sect. 7.5, Sect. 7.6 further delves into the phenomenon of new terrorism, encompassing its strategies and its reliance on technology. Continuing the analysis, Sect. 7.7 examines the impact of the Internet of Things (IoT) on terrorist activities. In Sect. 7.8, the key insights derived from this study are offered. Finally, the chapter concludes in Sect. 7.9.

7.2 Definitions and Concepts

Prior to delving further into this study, it is vital to underscore the intricate interplay between the definitions of several key concepts in the realm of terrorism and violent extremism. This emphasises the necessity for precision and clarity when studying these terms, while recognising their nuanced and multifaceted nature within the contemporary context of security and extremism discourse. However, despite this, the terminology within terrorism and extremist studies is a subject of ongoing and extensive debate, contributing to a lack of conceptual clarity surrounding the definitions of key terms. This lack of consensus is particularly evident in the definitions of 'radicalisation, 'terrorism' [56], 'violent extremism' and 'cyberterrorism'. Considering this complex context and for the purposes of this study, this chapter adopts broad definitions of the stated terms as discussed in the following subsections.

7.2.1 Propaganda

Online platforms have evolved into fertile grounds for the perpetuation of polarisation and dissemination of propagandistic content [68]. The term "propaganda" is defined as "dissemination of ideas, information, or rumour for the purpose of helping or injuring an institution, a cause, or a person" (Webster, 1961, as cited in [31, p. 96]). The advent of the internet has transformed the manner through which terrorists and extremists disseminate propaganda. This is characterised by their ability to seamlessly furnish content directly and persistently to a multitude of online platforms or users, circumventing intermediaries in the process [27]. Moreover, this medium facilitates instantaneous access to their target audience, enabling these malicious entities to tailor their messages contingent upon variables such as geographical location, gender, age, and other distinguishing attributes of the recipients. For example, ISIS adeptly tailors its messaging, differing substantially for Western Muslims in contrast to their counterparts residing in closer proximity to the Middle East [53].

7.2.2 Radicalisation

Radicalisation can be defined as the "the process by which a person comes to support terrorism and extremist ideologies associated with terrorist groups" (HM Government, 2015, p. 21, as cited in [56]). This definition centres on the emergence of extreme beliefs and principles. Importantly, radicalization is not intrinsically synonymous with violence, as not all instances of radicalisation result in harmful outcomes. An insignificant minority of radicalised individuals exclusively promote violent ideologies or engage in acts of terrorism. Nonetheless, for the purposes of this study, radicalisation pertains exclusively to the pernicious radicalisation of individuals engaging in or endorsing violent extremism [56].

7.2.3 Terrorism

Similarly, the definition of "terrorism" remains conceptually elusive. As stipulated by the Terrorism Act 2000, terrorism encompasses actions intended to influence the government, intimidate the public, or advance political, religious, or ideological agendas, often involving violence, harm to people or property, or disruption of electronic systems [59]. Therefore, for the purposes of this chapter, a 'terrorist' is defined as an individual or a member of an organisation resorting to acts of violence, intimidation, or coercion, often targeting civilians to evoke fear or further specific political, ideological, or religious agenda.

7.2.4 Extremism and Violent Extremism

The concept of violent extremism embodies a multifaceted construct, which currently lacks a definitive and universally accepted definition in both policy and academic circles [19]. The United Kingdom (UK) Government define extremism as "vocal or active opposition to fundamental British values, including democracy, the rule of law, individual liberty and mutual respect and tolerance of different faiths and beliefs"

(Home Office, 2015, as cited in [19]). Nonetheless, the aforementioned definition has encountered significant critique due to its marked ambiguity, stemming from the lack of precise characterisation of the underlying values, a factor that could potentially foster the stigmatisation of particular communities [19, 63]. This criticism also extends to the potential peril of eroding fundamental democratic tenets, as it could label individuals as extremists merely on the basis of possessing viewpoints that deviate from those prevailing in mainstream society [19, 26].

Stephens et al. [58] contend that conspicuous differences exist among extremism definitions, notably distinguishing between idealistic and behavioural facets. Idealistic extremism pertains to an ideology contravening a society's established values, while behavioural extremism centres on the methods employed by agents to attain a specific political objective. Within the spectrum of definitions for violent extremism, the prevailing emphasis converges upon the behavioural construct, wherein acts of violence function as instrumental means to achieve predetermined ends. In contrast, more idealistic characterisations of non-violent extremism predominantly pivot towards the extremist belief system itself [40]. The discourse has presented the viewpoint that distinguishing between non-violent and violent extremism could be characterised as "naive and dangerous" (Schmid, 2014, p. 20, as cited in [19]). This perspective is primarily based on the idea that individuals who espouse extremist ideologies might initially refrain from using coercive methods to bring about societal change, only to resort to violence when they deem it necessary (Ibid.). Nonetheless, it is important to emphasise that, within the scope of this chapter, a deliberate distinction between non-violent and violent extremism will be upheld. This distinction finds its rationale in the focused examination of radicalisation towards terrorism and violent extremism.

Notwithstanding the context presented, while disparities persist concerning the precise definitions of extremism, a prevailing consensus exists. This agreement acknowledges the establishment of a framework underpinned by an ideology endorsing forceful alterations to the societal and/or political fabric [7]. As a result, the operational definition of violent extremism (see below), as applied in this study, draws upon the definitions provided by both the Federal Bureau of Investigation (FBI) and the United Nations (UN). The FBI (n.d.) characterises violent extremism as actions that involve "encouraging, condoning, justifying, or supporting the commission of a violent act to achieve political, ideological, religious, social, or economic goals" (para. 1). Similarly, the UN [61] defines violent extremism as including individuals who participate in or support the use of violence to accomplish ideological, religious, political objectives, or bring about social change.

In light of the presented discussion, the author proposes the following definition which will be adopted in this study:

> Violent extremism refers to both actions that promote, justify, or support the use of violence to achieve political, ideological, religious, social, or economic objectives, and individuals who actively participate in or endorse acts of violence with the intent of advancing various goals.

Prior to concluding this section, it is important to reiterate that not every individual harbouring extremist beliefs will necessarily engage in violent actions. Therefore, as discussed, it remains imperative to formulate precise definitions for these ambiguous terminologies. In doing so, this will facilitate the distinction between non-violent extremists and their violent counterparts and effectively differentiate between ideological predisposition and actionable behavior [56].

7.2.5 Cyberterrorism

The digital realm harbours a conceptual parallel universe wherein the potential for harm and disruptive actions by terrorists is envisioned to extend into the online domain [21], a phenomenon often labelled as "cyberterrorism" (Conway, 2002, as cited in [21]). This term, frequently cited from Conway (2002, as cited in [21]), encapsulates the notion of leveraging online tools to inflict damage while concurrently advancing specific ideological agendas [21]. Denning [15], in an early exploration of this concept, defines cyberterrorism as the intersection of cyberspace and terrorism, encompassing politically driven hacking activities aimed at causing significant harm, including loss of life or substantial economic repercussions. An illustrative scenario involves infiltrating an air traffic control system to orchestrate a collision between two aircraft, thereby exemplifying the convergence of virtual and real-world consequences. Archer [1] expounds that descriptions of the "cyber terror threat" exhibit discernible congruence with conventional typologies of terrorism, as they involve the use of computer-based methods to coerce governments and instil fear within populations. Nevertheless, Holbrook [21] highlights a significant challenge in the concept of cyberterrorism, considering the lack of tangible instances beyond speculative constructs and theoretical frameworks.

Rosenfield [50, p. 77] categorises cyberattacks into two primary groups: those targeting data, such as website defacement and distributed denial-of-service (DDoS) attacks, and those focusing on control systems governing the usage of physical infrastructure [1]. Notably, the latter category, constituting a significant majority of cyberattacks, blurs the definitional boundaries of conventional terrorism, given its divergence from the established modus operandi. Conversely, the former category, comprising the bulk of cyber incidents, falls short of fulfilling the criteria requisite for a terrorism classification. Another complicating factor is the lack of empirical clarity in both typologies which presents a fundamental challenge in the discourse on cyberterrorism [21].

7.3   The Interplay Between Offline and Online Extremism

There exists an intricate interplay between online and offline extremist behaviours [66]. This perspective that finds empirical support in a study conducted by Baugut and Neumann [5], who conducted interviews with 44 convicted and former German and Austrian Islamists. Their research unveils a fascinating pattern wherein individuals consumed online propaganda materials and then engaged in face-to-face discussions with peers and Islamic preachers to deliberate upon the content. This exchange of ideas among peers, where specific online content was recommended for later viewing, illustrates a symbiotic relationship between the online and offline realms. The participants were also observed gathering for offline interactions while collectively engaging with online propaganda, followed by in-depth analyses of the content. These complex connections between online and offline communication within the study population challenge the traditional notion of a clear-cut dichotomy [66]. The evident continuity between the online and offline domains prompts a re-evaluation of the role of the internet in propaganda dissemination. While the internet undeniably serves as a transformative conduit for spreading extremist materials, it operates within a complex information landscape that transcends simplistic distinctions between the online and offline realms (Ibid.).

Building upon the insights from the complex connections between online and offline extremist behaviors discussed earlier, Situational Action Theory (SAT), as articulated by Wikström and Bouhana [67], offers a compelling framework that delves into an individual's susceptibility to radicalisation. By examining an individual's vulnerabilities, stressors, and interactions within their socio-environmental context, SAT illuminates the motivations that drive them to embrace extremist ideologies within the framework of prevailing norms. This approach investigates the reasons behind an individual's alignment with terrorism as a legitimate course of action and explores why certain individuals resonate with such ideologies while others do not. Importantly, SAT emphasises the influential role of contextual socialisation, whether it occurs offline or online [8]. This perspective underscores the need to consider the broader social context in which online propaganda operates, acknowledging that both online and offline factors contribute to an individual's radicalisation journey. This study primarily focuses on online factors and the impact of technology on terrorism and violent extremism.

Building upon this examination of the complex relationship between online and offline elements, Whittaker [66] contends that the binary dichotomy framework often attributes radicalising agency solely to the internet. This perspective overlooks essential factors such as vulnerabilities, stressors, and the intricate interplay between online and offline elements (Ibid.). In a similar vein, Reynolds and Hafez [49] conducted a study investigating three hypotheses that emphasised the significance of offline networks and online radicalisation in elucidating the recruitment of German foreign fighters. This study highlights the importance of recognising the influence of proximity to radical communities on individuals' tendencies to engage in online interactions. Notably, the mechanics of social media content-sharing algorithms, which prioritise interactions, are influenced by offline proximity dynamics [62]. As a result, individuals' propensity to communicate online is shaped by their proximity to radical communities [66]. Users are inclined to encounter or receive content recommendations from individuals within their immediate local network. This phenomenon underscores the intricate interplay between individuals' surroundings, where interactions seamlessly traverse both digital and physical domains, defying clear boundaries [66].

7.4   Filter Bubble and Echo Chambers

The UK Crown Prosecution Service (CPS) highlights the diverse nature of terrorist organisations, encompassing various ethnic, religious, political, and racial identities, each with distinct ideologies, objectives, and aims. Prominent examples include groups such as Al-Qaeda and the Islamic State (IS), known by various acronyms such as ISIL, ISIS, or Daesh, or right-wing extremist organisations. While not all of these organisations necessarily engage in physical attacks, their pervasive online presence presents a significant and ongoing threat (CPS, 2017, cited in [13]). This digital landscape of the internet provides an environment that can facilitate the dissemination and normalisation of deviant behaviours [57]. This phenomenon becomes particularly relevant within the context of contemporary terrorist organisations, where online platforms play a pivotal role in advancing their agendas. Neumann's [41] concept of an "echo chamber" further underscores this trend. Within these echo chambers, individuals become immersed in an environment where like-minded ideological cohorts dominate, leading to the reinforcement of their own beliefs. This amplification of viewpoints is explored in the work of Binder & Kenyon, demonstrating how this process can lead to the accentuation of ideological stances and perspectives [7]. Therefore, in understanding the use of the internet by extremist individuals for social networking and identity negotiation, it becomes imperative to explore the concepts of both filter bubbles and echo chambers in the following subsections.

7.4.1   Filter Bubbles

The increasing complexity of mechanisms built into mainstream platforms has raised scholarly concerns. Pariser [44] posits that recommendation systems, intended to enhance content discovery, inadvertently engender an escalation in exposure to extremist material, coining this phenomenon as a "filter bubble". The filter bubble phenomenon unfolds through the automated operation of algorithms, which curate and present content meticulously personalised to individual users based on their historical internet engagement. Consequently, these online mechanisms serve a curation function, limiting the array of content accessed by an individual, thereby shaping a bubble of tailored interests and personalised subjects. Pariser [44] contends that these pre-determined algorithmic selections effectively restrict users from venturing into uncharted territories of information and ideas, as they are systematically deprived of a diversified range of content beyond their preexisting political, religious, or ideological convictions. The algorithmic choices effectively reinforce users' "existing presumptions and consumption patterns" [56], thus creating an environment wherein individuals predominantly engage with like-minded adherents. Pariser [44] argues that these pre-determined algorithmic selections effectively restrict users from venturing into uncharted territories of information and ideas. These restrictions systematically deprive users of content beyond their existing political, religious, or ideological convictions. As a result, users' "existing presumptions and consumption patterns" are reinforced [56]. This creates an environment in which individuals primarily engage with like-minded adherents. Furthermore, scholarship indicates a propensity for recommendation algorithms to favour extremist content. The empirical study conducted by Reed et al. [47] pertaining to radical filter bubbles underscores the potential of social media algorithms to propagate extremist networks and discourses. The study reveals that, within the context of YouTube, instances where a user consumes violent extremist content trigger algorithmic recommendations for additional extremist material [47]. Consequently, individuals might find themselves ensnared within a network interwoven with extremist narratives. It is evident that algorithmically driven services play a pivotal role in shaping the online landscape, thereby yielding the potential to perpetuate extremist inclinations and substantively contribute to the process of radicalisation. As an integral component of the Internet's ecosystem, these algorithmic mechanisms possess the capacity to reinforce extremist attitudes, thus warranting meticulous scrutiny in efforts to comprehend and address the complex phenomenon of radicalisation.

7.4.2 Echo Chambers

Filter bubbles and virtual communities serve as platforms for connecting individuals who share similar viewpoints, thus facilitating the formation of echo chambers. Within an echo chamber, individuals consciously subject themselves to specific information sources that amplify and 'echo' their pre-existing beliefs, effectively disregarding contradictory information. This discerning exposure engenders the perception of a pervasive consensus and has the potential to distort one's understanding of underlying realities [64]. Notably, Von Behr et al. [64] conducted an investigation involving interviews with 15 radicalised individuals, culminating in the observation that the internet played a substantial role in the radicalisation process for all participants. Their findings indicate that cyberspace can magnify opportunities for accessing extremist content while concurrently serving as an echo chamber. Therefore, the internet assumes the role of an echo chamber, specifically fostering narratives of violent extremism by providing a platform that reinforces radical beliefs and contributes to the formation of identities. Consequently, it accelerates the process of radicalisation [64]. However, the study asserts that the internet consolidates rather than replaces the pre-existing offline trajectory of violent radicalisation. Notwithstanding the study's constraint of a relatively limited sample size and consequent generalisability challenges, it is vital to acknowledge the inherent complexities of obtaining participant access in the field of terrorism and extremist studies. Furthermore, it is important to note that the terminologies filter bubble and echo chamber are often utilised interchangeably. Yet, it is crucial to differentiate between them. An echo chamber, discussed in more detail below, occurs when individuals actively seek out information that aligns with their existing convictions, manifesting as a user-driven phenomenon. Conversely, a filter bubble refers to the automatic curation of news, solely driven by the platform, leading to the exclusion of diverse content. Currently, the extent of reciprocal impact between online recommendation systems and user preferences remains elusive. This differentiation assumes significance, considering that filtering technology itself could potentially assume a key role in catalysing violent extremism and subsequently offer a pivotal tool in countering such extremism [56].

Echo chambers are constructs within which individuals possess the ability to immerse themselves among cohorts who share similar dispositions, thereby fortifying their perspectives and culminating in the magnification of viewpoints [7]. This circumstance is further exacerbated by the increasing sentiment of collective identity that thrives within these virtual and often anonymous online interactions. As a result, these artificial and frequently anonymised virtual exchanges might potentially lead to a diminishment in individual inhibitions. This, in turn, fosters an increased reliance on the convictions held by others, thereby rendering individuals more susceptible to embracing exceedingly radical viewpoints [68]. An illustrative instance lies in the use of online forums by far-right extremists, dating back to the early 1980s, during which white supremacists established public bulletin board systems as platforms of engagement [2]. A prominent example of this phenomenon is embodied by Don Black, a former leader of the Ku Klux Klan (KKK), who established the white supremacist forum Stormfront in 1995—an entity that endures in operational capacity to the present day [2]. In a pertinent vein, Ducol et al. [16] posit that the dynamics of relationships cultivated in virtual spaces diverge from those originating in offline contexts. This divergence may potentially necessitate an extended period for their maturation. However, it concurrently fosters deeper levels of intimacy and connection.

7.5   The Impact of the Internet and Associated Technology on Radicalisation to Violent Extremism

The emergence of the internet has effectively transcended geographical barriers, facilitating unimpeded communication channels between terrorists and prospective recruits. This paradigm shift has significantly expanded the pool of potential recruits compared to previous generations. The digital realm, with its expansive capabilities, has created a mechanism through which individuals sharing similar interests are interconnected across vast distances, even in the absence of prior interpersonal affiliations. It thereby affords them the opportunity to connect with extremist entities [66]. This technological channel empowers terrorists to tailor propagandistic endeavours toward specific demographics, establishing direct lines of communication with individual recipients. Moreover, the interactive nature of online engagement can galvanise a sense of solidarity, consequently engendering a quasi-virtual community that binds terrorists and prospective recruits [31].

The following subsections will examine the impact of the internet and associated technology on radicalisation to violent extremism, delving into its multifaceted dimensions and implications.

7.5.1 Anonymity and Encryption

The notion of the internet, particularly the Surface Web, providing absolute anonymity to terrorist and extremist groups is challenged by Benson [6], who posits that users on the digital landscape are substantially less anonymous than commonly presumed. This is owing to the intricate architecture that underpins online interactions. The analogy of leaving "tracks in the snow" aptly captures the concept that actions undertaken on the internet, akin to footprints in the snow, inherently reveal the origin and destination of information [6]. It is the World Wide Web, a pivotal component of the internet, which notably accentuates this phenomenon by virtue of its widespread prevalence and enduring persistence [6]. Contrary to the notion of data irreversibility, several entities, including search engines and the Internet Archive, routinely download content posted by users, rendering the concept of deleted data being eradicated obsolete. Notably, companies such as Facebook and Myspace have gained a reputation for secretly collecting user data and retaining deleted content. This practice, as noted by Williams (2011) and cited in Benson [6], underscores the potential for extensive digital footprints that persist beyond the user's intentions while retaining their convenience. The illusion of anonymity further diminishes in light of the ease with which material posted online can be tracked and analysed, yielding a plethora of details about the originator, all within the boundaries of legal permissibility [6]. Notably, tools such as the program "Creepy" stand as testaments to the potential to track an individual's physical location through the aggregation of data sourced from SMPs (Ibid.). The legal and ethical implications of hacking cannot be overlooked, however, it is noted that unauthorised access to a website can potentially unveil a wealth of information about its users, given the willingness to violate legal frameworks that might apply in certain jurisdictions (Richmond, 2011, as cited in [6]). The comparative capacities for identifying real identities of sympathetic individuals between terrorist entities and governments introduce an asymmetry. While both possess a degree of technical capability, the disparities in financial and human resources skew the balance in favour of governments. Therefore, the limited veil of anonymity offered by the internet renders clandestine organisations vulnerable to contestation by governments with substantial resources. This compels these organisations to engage in a precarious competition to safeguard the secrecy of their operations against the threat of infiltration [6].

While addressing the challenge of the online anonymity of extremists and terrorists might not pose significant obstacles for law enforcement agencies (LEAs), the issue of encryption introduces a wide range of formidable problems. Encryption mechanisms significantly enhance the effectiveness of terrorists' communication networks, allowing them to evade potential interference from LEAs. This technology enables terrorists to secure the confidentiality of their communications, making it challenging for LEAs to access the information. This, in turn, facilitates clandestine activities such as terrorist recruitment campaigns while preserving an element of secrecy [66]. Furthermore, the growing proliferation of encryption on various websites and smartphone applications has raised concerns among LEAs and security experts. This trend is exemplified by major technology companies such as Apple and Google, which have embraced robust encryption mechanisms, including enhanced data encryption on smartphones. This adoption has potential implications for LEAs' ability to monitor and intercept communications [43]. Moreover, the use of dedicated communication applications further complicates matters for LEAs, amplifying the challenges they face in surveilling and decoding communications [31]. Within this context, the application Telegram has gained prominence as a communication tool, particularly among entities such as ISIS and other jihadist groups, enabling the dissemination of messages to an extensive membership base [65]. Telegram's introduction of "channels" marked a significant development, evident in the example of the ISIS-affiliated channel named "Nashir," translating to "Distributor" in English, which attracted attention shortly after the feature's implementation [65]. The proliferation of Telegram's usage among the Islamic State and Al-Qaeda is underscored by an ICT special report, revealing a substantial increase in associated channels. The Islamic State alone launched 700 new channels in March 2016 (ICT, 2016, as cited in [65]).

7.5.2   Exploitation of the Dark Web

In light of the limitations of the Surface Web in ensuring anonymity for terrorist groups, as discussed in the previous section, the Dark Web has emerged as their preferred communication channel which offers them an unprecedented level of security ([65], Montasari 2023[1]). Notably, conclusive evidence of this trend began to materialise only in 2013. To illustrate the significance of the Dark Web, consider the

---

[1] Refer to the following publications for more detailed information about the same and relevant technology:

Montasari R (ed) (2023) Applications for artificial intelligence and digital forensics in national security. Springer, Cham.

Montasari R, Carpenter V, Masys A (eds) (2023) Digital transformation in policing: the promise, perils and solutions. Springer, Cham.

Montasari R (ed) (2022) Artificial intelligence and national security. Springer, Cham.

intercepted communications between Ayman Al-Zawahiri, the leader of Al-Qaeda, and Nasir Al-Wuhaysi, the leader of Al-Qaeda in the Arabian Peninsula based in Yemen, by the U.S. National Security Agency [65]. This revelation underscores how global Al-Qaeda networks have exploited this clandestine segment of the internet interchangeably referred to as the deepnet, blacknet, or darknet (INSS, 2013, as cited in [65]). Furthermore, the Dark Web has provided terrorists with an avenue to procure funds, execute financial transfers, and acquire illegally manufactured explosives and weaponry, using digital currencies such as Bitcoin and other cryptographic forms of currency. An example of this is the "Fund the Islamic Struggle without Leaving a Trace" platform on the Deep Web, actively soliciting contributions for the jihad cause through discreet Bitcoin transfers [65].

For instance, the Sunni extremist organisation, Hizb al-Tahrir, has orchestrated a sophisticated network of interconnected websites spanning from Europe to Africa. These online domains serve the dual purpose of eliciting financial support for the pursuit of jihad from sympathisers and compelling prospective contributors to engage in similar acts. This multifaceted digital architecture is strategically designed to foster a sense of financial camaraderie and solidarity, with the end goal of marshalling resources for their jihadi cause [51]. An exemplar is found in a German website that extends an array of banking particulars, including the account numbers to which donations can be deposited. A similar scenario unfolds in Chechnya, where the internet, particularly its Dark Web segment, serves as a potent mechanism for advertising avenues to channel financial backing from supporters. These channels often intersect with financial institutions located in Sacramento, California, highlighting the globalized nature of the financial infrastructure supporting such activities [51].

7.6 New Terrorism: Strategies and the Use of Technology

Aligned with Rapoport's [46] conceptualisation of the fourth wave of terrorism, religion assumes a pivotal role in the ideologies of contemporary terrorism. The concerns surrounding and interpretations of religious attacks often stem from the perceived 'apocalyptic' dimensions inherent in these ideologies [48]. This perceptible shift of terrorism away from conventional political agendas [24] might signify a resurgence of an era in which religion was viewed by some as the "only accepted justification for terror" [45, p. 659]. This is particularly pertinent within Western societies where instances of political violence are met with stringent intolerance. Furthermore, instances such as the 1993 attack on the World Trade Centre could be interpreted as indications of subsequent events such as 9/11, indicative of gradual transformations

---

Montasari R, Carroll F, Mitchell I, Hara S, Bolton-King R (eds) (2022) Privacy, security and forensics in the Internet of Things (IoT). Springer, Cham.

Montasari R, Jahankhani H, Hill R, Parkinson S (eds) (2021) Digital forensic investigation of Internet of Things (IoT) devices. Springer, Cham.

Montasari R, Jahankhani H (eds) (2021) Artificial intelligence in cyber security: impact and implications. Springer, Cham.

that were underway as early as the 1990s [48]. The digital milieu witnessed during ISIS' online campaigns underscored the adept use of online platforms, particularly Twitter, for propagandistic ends. This highlights terrorist organisations' capability to harness contemporary technology for the propagation of their agenda. Noteworthy within the context of this Twitter campaign was the inception of the mobile application 'the Dawn of Glad Tidings', strategically devised to circumvent Twitter's spam detection algorithms (Ibid.).

Furthermore, the plausible integration of Internet of Things (IoT) devices (discussed in detail within Sect. 7.7) in a comparable manner emerges as a conceivable avenue (Ibid.). An additional observation that warrants consideration within the context of the aforementioned new era of terrorism pertains to its characterised 'apocalyptic' attributes, as expounded upon earlier. This characteristic has prompted assertions suggesting that the phenomenon of terrorism has intensified in its brutality, a trajectory attributed to a prevailing absolutism and the absence of willingness to engage in compromise. This argument raises a series of fundamental questions. For instance, does this contemporary landscape mark a departure from the traditional paradigm of the 'propaganda of the deed' [17] towards an era where the very "means have become an end in themselves" [14]? Alternatively, does the implication that the religious absolutism exhibited by 'new terrorists' represents an entirely unprecedented phenomenon neglect to acknowledge the resonating tendencies inherent in preceding waves of terrorist entities [12]?

Scholarly discourse has explored the idea that terrorism has evolved towards greater brutality over the years [24]. This evolution suggests that the comparatively 'less severe' forms of terrorism in the past might have been influenced by the relatively "reasonable" behaviours of earlier actors [14, p. 1]. A perspective articulated by Laqueur [29] posits that previous manifestations of terrorism adhered to a more restricted set of implicit rules. In comparison, the contemporaneous deployment of indiscriminate and ruthlessly executed strategies, exemplified by practices such as suicide bombings, aligns with the tenets of the new terrorism [14]. It is noteworthy, however, that tactics such as suicide bombing are not novel occurrences, for example, instances of suicide bombings were documented prior to the events of 9/11 [23]. Likewise, early adopters such as the Liberation Tigers of Tamil Eelam (LTTE)[2] regularly employed such violent strategies. Unlike conventional armed forces that depend on technology to establish dominance, terrorist organisations demonstrate a propensity to leverage any accessible technological means, regardless of their level of sophistication, if such tools serve their objectives. Consequently, even strategies involving attrition and employing rudimentary technologies can yield strategic value for terrorist entities. To illustrate, the basic use of the Casio F-91W digital watch, employed in the construction of explosive devices by al Qaeda, was so pervasive that it earned the moniker "sign of al-Qaida" [60].

In light of the presented discussions, the following subsections examine strategies of a number of terrorist and extremist organisations along with ways in which these organisation exploit technology.

---

[2] Refer to Chap. 4 for a more in-depth discussion regarding the Liberation Tigers of Tamil Eelam.

7.6.1   ISIS

The strategy employed by ISIS to attract and mobilise young individuals into its ranks involves a concerted effort to glorify and normalise life within the organisation. This constitutes a powerful form of propaganda that warrants significant attention [52]. This enterprise is particularly perilous due to its potential to ensnare vulnerable minds. One of the pivotal facets of this propaganda campaign is the romanticisation of existence within ISIS, which is strategically calibrated to appeal to prospective jihadi brides. This aspect of the campaign positions ISIS as an advocate for gender equality, projecting an image of inclusivity and familial support. By showcasing visual depictions of ISIS members partaking in conventional familial pursuits and engaging in recreational activities with children, a calculated attempt is made to convey a perception of ISIS as a pro-family institution [31]. A critical turning point is observed in the recruitment landscape, evident from the substantial influx of approximately 100 French women who affiliated themselves with ISIS in 2014. This recruitment included roles ranging from suicide bombers to jihadi brides. To a significant extent, this phenomenon can be attributed to the compelling influence of gender-targeted propaganda. The appeal is strengthened by carefully crafted narratives that portray life within ISIS ranks as ideal, creating a utopian facade designed to captivate male aspirants. This is exemplified through instances such as the widespread dissemination of images featuring joyful camaraderie amongst young male ISIS fighters, bolstering the portrayal of a blissful existence [52]. A key element underpinning this propaganda strategy is the deliberate effort to depict the lifestyle of militants and their daily routines as ordinary and unremarkable. This calculated approach aims to create a positive perception of life within ISIS, particularly appealing to potential recruits from Western countries who might seek familiarity in their envisioned transition. Such proclivity is rooted in the desire for continuity with certain aspects of their previous lives. Notably, the propaganda apparatus also exhibits a more practical and materialistic dimension, designed to attract individuals who might not necessarily embrace the core ideological or theological foundations being promoted (Ibid.).

7.6.2   Al-Qaeda

Terrorist organisations have harnessed the capabilities of the internet to engage in fundraising activities, mirroring the practices employed by various political entities. A notable illustration of this paradigm is evident in the operations of Al-Qaeda, an entity that has consistently accorded paramount importance to the acquisition of financial resources. The modus operandi of Al-Qaeda's global network is intricately centred on websites, chat rooms, and forums that serve as conduits for transactions involving non-governmental organisations, charitable entities, and other financial institutions [51]. Considering that Al-Qaeda terrorist organisation has been extensively discussed in Chap. 2 of the book, this chapter does not aim to delve further into this organisation.

7.6.3 Right-Wing Extremists

Similarly, the internet has emerged as a critical medium through which right-wing extremist factions have established low-cost, effective, and secure avenues for networking and communication. This paradigm shift is characterised by an ostensibly extensive support base, facilitated by online platforms, which amplifies the semblance of a sizable following. The strategic deployment of this virtual landscape enables far-right activists to efficiently identify and recruit potential supporters for their causes. A prominent strategy employed by these groups is the dissemination of online content, including racist biographies, manifestos, operational directives, and training materials. All of these materials are meticulously designed and arranged to substantiate the purported legitimacy of their objectives [68]. Far-right networks adeptly leverage online platforms to fulfil a diverse array of objectives, analogous to the strategies employed by their extremist counterparts. An integral facet of their digital footprint is the solicitation of funds, pivotal for both offline and virtual undertakings (Ibid.). This financial mobilisation unfolds across an intricate array of virtual spaces, consisting of websites, SMPs, email distribution lists, messaging applications, and various online forums. These channels offer a dynamic platform to articulate their financial requirements, direct potential contributors towards avenues for both offline and online donations, and offer merchandise for sale, thus constituting a multifaceted approach (Ibid.).

The process of resource accumulation unfolds through manifold trajectories. A straightforward approach involves the issuance of direct appeals for donations within arenas where their support base congregates. Simultaneously, crowdfunding platforms and donation applications have emerged as practical tools, facilitating streamlined financial engagement. However, a notable distinction arises within the far-right landscape, marked by a deliberate avoidance of mainstream crowdfunding platforms such as Indiegogo and GoFundMe. Instead, there is a strategic shift towards specific platforms purposefully designed to accommodate their perspectives, including platforms such as GoyFundMe, Hatreon, and WeSearchr. These platforms offer a more receptive environment conducive to their ideological narratives (Ibid.). Far-right groups have adopted an additional source of income through e-commerce activities. This involves the selling of various products utilising a range of methods including direct sales on websites that are part of online retail platforms and payment systems, as well as using well-known platforms such as eBay, Amazon, and Etsy as intermediaries. This economic activity, centred around the exchange of goods, serves as a significant means of financial gain (Ibid.). The spread of radical ideologies is also evident within the music streaming services, facilitated by platforms such as Spotify and iTunes. Furthermore, self-publishing platforms such as Amazon's CreateSpace function as avenues for the propagation of radical textual material, thus expanding the reach of far-right narratives [25].

7.6.4 The Internet and Social Media Platforms

The internet is also used to enable the transfer of knowledge and facilitate operational coordination [68]. Geospatial imagery, such as Google Earth, can be employed to research areas and formulate prospective attacks [31]. Extremists can swiftly and easily transmit information around the world through free or inexpensive streaming services, file storage systems, and end-to-end encrypted communication applications [3]. They can connect geographically dispersed users and share information amongst them [28]. Social media, encrypted messaging channels, and other similar platforms can also connect individuals living in proximity to each other, enabling offline activities by helping them connect, coordinate, and arrange meetings with one another [54]. Encryption conceals interactions and messages, which is a concern due to its growing adoption on websites and smartphone applications [68]. Apple and Google revealed in September 2014 that data saved on smartphones would have stronger encryption [43]. Compared to using Twitter or Facebook alone, these apps greatly increase the difficulty for law enforcement to intercept communications [31].

Expanding upon the digital landscape, terrorist groups also leverage SMPs to significantly amplify their reach and impact. The accessibility and anonymity offered by SMPs enable these groups to create numerous anonymous accounts effortlessly, evading detection by LEAs and establishing cost-effective, uncomplicated, and interception-resistant communication networks, as previously discussed (Ibid.). Moreover, the utilisation of these platforms for propagandistic purposes leads to the inadvertent exposure of individuals to terrorist content. This is a departure from traditional terrorist websites, where users must actively initiate searches to access such material (Ibid.). Furthermore, the widespread availability of smartphones, providing constant and ubiquitous internet access, has effectively dissolved the temporal and spatial constraints that once limited an individual's exposure to extremist content through SMPs. However, this unregulated dissemination of information on SMPs also facilitates the spread of inaccurate and misleading information, which can have far-reaching consequences [32]. In this dynamic online environment, terrorist groups such as ISIS harness advanced technological tools to craft propaganda materials ranging from short videos to elaborate films. These materials often feature sophisticated digital artistry and Hollywood-style special effects, portraying ISIS militants as heroic and valorous figures [31]. Addressing the multifaceted challenge of combating this misinformation proves to be a daunting task. While efforts to suspend pro-ISIS social media profiles have been undertaken, their effectiveness is limited as new pro-ISIS accounts swiftly replace the suspended ones in a continuous cycle. Similarly, attempts by LEAs to deter potential ISIS recruits at airports face inherent limitations, as they cannot entirely prevent all prospective recruits from departing to join the ranks of the terrorist organisation (Ibid.).

The Plymouth incident, a tragic event in which Jake Davison fatally took the lives of five individuals, serves as a recent and alarming example that vividly illustrates the pervasive reach of online terrorism propaganda [4]. Davison's deep involvement with the prominent YouTube channel Incel TV, boasting a substantial subscriber base of 18,000, highlights the significant influence of this digital platform in propagating the Incel (Involuntary celibate) philosophy (Ibid.). Through his digital presence on YouTube, Davison actively engaged in espousing misogynistic hate speech, promoting acts of violence, and aligning himself with the Incel ideology. Notably, Incel is not officially designated as a proscribed terrorist organisation within the UK. However, categorising Davison's reprehensible actions as a terrorist attack implies an acknowledgment of his online conduct leading up to the incident as constituting terrorist behaviour [11]. This classification further accentuates the immense impact of online terrorist propaganda [13]. The resonance of support garnered by Davison, along with the endorsement received by Incel TV, serves as a stark reminder of the clandestine yet potent promotion of terrorism in the online realm, both within the UK and on an international scale (Ibid.).

7.6.5   Other Strategies

The interplay between terrorist organisations and the digital realm extends beyond communication to include operational coordination, as stipulated by Rusumanov [51]. This symbiotic relationship enabled the orchestration of the September 11 attacks by Al-Qaeda operatives, with a trove of encrypted messages discovered on the computer of Abu Zubaydah, a detained Al-Qaeda terrorist linked to the attacks. These messages, encrypted and concealed within a password-protected website section, spanned the period from May to September 2001, coinciding with the operational timeline. To ensure anonymity, the attackers utilised public computers and open e-mail accounts, illustrating the extent to which free web-based email services were employed in their communication strategies [51]. Extending this paradigm, another method used is steganography, a technique that involves embedding messages within graphic files, effectively concealing crucial instructions, such as maps, images, and technical details pertaining to the deployment of explosives [20].

7.7   The Impact of IoT on Terrorist Activities

In the wake of the emergence and widespread adoption of the internet, there has been an accelerated advancement in the realm of information and communication technology. One notable exemplar of such technological advancements is the IoT. Coined by Kevin Ashton in 1999 with the original intent of promoting radio-frequency identification (RFID) technology, the term "Internet of Things" has become emblematic of a broader concept [30]. However, the inception of interconnected devices traces its origins back to 1832, a pivotal year marked by the development of the first electromagnetic telegraph. This technological innovation facilitated direct interaction between two machines through the transmission of electrical signals. The true history of the IoT, however, commenced with the establishment of the internet in the late 1960s. During this period, the nascent idea was often referred to as "embedded internet" or "pervasive computing". It was not until around 2010/2011 that the term "Internet of Things" gained substantial recognition, and its mainstream integration was realised in early 2014 [30, 56].

The IoT functions as a network that consists of a diverse array of entities, connecting them through online channels. This involves the interlinking of distinctly identifiable embedded computing devices within the existing internet infrastructure [38]. While some IoT devices are conventional items embedded with internet connectivity, others are sensing devices designed exclusively with IoT principles in mind (Ibid.). The IoT consists of a wide spectrum of technologies, including unmanned aerial vehicles (UAVs), smart swarms, the smart grid, intelligent buildings, domestic appliances, autonomous cyber-physical and cyber-biological systems, wearable technology, embedded digital entities, machine-to-machine communication, RFID sensors, and context-aware computing, among others [38, 56]. Montasari and Hill [38] assert, "Each of these technologies has become a specific domain in its own merit. With the new types of devices constantly emerging, the IoT has almost reached its uttermost evolution."

The IoT has brought forth a multitude of societal advantages, effecting a revolutionary transformation in the daily lives of numerous individuals within these societies [35–38]. To illustrate, IoT-connected sensors have been instrumental in aiding agricultural practitioners in the oversight of their crops and livestock, thereby augmenting production efficiency and livestock health monitoring. Additionally, intelligent health-connected devices have been harnessed to improve or preserve patients' well-being through wearable technologies. However, notwithstanding these manifold benefits, it is imperative to recognise that IoT devices concurrently give rise to a host of challenges pertinent to national security (Ibid.). Despite the semblance of shared monitoring requirements with cloud computing, the decentralised nature of IoT, coupled with factors such as volume, variety, and velocity, engenders a distinct and elevated spectrum of security quandaries. This manifests in scenarios where malicious actors, including sophisticated terrorists, could potentially turn IoT nodes into zombies, intercept and manipulate cardiac devices, orchestrate Distributed Denial of Service (DDoS) attacks, and exploit vulnerabilities in In-Vehicle Infotainment systems, Closed-Circuit Televisions (CCTVs), and Internet Protocol (IP) cameras, among other manifestations (Ibid.; [56]).

Similarly, Unmanned Aerial Vehicles (UAVs), commonly referred to as drones, have garnered particular appeal not only among state actors but also among terrorist entities, insurgents, and other non-state militant actors. UAVs represent a cost-effective modality for executing targeted assaults, devoid of personnel risk [18]. This capacity is highlighted in the potential for terrorists to employ commercially available UAVs laden with explosives, thereby executing devastating attacks against civilian, military, or political targets, while evading the emotive responses (such as disgust, shock, anger and other emotions from the public) often provoked by conventional suicide attacks. As the prevalence of commercially accessible UAVs continues to expand, the proliferation of drone-based attacks is anticipated to parallel the trajectory of suicide bombings and other forms of terrorism (Ibid.).

The noticeable trend towards drone-facilitated terrorism and insurgent deployment is evident in instances such as the attempted assassination of the Venezuelan president in 2018 utilising a DJI Matrice 600, capable of carrying payloads exceeding 5 kg. In conflict zones such as Iraq and Syria, ISIS has harnessed drone footage to coordinate assaults on military installations, employing drone-captured imagery in recruitment activities targeting young individuals ([18], as cited in [56]). Western intelligence agencies have responded proactively, pre-empting several drone-related attacks during the planning stages. Notable instances include the discovery of drone manuals and blueprints of prominent London shopping areas during a counterterrorism operation in the UK in 2016 ([18], as cited in [56]). Analogously, in 2012, a US-based individual faced charges for attempting to deploy explosives-laden drones to target buildings in Washington, D.C., with purported motives of avenging U.S. drone strikes in Iraq. Likewise, a drone successfully approached the vicinity of the German Chancellor in 2013, illustrating the vulnerabilities posed by such technology ([18], as cited by [56]). These prevailing threats are poised to be further exacerbated in light of the burgeoning prevalence of 5G technology and its related technologies, which increasingly pervade the security landscape of the IoT and its edge ecosystem. By leveraging the velocity, scale, and computational capabilities afforded by 5G-enabled devices, terrorists could potentially orchestrate catastrophic swarm attacks or expedite voluminous data processing to facilitate covert cyber activities [56].

7.8   Discussion

In light of the transformation brought forth by the evolution of Web 2.0, the internet has now transformed into an interactive platform that has assumed a preeminent role in facilitating interpersonal communication. Consequently, governmental entities have, to a certain extent, underestimated the considerable influence wielded by social media and the internet as channels for disseminating violent norms and values. Furthermore, increased interactivity and anonymity online facilitate the seamless dissemination of propaganda, thereby contributing to an amplified exposure to provocative and emotionally charged content. This increased exposure, for certain individuals, can instigate moral indignation and engender the phenomenon of othering. In exploiting this perceptual construct of the 'other', extremist factions strategically seclude susceptible individuals from their offline social circles, entrenching them within an echo chamber devoid of counter-narratives and dissenting viewpoints. Consequently, the online environment functions as a reinforcing crucible, upholding and affirming extremist identities, thereby impacting the trajectory of radicalisation.

Moreover, the process of radicalisation, wherein individuals transition into extremists, is a socio-cultural phenomenon that bears substantial real-world consequences. This transformation is closely linked to the internet, which serves as a channel enabling individuals to establish networks, forge connections, and nurture a sense of collective identity—an interactive dynamic unattainable through traditional mediums such as one-way pamphlets, magazines, and broadcasts [56]. This interactivity underscores what Bowman-Grieve [10] describes as a "virtual community", a digital space allowing relatively unregulated communication and the sharing of values, norms, and beliefs. Virtual communities transcend geographical boundaries, enabling individuals to establish relationships and foster a sense of community devoid of spatial and temporal constraints [56]. Within this context, extremists find a central domain and a secure sanctuary where they can openly embrace and promote violent viewpoints without encountering counter-extremist resistance [10].

These extremist virtual networks facilitate a profound sense of belonging, which tends to attract individuals lacking robust social ties in their offline surroundings. Consequently, extremist groups guide individuals toward alternative technology (alt-tech) platforms, resulting in their social isolation within an echo chamber (Ibid.). The process of reinforcing and legitimising extremist ideologies within such an interconnected virtual community can significantly accelerate the trajectory of radicalisation. However, the academic discourse pertaining to the degree of influence exerted by the internet and associated technology in the radicalisation of individuals is characterised by its contentious nature. For instance, in comparison with previous research, Bouhana and Wikström [9] conducted an extensive examination of the literature on extremism and ascertained that the internet does not function as a causative factor in driving radicalisation towards violent extremism. Their analysis led them to the conclusion that, instead, the online environment impedes the formation of close-knit relationships and communal ties. Consequently, the field of extremist studies is marked by a lack of consensus on this matter. Such oversimplified narratives, however, possess the potential to ingrain themselves within public policy frameworks. As an illustration, the UK Home Affairs Committee [22], in a notable assertion, posited that the internet "will rarely be a substitute for the social process of radicalisation" (p. 7) thereby implying that the internet should not be regarded as a social structure in itself.

Finally, the dynamics of radicalisation are propelled by distinctive mechanisms intrinsic to information and communication technology, comprising elements such as algorithmic filtering and interactive platforms. This, in turn, engenders a proliferation of avenues through which radicalisation can be fostered. As a result, the internet becomes a fertile ground affording multifarious opportunities for the process of radicalisation to transpire [56].

7.9   Conclusion

There exists a growing recognition of the pivotal role that the internet plays in the process of individuals' radicalisation. This recognition stems from the transformative evolution of the internet, which has assumed the role of a dynamic domain wherein individuals engage, partake in communal experiences, and foster profound affiliations with their peers. Within this digital environment, the formation of virtual communities emerges as a notable phenomenon. Such communities possess the potential to fortify or amplify preexisting extremist ideologies, initially incubated through sentiments of moral indignation and grievances. The resultant milieu offers extremists an unprecedented platform to cultivate a sense of communal belonging within the online sphere, a possibility that was hitherto implausible during previous decades [56]. Furthermore, the internet has ushered in a transformative paradigm for the dissemination of terrorist and extremist ideologies. In this digital landscape, an environment conducive to the propagation of deviant behaviors has emerged, often festering within echo chambers. This phenomenon contributes to the perpetuation of a self-reinforcing cycle of propaganda. In this digital context, the inherent attributes of the internet, including its effortless communication mechanisms and cost-effectiveness, have streamlined the sharing of information, facilitating the orchestration of operational planning and coordination among these groups. Moreover, the enlarged recruitment pool facilitated by the internet extends the reach of these ideologies to a broader and potentially untapped audience. Crucially, the interplay between the online and offline realms, as underscored by Whittaker [66], serves as a dynamic interaction that shapes individuals' decision-making processes.

Therefore, understanding the interconnectedness of these domains is vital when analysing the broader impact of digital platforms. The transformative power of the internet and associated technologies is not exclusive to terrorist organisations; rather, it extends to diverse entities seeking to advance their agendas within the digital realm. It is therefore imperative to recognise that governments also harness the potential of the internet and related technologies for their own strategic objectives. In essence, the internet and associated technology have profoundly reshaped the landscape of extremist activities, creating a complex interplay between virtual and real-world domains. While it enhances the reach and capabilities of these groups, its effects are multifaceted and extend to a spectrum of actors, including governments. As a result, comprehensive understanding and response strategies are imperative in the face of these evolving dynamics.

References

1. Archer EM (2014) Crossing the Rubicon: understanding cyber terrorism in the European context. Eur Leg 19(5):606–621
2. Bartlett J (2017) The far right: from hope to hate: how the early internet fed the far right. The Guardian. Available at: https://www.theguardian.com/world/2017/aug/31/far-right-alt-right-white-supremacists-rise-online. Accessed 27 Sept 2023
3. Baele SJ, Brace L, Coan TG (2020) Uncovering the far-right online ecosystem: an analytical framework and research agenda. Stud Confl Terror 46(9):1599–1623
4. Bancroft H, Mathers M, Tidman Z (2021) Jake Davison named as Plymouth Shooter: what we know so far. Available at: https://www.independent.co.uk/news/uk/home-news/jake-davison-plymouth-shooting-b1901948.html. Accessed 27 Sept 2023
5. Baugut P, Neumann K (2020) Online propaganda use during Islamist radicalization. Inf Commun Soc 23(11):1570–1592
6. Benson DC (2014) Why the internet is not increasing terrorism. Secur Stud 23(2):293–328
7. Binder JF, Kenyon J (2022) Terrorism and the internet: how dangerous is online radicalization? Front Psychol 6639
8. Bouhana N (2019) The moral ecology of extremism: a systemic perspective
9. Bouhana N, Wikstrom PH (2011) Al Qa'ida-influenced radicalisation: a rapid evidence assessment guided by Situational Action Theory. Home Office. Available at: https://assets.publishing.service.gov.uk/government/uploads/system/uploads/attachment_data/file/116724/occ97.pdf. Accessed 27 Sept 2023
10. Bowman-Grieve L (2013) A psychological perspective on virtual communities supporting terrorist & extremist ideologies as a tool for recruitment. Secur Inform 2(1):1–5
11. Casciani DC, De Simone DDS (2021). Incels: a new terror threat to the UK? https://www.bbc.co.uk/news/uk-58207064
12. Copeland T (2001) Is the "new terrorism" really new?: an analysis of the new paradigm for terrorism. J Confl Stud 21(2):7–27
13. Correia VA, Sadok M (2021) Governing online terrorist propaganda: a societal security issue. In: Proceedings of the 7th international workshop on socio-technical perspective in IS development (STPIS 2021), Trento, Italy, pp 232–242
14. Crenshaw M (2003) Is today's "new" terrorism qualitatively different from pre-September 11 "old" terrorism? Palestine-Israel J 10(1)
15. Denning D (2001) Activism, hacktivism, and cyberterrorism: the internet as a tool for influencing foreign policy. In: Arquilla J, Ronfeldt D (eds) Networks and netwars: the future of terror, crime and militancy. RAND Corporation, pp 239–288
16. Ducol B, Bouchard M, Davies G, Ouellet M, Neudecker C (2016) Assessment of the state of knowledge: connections between research on the social psychology of the Internet and violent extremism, 16-05. TSAS—The Canadian Network for Research on Terrorism, Security, and Society, Waterloo
17. Fleming M (1980) Propaganda by the deed: terrorism and anarchist theory in late nineteenth-century Europe. Stud Confl Terror 4(1–4):1–23
18. Grossman N (2018) Are drones the new terrorist weapon? Someone tried to kill Venezuela's president with one. The Washington Post. Available at: https://www.washingtonpost.com/news/monkey-cage/wp/2018/08/10/are-drones-the-new-terrorist-weapon-someone-just-tried-to-kill-venezuelas-president-with-a-drone/. Accessed 27 Sept 2023
19. Gunton K (2022) The impact of the internet and social media platforms on radicalisation to terrorism and violent extremism. In: Montasari R, Carroll F, Mitchell I, Hara S, Bolton-King R (eds) Privacy, security and forensics in the Internet of Things (IoT). Springer, Cham, pp 167–177
20. Haig Z, Kovács L (2007) New way of terrorism: Internet- and cyber-terrorism. Acad Appl Res Milit Public Manag Sci 6(4):659–671
21. Holbrook D (2015) A critical analysis of the role of the internet in the preparation and planning of acts of terrorism. Dyn Asymmetr Confl 8(2):121–133

22. Home Affairs Committee (2014) Counter-terrorism: seventeenth report of session 2013–14. The Stationery Office. Available at: https://publications.parliament.uk/pa/cm201314/cmselect/cmhaff/231/231.pdf. Accessed 27 Sept 2023
23. Horowitz MC (2015) The rise and spread of suicide bombing. Annu Rev Polit Sci 18:69–84
24. Jenkins BM (2006) The new age of terrorism. In: The McGraw-Hill homeland security handbook, pp 117–130
25. Keatinge T, Keen F, Izenman K (2019) Fundraising for right-wing extremist movements: how they raise funds and how to counter it. RUSI J 164(2):10–23
26. Khan S (2019) Challenging hateful extremism. Commission for Countering Extremism
27. Knox EG (2014) The slippery slope of material support prosecutions: social media support to terrorists. Hastings LJ 66:295
28. Koehler D (2014) The radical online: Individual radicalization processes and the role of the Internet. J Deradical 1:116–134
29. Laqueur W (1998) The new face of terrorism. Wash Q 21(4):167–178
30. Lueth KL (2014) Why the Internet of Things is called Internet of Things: definition, history, disambiguation. IoT Analytics. Available at: https://iot-analytics.com/internet-of-things-definition/. Accessed 27 Sept 2023
31. Lieberman AV (2017) Terrorism, the internet, and propaganda: a deadly combination. J Natl Secur Law Policy 9:95
32. McNeal GS (2006) Cyber embargo: countering the internet jihad. Case W Res J Int L 39:789
33. Meleagrou-Hitchens A, Alexander A, Kaderbhai N (2017) The impact of digital communications technology on radicalization and recruitment. Int Aff (Royal Institute of International Affairs 1944-) 93(5):1233–1249
34. Montasari R (ed) (2023) Applications for artificial intelligence and digital forensics in national security. Springer, Cham.
35. Montasari R, Jahankhani H, Hill R, Parkinson S (eds) (2020) Digital forensic investigation of Internet of Things (IoT) devices. Springer
36. Montasari R, Hill R, Montaseri F, Jahankhani H, Hosseinian-Far A (2020) Internet of things devices: digital forensic process and data reduction. Int J Electron Secur Digit Forens 12(4):424–436
37. Montasari R, Hill R, Parkinson S, Peltola P, Hosseinian-Far A, Daneshkhah A (2020) Digital forensics: challenges and opportunities for future studies. Int J Organ Collect Intell (IJOCI) 10(2):37–53
38. Montasari R, Hill R (2019) Next-generation digital forensics: challenges and future paradigms. In: 2019 IEEE 12th international conference on global security, safety and sustainability (ICGS3), London, UK, pp 205–212
39. Neumann PR (2013) Options and strategies for countering online radicalization in the United States. Stud Confl Terror 36(6):431–459
40. Neumann PR (2013a) The trouble with radicalization. Int Aff 89(4):873–893
41. Neumann PR (2013b) Options and strategies for countering online radicalization in the United States. Stud Confl Terror 36(6):431–459
42. O'Callaghan D, Greene D, Conway M, Carthy J, Cunningham P (2015) Down the (white) rabbit hole: the extreme right and online recommender systems. Soc Sci Comput Rev 33(4):459–478
43. Oremus W (2015) Obama is annoyed the government can't spy on your iMessages and Snapchats. Slate Magazine, 19 Jan. https://slate.com/technology/2015/01/obama-wants-backdoors-in-encrypted-messaging-to-allow-government-spying.html
44. Pariser E (2011) The filter bubble: what the internet is hiding from you. Penguin Press
45. Rapoport DC (1983) Fear and trembling: terrorism in three religious traditions. Am Polit Sci Rev 78(3):658–677
46. Rapoport DC (2001) The fourth wave: September 11 in the history of terrorism. Curr Hist 100(650):419–424
47. Reed A, Whittaker J, Votta F, Looney S (2019) Radical filter bubbles: social media personalization algorithms and extremist content. Global Research Network on Terrorism and Technology

48. Rees J (2022) The internet of things and terrorism: a cause for concern. In: Montasari R, Carroll F, Mitchell I, Hara S, Bolton-King R (eds) Privacy, security and forensics in the internet of things (IoT). Springer, Cham, pp 197–202

49. Reynolds SC, Hafez MM (2019) Social network analysis of German foreign fighters in Syria and Iraq. Terror Polit Violence 31(4):661–686

50. Rosenfield DK (2009) Rethinking cyber war. Crit Rev 21(1):77–90

51. Rusumanov V (2016) The use of the Internet by terrorist organizations. Inf Secur 34(2):137–150

52. Saul H (2015) Is this the most dangerous ISIS propaganda? The Independent, 20 Mar. https://www.independent.co.uk/news/world/middle-east/the-most-dangerous-isis-propaganda-yet-jihadi-brides-with-m5s-fighters-relaxing-and-children-playing-used-to-present-caliphate-as-a-utopia-10121653.html

53. Shane S, Hubbard B (2014) ISIS displaying a deft command of varied media. New York Times, 30

54. Simi P, Futrell R (2006) Cyberculture and the endurance of white power activism. J Polit Milit Sociol 34(1):115–142

55. Speckhard A, Ellenberg M, Morton J, Ash A (2021) Involuntary celibates' experiences of and grievance over sexual exclusion and the potential threat of violence among those active in an online incel forum. J Strateg Secur 14(2):89–121

56. Sullivan A, Montasari R (2022) The use of the Internet and the internet of things in modern terrorism and violent extremism. In: Montasari R, Carroll F, Mitchell I, Hara S, Bolton-King R (eds) Privacy, security and forensics in the internet of things (IoT). Springer, Cham, pp 151–165

57. Sutherland EH, Cressey DR, Luckenbill DF (1992) Principles of criminology. Altamira Press

58. Stephens W, Sieckelinck S (2020). Being resilient to radicalisation in PVE policy: a critical examination. Critical Studies on Terrorism 13(1):142–165

59. Terrorism Act 2000, c. 11. Available at: https://www.legislation.gov.uk/ukpga/2000/11/contents. Accessed 27 Sept 2023

60. The Guardian (2011) Guantánamo files: how interrogators were told to spot al-Qaida and Taliban members. Available at: https://www.theguardian.com/world/interactive/2011/apr/25/guantanamo-files-interrogators-al-qaida-taliban. Accessed 14 Sept 2023

61. United Nations, General Assembly, Human Rights Council (2016). Report on best practices and lessons learned on how protecting and promoting human rights contribute to preventing and countering violent extremism: Report of the United Nations High Commissioner for Human Rights. A/HRC/33/29.

62. Valentini D, Lorusso AM, Stephan A (2020) Onlife extremism: dynamic integration of digital and physical spaces in radicalization. Front Psychol 11:524

63. Vincent C, Hunter-Henin M (2018) The trouble with teaching 'British values' in school. Independent. https://www.independent.co.uk/news/education/british-values-education-what-schools-teach-extremism-culture-how-to-teachers-lessons-a8200351.html. Accessed 27 Sept 2023

64. Von Behr I, Reding A, Edwards C, Gribbon L (2013) Radicalisation in the digital era: the use of the Internet in 15 cases of terrorism and extremism. RAND

65. Weimann G (2016) Terrorist migration to the dark web. Perspect Terror 10(3):40–44

66. Whittaker J (2022) Rethinking online radicalization. Perspect Terror 16(4):27–40

67. Wikström POH, Bouhana N (2016) Analyzing radicalization and terrorism. In: LaFree G, Freilich JD (eds) The handbook of the criminology of terrorism, pp 175–186

68. Williams HJ, Evans AT (2022) Extremist use of online spaces. RAND Corporation, CT-A1458-1

Chapter 8
Machine Learning and Deep Learning Techniques in Countering Cyberterrorism

Abstract In recent years, artificial intelligence (AI) has emerged as a critical tool in automating data detection and acquisition processes, significantly contributing to bolstering national security efforts. The integration of human-like attributes into machines enables AI to address challenges akin to human cognitive faculties, thus becoming a preferred approach in decision-making. Notably, AI has gained substantial traction in identifying criminal and radical activities online. Within the domain of AI, machine learning (ML) plays a pivotal role in automating data detection, particularly in detecting radical behaviours online. However, despite its potential, ML encounters challenges arising from the complexity and diversity of cyber-attacks, which might impact the accuracy and applicability of the algorithms. The primary aim of this chapter is to delve into the technical dimensions of AI concerning cyber security and cyber terrorism. Particularly, the chapter focuses on investigating the pivotal role ML and deep learning (DL) play in identifying extremist content and activities on online platforms. Through a comprehensive examination of these two domains, the chapter endeavours to illuminate the potential pathways through which AI can bolster cyber security measures and effectively combat cyber terrorism.

Keywords National security · Cyberterrorism · Extremism · Internet · Social media · Artificial intelligence · Machine learning · Deep learning · Digital forensics · Algorithms · Counter terrorism policing · Privacy

8.1 Introduction

In recent years, AI has played a pivotal role in automating data detection and acquisition processes, particularly in bolstering national security efforts [56]. AI equips machines with human-like attributes, enabling them to approach challenges in ways akin to human cognitive faculties [24]. As a preferred approach in decision-making, AI has gained traction in identifying criminal and radical activities online, as evidenced in various surveillance techniques such as wiretapping and dragnet surveillance (Ibid.) [56]. Similarly, ML, a subfield of AI, plays a key role in automated data

© The Author(s), under exclusive license to Springer Nature Switzerland AG 2024
R. Montasari, *Cyberspace, Cyberterrorism and the International Security in the Fourth Industrial Revolution*, Advanced Sciences and Technologies for Security Applications, https://doi.org/10.1007/978-3-031-50454-9_8

detection, especially in detecting radical behaviours online.[1] However, despite their significant potential, AI and ML encounter various challenges due to the complexity and diversity of cyber-attacks, potentially limiting the accuracy and applicability of their algorithms [5, 15]. Moreover, the broad applications of AI introduce a multitude of technical, operational, ethical, legal, security, and privacy challenges which will be comprehensively addressed in Chap. 9 of this book.

In contrast, the focus of this chapter is to explore the technical aspects of AI within the context of cyber security and cyber terrorism. Specifically, the chapter will delve into the relevance of ML and DL in these fields, with a specific emphasis on detecting extremist content and activities online. Filling this research gap is particularly important as while ML has found extensive applications in various cybersecurity domains, there has been relatively limited research on its use for examining radical behaviours and countering terrorist activities in cyberspace. Therefore, by exploring the technical aspects of AI, this chapter will illuminate the potential pathways through which AI can bolster cyber security measures and effectively combat cyber terrorism. Furthermore, by delving into these critical technical aspects of AI application, this study contributes to a broader understanding of its role in safeguarding the digital landscape against emerging threats, such as radical behaviours and terrorist activities in cyberspace.

The subsequent sections of this chapter are structured as follows. Section 8.2 provides a contextual background to the subject matter while Sect. 8.3 delves into a comprehensive examination of various ML techniques applicable in combatting cyberterrorism. Section 8.4 investigates the topic of feature extraction and selection, whereas Sect. 8.5 explores the intersection of DL and intrusion detection systems. Section 8.6 examines counter-terrorism policing, aiming to enhance understanding and address the issue under investigation by analysing the recruitment tactics employed by cyber terrorists. Section 8.7 discusses the key findings. Finally, the chapter concludes in Sect. 8.8.

8.2 Background

The manual identification of radical posts and chat rooms online aimed at radicalising young individuals is practically infeasible, necessitating the implementation of an automated system capable of efficiently extracting such data [15]. Within this context,[2] ML emerges as a potent AI solution for the automated detection, collection and processing of vast and diverse textual data online.[3] Through diverse training

---

[1] For instance, ML algorithms can learn from training datasets to identify patterns and classify data as radical or non-radical.

[2] Covering both radical and extremist behaviour.

[3] NOTABLY, Twitter has emerged as a prominent platform used by violent extremist groups, prompting the application of ML and data mining techniques to detect hashtags employed in these posts [5].

techniques, ML effectively identifies patterns and recurrences in vast datasets [57], thereby streamlining and expediting the process of digital forensic investigations for law enforcement purposes [37]. The extracted data is subsequently subjected to classification techniques within the ML framework to distinguish between radical and non-radical content. This classification process involves examining writing styles, structural characteristics, and temporal patterns [15].

To facilitate the automated detection of such content, ML algorithms are fed data sets, commonly known as training sets, containing pertinent patterns that enable them to learn and adapt to real-life scenarios [5]. Within this context, ML establishes mathematical models based on sample data, commonly referred to as training data, to enable predictions without requiring manual programming [3, 24]. Specifically, the data is categorised into two classes: radical and non-radical [15]. Similarly, the process of employing ML techniques involves two distinct phases: the training phase and the testing phase. During the training phase, mathematical computations are applied to the training dataset to facilitate the learning of behavioural patterns over time. Subsequently, the acquired knowledge is tested to ascertain its ability to classify data as normal or intrusive [34]. It is essential to acknowledge that the data utilised in ML algorithms to combat cyberterrorism differs from the data employed by law enforcement agencies (LEAs). ML techniques are also used for surveillance purposes in the context of counterterrorism and digital policing online. Within this context, surveillance methods, such as Dragnet surveillance, involve the comprehensive collection of data concerning all individuals, as opposed to focusing solely on those under suspicion. However, it is believed to minimise intrusiveness by primarily focusing on metadata,[4] which provides governments with a basis for justifying these practices [56].

In the context of counterterrorism methods, the integration of AI-driven techniques, irrespective of their specific application such as AI-driven cybersecurity, automated detection measures, or surveillance, poses a myriad of challenges. As alluded to briefly in the preceding section, these challenges, while elaborated upon in Chap. 9, merit preliminary discussion. Notably, among the challenges demanding attention, privacy and security concerns loom large, considering their intrinsic potential to infringe upon the fundamental human rights of individuals. The privacy-security debate revolves around the complex balance between individuals' privacy entitlement and their right to security [56]. Navigating this delicate balance constitutes a central concern in the deployment and regulation of AI-based mechanisms aimed at safeguarding national security while upholding citizens' fundamental rights.

The ongoing debate regarding the prioritisation of security over privacy suffers from inherent flaws, with an increasing inclination toward favouring security [46]. This discourse holds substantial importance, considering its far-reaching implications, including its impact on the very "foundations" of "freedom and democracy" (Ibid., p. 1). Furthermore, the prevailing perception suggests that privacy has become increasingly elusive, leading individuals to believe that sacrificing privacy may

---

[4] In its simplest term, metadata refers to data about the data.

enhance the security of their personal information [46]. While governments naturally prioritise security to safeguard national interests and critical infrastructures, some scholars contend that policy discussions should grant greater prominence to the privacy and rights of citizens (Ibid.). A pivotal aspect of this debate centres around data collection methods, distinguishing between non-intrusive and invasive approaches. Non-intrusive data collection, though conducive to safeguarding individuals' privacy and their devices, can be time-consuming, prompting agencies to opt for quicker yet more invasive methods such as hacking [2, 55]. As previously mentioned, Chap. 9 will provide a comprehensive exploration of the ethical, legal, privacy, and technical challenges associated with the use of AI and ML in this context.

8.3   Techniques and Methods

ML techniques offer a diverse range of benefits and applications, encompassing risk indicators that detect warning signs and facilitate the identification of linkages between threats, ultimately leading to the apprehension of perpetrators involved in cyber-attacks [33]. In the context of ML, four primary learning methods are widely employed, namely supervised, unsupervised, semi-supervised, and reinforcement learning. Supervised learning necessitates the labelling of all input data as either "good" or "bad" to facilitate the model's training. For instance, in fraud detection cases, supervised learning relies on predictive data analysis [25]. In the domain of cyberterrorism, supervised learning classifiers are utilised within data mining approaches to generate predictive models capable of pre-empting terrorist attacks [57]. This approach enables the processing of vast and diverse datasets, which would be impractical to manage manually, allowing the classification of data as either radical or non-radical [15].

While supervised learning yields accurate results using known and labelled input data, it entails offline analysis and computationally complex operations [18]. In contrast, unsupervised learning does not require a training set. This model autonomously updates itself through continual processing of new information and data, thereby adapting and evolving accordingly (Ibid.). An exemplar technique employed in the context of cyberterrorism is the Support Vector Machine (SVM), known for its highly successful classification capabilities [57]. Unsupervised learning involves input data lacking classification, allowing the system to autonomously update and evolve based on the continuous influx of new information and data [25]. Unsupervised learning, while not yielding results as reliable as those obtained through supervised learning, offers the advantage of employing unknown input data and conducting real-time analyses, thereby reducing the need for extensive human intervention [18].

Semi-supervised learning is particularly suitable for situations where obtaining information proves costly, impractical, or necessitates human intervention, and it facilitates the storage of data in group variables, even when the identity of the data is undefined [25]. On the other hand, reinforcement learning techniques empower machines to autonomously ascertain optimal actions within a given environment (Ibid.). In this context, reinforcement learning entails the mapping of situations to determine which actions lead to the highest rewards in practice [49]. Effectively, it represents a process of trial-and-error wherein an "agent" interacts "with the environment", progressively "learning an optimal policy" [26, p. 5]. The identification of ML techniques necessitates an examination of the essential functions that these algorithms serve, as they play a crucial role in detecting and analysing significant security risks. Among these functions, regression, clustering, and classification stand out as three particularly effective methods frequently employed within the ML landscape [25]. In fact, both supervised and unsupervised techniques heavily rely on these learning algorithms, as they form the essential framework for the learning process [40].

Regression, a fundamental ML technique, involves the automated prediction of recurring attributes present in a given dataset [15]. Within the cybersecurity domain, regression finds substantial utility in fraud detection, where it aids in understanding existing data patterns and making predictions about future changes [25]. Three specific types of regression are particularly relevant in this context, including linear regression, polynomial regression and ridge regression. Linear regression involves borrowing methods from various domains and applying them to data values within a constant scope, such as logistics and statistics, rather than categorising them into discrete groups (Ibid.). Polynomial regression, on the other hand, is employed to identify and analyse non-linear situations, such as the behaviour of viruses, with Python serving as a prevalent tool for implementation. Finally, ridge regression addresses the issue of multicollinearity, which involves the combination of multiple regression datasets. By incorporating a degree of inclination, this method reduces standard errors in the analysis (Ibid.).

In contrast, clustering in the context of ML involves the grouping of objects possessing similar characteristics [15]. This algorithm effectively categorises data into groups with shared attributes, such as specific text or objects associated with radical characteristics (Ibid.). Notably, clustering is prominently utilised in forensic analysis, where its objective is to detect anomalies by classifying activities into distinct categories, separating those deemed benign from potentially malicious [25]. It is primarily employed as a preventive measure, aiming to avert issues before they arise. In cybersecurity, clustering is instrumental in secure email gateways, where it aids in identifying outliers and segregating legitimate files from potentially harmful ones, particularly during malware analysis (Ibid.).

Similarly, classification involves identifying "common properties among different crime entities" and organising "them into predefined classes" [7, p. 52]. This technique finds application in various domains, including email spam filtering and predicting crime trends, as it expedites the identification of criminal content. Nonetheless, it is essential to note that, despite its efficacy, some advocate for complementary manual checking, given that no classifier attains complete accuracy [15]. Notably, one of the initial ML methods utilised to address cybersecurity concerns was spam filters, which were predominantly employed when the nature of classification was already known to the computer [25]. Spam filter applications often leverage the Naive Bayes algorithm, a statistical approach that employs a dataset for sample identification. In mail services, this algorithm detects word frequency within the text and employs clustering to classify the data [25]. In the context of cyberterrorism, classification plays a pivotal role in determining whether a given text exhibits radical or non-radical characteristics. As defined by The Federal Bureau of Investigation (FBI), radical characteristics comprise violent acts or threats to harm human life, mass intimidation, and attempts to influence government policy and actions related to counterterrorism [15]. Several classification techniques, including Support Vector Machines (SVM), Decision Trees, and Naive Bayes, are employed in ML, contributing significantly to the analysis of cybersecurity issues [25].

Ch et al. [5] developed a comprehensive methodology for analysing cybercrime incidents, encompassing four distinct phases applied to the data. These phases involve reconnaissance, pre-processing, data clustering, and classification. During the reconnaissance phase, structured and unstructured data are collected and preserved in their raw form. Subsequently, the pre-processing phase facilitates feature extraction, which not only reduces data dimensions from high to low but also involves the utilisation of the Term Frequency-Inverse Document Frequency (TF-IDF) vectorisation process [58]. TF-IDF serves the purpose of identifying relevant terms in textual data and consists of two main components: Term Frequency (TF), which assesses the frequency of a word's occurrence in a given text and gauges its significance within the entire corpus through a specific equation [47]. Upon pre-processing the data, the organisation is improved, leading to enhanced data visualisation as the dimensions are reduced.

One of the most effective feature extraction methods involves employing the bag-of-words model, which associates a model with each feature, thereby contributing to increased accuracy [5]. The initial step in this approach involves tokenising the sentence, essentially breaking down the text into individual components, and subsequently applying weights to the words (Ibid.). Tokenisation, a widely utilised technique in numerous ML methods, is essential for handling text data by decomposing it into discrete characters [45]. Following tokenisation, the term frequency (TF) is evaluated based on the statistical presence of each term in the text, and the TF values are stored in a matrix format known as the TF matrix (8.1) [5].

$$\text{tf-idf} = \log \frac{N}{1 + N_w}$$
$$\text{tf(term)} = \frac{N_{term}(Record)}{T_{term}(Record)} \tag{8.1}$$

where N_{term} denotes the number of occurrences of a specific term within a record, while T_{term} represents the total number of terms present in that same record (Ibid.).

The subsequent step in the TF-IDF methodology involves the assessment of the Inverse Document Frequency (IDF), a process of downscaling or simplifying words that occur multiple times within the text. This step results in the construction of an IDF matrix (8.2) (Ibid.).

$$\sum_{i=1}^{N} x^2 = \frac{(O-E)^2}{E} idf(term) = \log \frac{N}{1+N_{term}} \qquad (8.2)$$

Consequently, the subsequent phase involves the computation of the final TF-IDF score by performing the multiplication of both the TF and the IDF frequencies (8.3) (Ibid.).

$$tf\text{-}idf(term) = tf(term) \times idf(term) \qquad (8.3)$$

Subsequently, the sentences are scored and allocated to their respective texts, with the threshold of TF-IDF determined based on the scoring of all sentences within the text (Ibid.). This process serves not only to assess the correlation of terms within the data but also to ascertain if the relationship between the two variables accurately represents the dataset (Ibid.). It is important to note that this is just one approach among many, illustrating how ML techniques are employed for the collection and detection of radical characteristics in textual data.

An alternative approach for detecting radical characteristics in textual data is the application of Naïve Bayes, which constitutes a supervised learning method aiming to assign the most probable category, such as radical characteristics, to a given example based on its feature vector [43]. This technique has demonstrated considerable success in practical applications, outperforming some more advanced methods due to its effectiveness in text classification and system performance management (Ibid.). Naïve Bayes functions as a probabilistic classifier, operating on the assumption of object probabilities, thereby making it highly adaptable within the context of ML [6]. In text classification scenarios, this method relies on training datasets containing text categories, with word frequencies utilised as features, and crucially, the assumption of independence among these features (Ibid.). The effectiveness of Naïve Bayes is enhanced when the features are entirely independent and does not significantly depend on the variety of feature dependencies [43]. Bayes' theorem serves as a fundamental tool in this realm of research, as it enables the determination of hypothesis probabilities based on conditional probabilities (Ibid.). The formula for Bayes' theorem is as follows:

$$P(A|B) = \frac{P(B|A).P(A)}{P(B)}$$

In this context, the notation $P(A|B)$ denotes the posterior probability, which represents the probability of hypothesis A given the observed event B. On the other hand, $P(B|A)$ stands for the likelihood probability, indicating the probability of the evidence given that the hypothesis is true. Furthermore, $P(A)$ corresponds to the prior probability, signifying the probability of the hypothesis before the evidence is observed. Finally, $P(B)$ refers to the marginal probability, representing the probability of the evidence itself (Ibid.).

As previously mentioned, Naïve Bayes represents a widely employed method for text classification challenges and has gained popularity in the detection of radical text online. This algorithm leverages word frequency to create training datasets, comprising terms and corresponding class details for each sample [25]. Concerning its application in automatically detecting radical characteristics, Naïve Bayes attempts to classify posts on platforms such as Twitter by analysing their structure, style, and temporal patterns [15]. However, an inherent limitation of this approach is its assumption of feature independence, which hinders its ability to learn the relationships between the suggested features [59].

Moreover, an additional tool employed in data extraction, particularly focusing on the psychological aspects of text, is the Linguistic Inquiry and Word Count (LIWC). Recent research has utilised LIWC to detect extremist behaviours and gain insight into the psychological dynamics behind the recruitment process [38]. Given the evolving and increasingly sophisticated strategies of terrorist organisations, the analysis of their texts has become more intricate and nuanced. For instance, the use of code words during conversations has become prevalent, complicating text deciphering and analysis. Consequently, counterterrorism agencies have conducted extensive research into the language employed by terrorists and the underlying thought processes (Ibid.). This pursuit has become particularly crucial in contemporary times as terrorist organisations adapt and employ more intelligent and covert methods in their operations. The LIWC program comprises two primary components: the dictionaries it references and the processing component [51]. The processing component serves as the core of the program, where all texts are stored and analysed word-by-word. Once the analysis of the entire text corpus is completed, the results are compared with the dictionary. The dictionary plays a pivotal role in this process, as it encompasses a collection of words and phrases associated with specific categories. A fundamental aspect of this process is the determination of whether a word conveys negative or positive emotion (Ibid.). Various linguistic elements are examined through LIWC analysis, such as word count, which can provide insights into the level of engagement between two individuals engaged in a conversation and identify conversational dominance (Ibid.). Additionally, pronouns might reveal the parties involved in the conversation or whether individuals outside of the conversation are referenced. Furthermore, LIWC analysis takes into account the presence of assents and emojis, which can indicate the positivity of the conversation and the level of agreement and understanding (Ibid.). It is important to recognise that individuals respond differently to different situations, and some might not overtly express emotions in written communication, or not at all. Nevertheless, extensive studies and research have demonstrated the accuracy of LIWC in detecting emotions in language through text analysis (Ibid.).

8.4   Feature Extraction and Selection

ML techniques have been developed to automatically identify patterns in data and unveil concealed knowledge within the vast amounts of online information. Nevertheless, such data often contains a substantial level of noise [50]. Additionally, the content found online is generally characterised by informal writing styles, encompassing misspellings, poor punctuation, and grammar errors. The primary approach for mitigating these noisy features is referred to as dimensionality reduction, which comprises two sub-parts: feature selection and feature extraction (Ibid.). Both these processes hold the potential to enhance the performance of ML by enabling the creation of more robust models, reducing storage requirements, and lowering data complexity. The objective of feature selection is to "choose a subset of features for improving prediction accuracy or decreasing the size of the structure without significantly decreasing prediction accuracy of the classifier built using only the selected features" [10, p. 132]. Features are selected based on a predetermined set of criteria, often aiming to preserve the closeness of the resulting class distribution to the original data (Ibid.). Feature extraction refers to the process of projecting features into a novel "feature space with lower dimensionality", wherein the newly constructed features typically comprise "combinations of the original features" [50, p. 1]. It accomplishes two crucial tasks: mapping the original feature space to a new space with reduced dimensions and amalgamating the original feature space, which is a complex operation. On the other hand, feature selection is regarded as a superior dimensionality reduction approach in terms of interpretability and readability compared to feature extraction even though the issue of combining both techniques is not commonly encountered [50]. In the context of this chapter, with a focus on ML and classification challenges, feature selection emerges as the most effective tool for employment.

As previously mentioned, feature selection algorithms can be categorised into supervised, unsupervised, and semi-supervised feature selection, depending on whether the training set is labelled or not. Supervised methods in feature selection can be further distinguished into distinct models: filter models, wrapper models, and embedded models. The filter model serves the purpose of isolating the classifier from the feature selection process to prevent bias from transferring between the two, which could lead to significant bias downstream. This model relies on consistency, distance, and dependency within the training data (Ibid.). On the other hand, the wrapper model employs predictive accuracy to assess the quality of the selected features. However, a drawback of this model is its typically time-consuming and costly nature for organisations. In contrast, the embedded model is specifically designed to establish the connection and relationship between the filter and wrapper models, combining the efficiency of the filter model with the accuracy of the wrapper model, and performing feature selection during learning time (Ibid.). Critics of the supervised method argue that while this approach evaluates the relevance of data in reference to labelled information, it necessitates a large volume of labelled data, which may not always be feasible (Ibid.). Consequently, researchers turn to unsupervised feature selection, which involves unlabelled data. However, the challenge lies in understanding the relevance of these features. The goal of unsupervised methods is to incorporate the natural classification of the data and enhance clustering accuracy based on given criteria [4]. Furthermore, semi-supervised feature selection aims to improve the learning performance of the model, taking into account the model's score (Ibid.).

In many classification problems, the machine might struggle to learn the good classifiers before removing unwanted features due to the large size of the data sample [50]. It is important to note that feature selection primarily impacts the training stage of classification. As Tang et al. elucidate, after generating features,

> Instead of processing data with the whole features to the learning algorithm directly, feature selection for classification will first perform feature selection to select a subset of features and then process the data with the selected features to the learning algorithm. (2014, p. 5)

Researchers aspire to find the most accurate feature subsets by searching through all possible combinations. However, such an exhaustive approach can be time-consuming and impractical [50]. To address this, a recommended solution is the adoption of a stopping criterion, which enables the search process to terminate when prompted by the machine. Nevertheless, limitations in algorithm scalability might arise when the data exceeds the machine's memory capacity, necessitating further processing that could be costly or even infeasible. The use of feature selection techniques could potentially mitigate this scalability issue, but it would necessitate the retention of full dimensionality, leading to additional complexities (Ibid.) [4]. Addressing feature selection as "the optimal solution can only be obtained by an exhaustive search given an evaluation or search criterion" [4, p. 75], it is crucial to recognise the dynamic nature of features. Consequently, awareness of emerging areas in feature selection, such as online feature selection, is essential. This domain pertains to network data streams and videos, offering applicability in dynamic situations and feature spaces. However, incorporating online feature selection raises concerns about its practicality and relevance in certain contexts [4], the discussions of which falls outside the purview of this chapter.

8.5   Deep Learning and Intrusion Detection Systems

Research has highlighted the necessity for increased training duration to effectively process the vast number of datasets involved in the detection of radical behaviour online [23]. To address this concern, a promising solution has emerged in the form of DL, a recently developed technology that employs a novel learning approach (Ibid.). As a subfield of ML, DL emulates human thought patterns and, notably, the decision-making process [29]. Furthermore, it significantly reduces training time and ensures precise system operation [23]. The underlying structure of DL involves interconnected layers, with each layer receiving output from the preceding one (Ibid.). Within the realm of DL, it is more advantageous to employ algorithms that autonomously extract features that best represent the data, rather than relying on manual feature identification. Consequently, various domains within cybersecurity, such as vehicle autonomous systems, natural language processing (NLP), and image processing, have embraced DL as a promising methodology (Ibid.). In comparison to traditional ML, which addresses problems by dividing them into manageable pieces and solving them independently and efficiently, DL exhibits a longer training time for algorithms. However, this extended duration ultimately results in faster training times due to its comprehensive and accurate learning process, ultimately benefiting the overall performance (Ibid.). It is important to note that ML is adaptable to various computing systems and requires less data to construct the training dataset. Conversely, DL operates optimally on complex and high-performance machines, but it demands substantial amounts of data to complete the training set. This can be attributed to its holistic approach in addressing the issue across a spectrum, ensuring thorough and accurate learning (Ibid.).

In response to the evolving landscape of cyber threats, conventional and traditional methods alone do not suffice to ensure national security. This has prompted the adoption of dynamic approaches such as intrusion detection systems (IDS) to bolster cybersecurity [20]. The concept of intrusion detection was first introduced in approximately 1980 by security expert James Anderson, defining it as a mechanism or activity capable of detecting attacks and preventing unauthorised access to a system through continuous monitoring [28]. Consequently, scholars have investigated the recent advancements in cyberterrorism and intrusion detection, recognising the need for a system capable of serving as a preventive tool against these cybersecurity threats [23]. IDSs serve the purpose of identifying potential threats or malicious activities, constituting a network-level defence mechanism to safeguard computer networks [36]. To effectively counter such attacks, DL is integrated into the detection system [23]. By analysing network traffic, DL aids in the prevention of unauthorised access and the elimination of threats that exploit vulnerabilities in the network infrastructure [54]. However, due to the lack of structured formats in the generated traffic, classifying activities into distinct categories proves to be a challenging task [40]. Research suggests that implementing the detection system as an anomaly-based approach with a learning system would be more effective [23]. Anomaly-based detection methods focus on identifying deviations from normal network behaviour, in contrast to signature-based methods where attacks are detected based on pre-defined patterns stored in a database as suspicious activities [28].

In essence, intrusion detection methods can be broadly classified into two categories: signature-based and anomaly-based approaches. The former relies on a configuration of known signatures corresponding to previously identified attacks [36]. This method proves accurate and effective in detecting known attacks and threats. However, it might encounter challenges when faced with novel or diverse attacks, as it struggles to match patterns and subsequently fails to detect intrusions [28]. On the contrary, anomaly-based methods aim to assess network activity in relation to pre-established thresholds, and an alert is triggered if the behaviour surpasses these predefined thresholds (Ibid.). This approach tends to generate logs to record malicious activities occurring in the system [36]. Nevertheless, one notable drawback of anomaly-based detection lies in its tendency to produce a considerable number of false positives, which remains a persistent concern among detection methods [28]. Despite this challenge, the anomaly-based detection systems offer several advantages, particularly in its capacity for early detection, which is conducive to detailed and high-quality information and evidence collection [23]. Within IDSs, there are three primary classifiers, namely single, hybrid, and ensemble classifiers, that constitute the main techniques employed (Ibid.). The following subsections examine these three classifiers in detail.

8.5.1   Single Classifiers

An ML technique utilised in the development of IDSs is commonly referred to as a single classifier [16]. These classifiers, also known as Single Machine Learning Classifiers, are characterised by the adoption of a sole algorithm responsible for the classification process [36]. Within the scope of this chapter, particular attention will be given to four techniques exclusively applied and employed in this classification type: Fuzzy Logic, Support Vector Machines (SVM), Decision Trees, and K-Nearest Neighbour (K-NN).

A.   Fuzzy Logic

Fuzzy Logic, also referred to as fuzzy set theory, holds substantial promise as one of the most potent techniques in single classification [44]. Besides its application in anomaly-based detection, this approach finds extensive use in various engineering domains (Ibid.). Fuzzy Logic plays a significant role in decision-making and reasoning within the classification process, making it an appealing choice for corporations and organisations due to its human-like characteristics (Ibid.). However, unlike conventional binary decision-making, where outcomes are limited to either false (0) or true (1), fuzzy set theory relaxes these boundaries, allowing the degree of truth to assume values between 0 and 1 [16].

B.   Support Vector Machines (SVM)

Introduced in the mid-1990s, Support Vector Machines (SVMs) have garnered extensive research attention within the domain of ML, particularly concerning IDSs [16]. SVMs utilise the training data to learn the distinguishing characteristics of normal behaviour, enabling them to classify remaining data instances as anomalies (Ibid.). By leveraging a hyperplane, SVMs categorise collected data into two distinct groups: the anomalies or programming problem, and the supporting vectors [44]. SVMs are noteworthy for their ability to transform "the input vector into a higher dimensional feature space", enabling them to determine the optimal separating hyperplane. This quality makes SVMs particularly attractive for binary classification tasks [52, p. 11995]. Furthermore, SVMs incorporate a user-specified parameter, known as the penalty factor, which allows users to strike a balance between misclassified samples and decisions pertaining to the decision boundary [52].

C. Decision Trees

As per research findings, Decision Trees constitute one of the simplest yet highly powerful tools in text classification [16, 44]. This algorithm holds significant importance due to its capacity to pre-establish a set of attributes, enabling efficient data classification [44]. However, to classify a sample, a series of predetermined decisions must be made. Subsequently, each decision influences the subsequent one, culminating in a tree-like structure (Ibid.). Notably, in the context of cyberterrorism, there arose a requirement for a classifier capable of predicting "the value of a target class for an unseen test instance, based on several already known instances," a task for which Decision Trees were deemed remarkably effective [16, p. 10].

D. K-Nearest Neighbour

K-Nearest Neighbour (K-NN), despite being one of the earliest classification techniques, retains its efficacy and continues to find consistent usage [44]. In K-NN, the number of samples (k) in the training dataset nearest to the stored sample is determined, and subsequently, the class label is assigned based on the most frequent occurrence among these k samples for the test samples [16]. Altering the value of k yields different performances, and the K-NN method identifies nearest neighbours by searching input vectors and classifying new instances to arrive at the result [44]. It possesses the ability to calculate the approximate difference between two points, rendering it the most straightforward and nonparametric approach within IDSs (Ibid.).

8.5.2 Hybrid Classifiers

The concept underlying Hybrid Classifiers involves the amalgamation of two or more ML techniques to achieve a substantial enhancement in system performance [52]. Employing a combination of multiple algorithms, whether supervised or unsupervised, yields better outcomes compared to a single classifier [28]. This procedure comprises two functional components: the first component processes raw data and provides immediate results, while the second component utilises these results to generate the final outcome [44, 52]. In this approach, certain clustering-based techniques are applied during the pre-processing step to identify and eliminate anomalies. Subsequently, the results of the clustering process serve as training samples for pattern recognition, facilitating the design of the classifier [16]. The adoption of this hybrid approach substantially enhances efficiency, a fact supported by extensive research demonstrating the superiority of hybrid methodologies when compared to individual learning approaches [36].

8.5.3 Ensemble Classifiers

The scope of literature concerning Ensemble Classifiers has garnered increasing interest, particularly due to its current status as the least explored type of classifier among the three [16]. Introduced in the late 1980s, Ensemble Classifiers are employed when multiple weak learners are combined to enhance classifier performance (Ibid.). The individual classifiers, such as K-NN, SVM, or Decision Trees, are considered weak learners due to their limited decision-making capabilities, however, their combination results in a more robust and efficient performance [36]. Ensemble Classifiers differ from a set of base classifiers, with these baselines utilised for validation and performance evaluation by aggregating their predictions [20, 44]. Integrating complementary classifiers augments robustness and accuracy of performance while enhancing representativeness in statistical and computational terms [16]. Additionally, this classifier provides insights into the machine's capacity and its proficiency in identifying attacks, along with the extent of misclassifications or inaccuracies in the process [44]. Various ensemble methods and strategies exist to combine weak learners, including bagging, boosting, and weighted majority voting (WMV) [16, 20, 36]. Among these approaches, WMV is widely recognised as particularly successful due to its simplicity, effectiveness in practice, and intuitive nature [20].

As depicted in Fig. 8.1 [42], feature selection plays a pivotal role in alleviating the system's burden and identifying a relevant subset that collaborates with the classifier to achieve high classification accuracy.

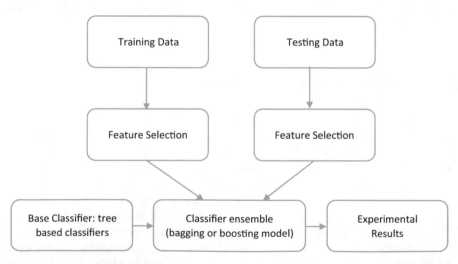

Fig. 8.1 The proposed ensemble model. Adapted from "Improving performance of intrusion detection system using ensemble methods and feature selection," by Pham et al. [42], *Proceedings of the Australasian computer science week multiconference*, p. 4. Copyright 2018 by the Association for Computing Machinery (ACM)

8.6 Counter Terrorism Policing

This section aims to enhance understanding of the pressing issue of cyber terrorism by analysing the recruitment methods employed by cyber terrorists, with a particular focus on delineating their pre-terrorist profiles. Additionally, the section delves into the efforts made by the Institute for Strategic Dialogue in countering online violent extremism and providing guidance to governmental entities [21]. Furthermore, the discussion encompasses the Global Terrorism Database, which is overseen by the Study of Terrorism and Responses to Terrorism and serves as an accessible open-source repository containing comprehensive information on international terrorist attacks and incidents worldwide since 1970 [48].

8.6.1 Pre-terrorist Profile

Warning signs in the context of targeted violence are defined as behavioural indicators that precede and are related to acts of violence, with potential predictive capabilities [30, 32]. Meloy [30] and Meloy and O'Toole [32] have identified eight distinct warning behaviours that indicate a proclivity for violence. These warning behaviours consist of identification, fixation, pathway warning behaviours, novel aggression warning behaviours, last resort warning behaviours, energy burst, leakage warning behaviours, and directly communicated threat warnings [22]. The behaviour of identification is closely associated with an individual's affiliation with a specific group or person, particularly with close ties to terrorist organisations or engagement with terrorist propaganda [31]. Conversely, 'fixation' shares similarities with identification, involving an individual's intense interest and passion. However, when it becomes pathological and disrupts social functioning, it becomes a potential danger and societal threat. Unlike identification, fixation is characterised by a pathological admiration and obsession with a particular cause or group, while identification warning signs originate from the individual [22]. Additionally, 'leakage' represents a warning behaviour where information about the planning and execution of an attack is inadvertently or intentionally communicated to a third party, a cue that can be detected by certain automated systems [41]. One such system is the LIWC, as discussed in Sect. 8.5. LIWC is capable of extracting psychological insights from vast amounts of text [38]. In this context, linguistic markers from LIWC are used as inputs to algorithms to identify radical behaviours and characteristics [22]. However, these algorithms are not designed for automatic detection but rather to narrow down and focus on pertinent information within the text (Ibid.). The initial step in detecting radical language involves defining a pre-established set of words through databases such as WordNet, which are relevant to identifying radical phrases and violent threats. Subsequently, posts flagged with these markers are compared with the word list for further analysis (Ibid.).

Cohen et al. [8] characterise warning signs as readily identifiable behaviours on SMPs. However, it is crucial to acknowledge that a substantial number of posts on these platforms might generate false positives, as individuals often engage in discussions concerning sensitive issues without any intent for violent action or espousing radical beliefs. Therefore, in order to minimise the occurrence of false positives, it is essential to concentrate efforts on forums, chats, and websites recognised for disseminating content associated with violent extremism [22]. Moreover, language usage on social media is highly informative about the nature of conversations. Pronouns used in discourse can reveal the dynamics between participants, while the use of positive emotion words and expressions of agreement can indicate the level of consensus within the conversation [51]. Additionally, the word count serves as a revealing metric, shedding light on the dominance of conversational participants and their respective levels of engagement (Ibid.).

Moghaddam's [35] conceptual framework, known as the "Staircase to Terrorism", provides insight into the individual journey within the radicalisation process [39]. The framework consists of distinct stages, starting with the 'ground floor', which emphasises the psychological interpretation of material conditions, with a particular focus on the Relative Deprivation Theory [27]. At this initial stage, individuals who perceive their material circumstances as unfair or experience a sense of injustice might be motivated to progress further in their search for a stronger identity or an improved quality of life (Ibid.). The Relative Deprivation Theory highlights the impact of unfulfilled expectations, which can negatively affect individuals and potentially lead to radical thoughts and behaviours (Ibid.).

Moving up to the 'first floor' of the model, one encounters the concept of "perceived options to fight unfair treatment". Here, individuals identify potential opportunities to enhance their mental well-being and living conditions [39]. Moghaddam [35] examines the impact of this stage on an individual's identity through the lens of the Terror Management Theory, elucidating how it influences their perspective. During this phase, individuals might strive for social mobility to improve their life satisfaction. However, if these efforts prove futile, some individuals may turn to alternative means to fulfil their needs, driven by a desire for a sense of community [27]. One influential perspective is the Rational Choice Theory, which suggests that radical behaviours and threats can emerge from the aspiration to achieve socio-political goals (Ibid.) [9]. Additionally, individuals on this 'first floor' might experience a sense of low self-efficacy, as defined by Bandura [1], regarding their belief in their capacity to attain desired objectives within society. If an individual perceives a lack of meaningful contribution within society, they are likely to transition to the subsequent floor in the model [27].

The second floor of Moghaddam's model focuses on the concept of the displacement of aggression [39]. Moghaddam draws upon Freud's theory of displaced aggression to suggest that anti-Americanism is employed as a means to divert criticism away from Middle Eastern governments [27]. In this stage, individuals channel their anger and frustration towards entities that might not necessarily be responsible for the initial aggression. Nevertheless, the display of aggression displacement facilitates the individual's progression to the next floor. On the third floor, individuals become integrated into a terrorist group and become increasingly involved in the moral engagement and reasoning behind the organisation's goals (Ibid.). The sense of belonging and attachment established in the preceding stages grants distinct identities to these individuals, with terrorists achieving this through instilling fear and isolation [35]. Understanding the use of violence in this context also involves recognising the disengagement from inhibitory mechanisms. These mechanisms, seen as crucial psychological barriers that typically deter individuals from inflicting harm upon themselves or others, assume significance [27] in such understanding.

Floor four of the model corresponds to the solidification of thoughts and beliefs, resulting in an intensified division between "us" and "them". This reinforces the individual's perception of the terrorist organisation as increasingly legitimate [39]. Additionally, on this floor, individuals assume specialised roles within the group, further accentuating their sense of identity. These roles might stem from their talents, motivations, and the terrorist group's needs [27]. This adoption of a role not only indicates the presence of indoctrination but also signifies a gradual assimilation of the specific morals held by the group (Ibid.). The final floor, floor five, encompasses the translation of thoughts and beliefs into actual action [39]. Through social categorisation and psychological distancing from the intended victims, individuals can be trained to carry out acts of violence without experiencing guilt or empathy [27]. This floor primarily revolves around theories of obedience and conformity, while also considering how all the preceding floors contribute to this ultimate stage (Ibid.). The transition between different floors within the model is associated with inflicting harm upon others, signifying that once an individual advances to the next level, a point of no return is reached. Consequently, the progression between floors suggests that the individual seeks further solutions and alternative perspectives to address their existing challenges [39].

8.6.2   Institute for Strategic Dialogue

The Institute for Strategic Dialogue (ISD) is an independent non-profit organisation specialising in countering violent extremism online and providing governments with guidance on defending and addressing such threats [21]. Since approximately 2006, ISD has been a key leading organisation in combating various forms of extremism. It actively collaborates with governments, communities, and businesses to mobilise concerted efforts in response to this issue. One of ISD's core components is its Digital Analysis Unit, which tracks hate and extremism online, identifies disinformation, and scrutinises online manipulation [19]. This unit boasts one of the most advanced open-source data analytics and risk identification capabilities within this research domain. ISD is claimed to have successfully removed numerous Facebook pages, Twitter accounts, Telegram channels, white supremacist content, and Islamist disinformation (Ibid.). Moreover, the Digital Analysis Unit endeavours to pre-emptively detect extremist content and behaviours before they manifest as threats and become public. The collected data are intended to inform policymakers and companies in devising effective strategies to counter violent and threatening extremism online. This, in turn, aims to foster safer communities and enhancing the resilience of young people against extremist and hate speech both online and offline (Ibid.).

Beyond its analytical efforts, ISD also designs and implements primarily educational programs aimed at empowering practitioners and governments to proactively prevent extremism. These initiatives have reached over 12 countries, making a substantial impact in the field. Notably, ISD's contributions extend to policy development, providing training and support to more than 40 governments and 150 cities worldwide (Ibid.). One of the notable and effective collaborations involves the partnership between ISD and the Global Counterterrorism Forum (GCTF), an informal platform that brings together policymakers and practitioners worldwide to exchange strategies and tools for preventing and mitigating the terrorist threat (Ibid.). Through this alliance, these platforms have pooled their expertise to present multiple recommendations, exemplified by the "Zurich-London Recommendations on Preventing and Countering Violent Extremism and Terrorism Online" [12]. This document emphasises the importance of a balanced approach for governments when countering the use of the internet for extremist behaviour and terrorist threats, taking into account privacy, religious sensitivities, and freedom of expression (Ibid.). In the context of countering terrorism online, the document outlines two main approaches: content-based and communications-based responses. A content-based response involves collaboration with international agencies and private companies to remove radical content and filter information to ensure its accuracy and non-threatening nature. On the other hand, a communications-based response involves the strategic challenge of terrorist organisations through communicative methods and counter-narratives (Ibid.). This chapter underscores the significance of regularly employing both of these responses in the mitigation efforts and argues that incorporating them within policy is essential for effectively countering violent extremism (Ibid.).

8.6.3   Global Terrorism Database

The Global Terrorism Database (GTD), managed by the Study of Terrorism and Responses to Terrorism (START), serves as an accessible open-source repository containing comprehensive information on international terrorist attacks and incidents worldwide since 1970. As elucidated by START [48], an organisation primarily funded by various US agencies and departments, the database comprises over 200,000 cases, encompassing details such as the date, location, number of casualties, responsible group, and the types of weapons utilised. Established in 2001 under the guidance of Dr. Gary LaFree and his team, the GTD was devised with the purpose of potentially forecasting the movements and dynamics of terrorist organisations. In essence, while AI has long been employed to predict potential attacks through classification and analysis of language patterns in remote corners of the online realm, this novel method represents a more pragmatic approach to monitoring, thereby enhancing the accuracy of preventive measures (Ibid.). According to START, the database gathers information on a minimum of 45 diverse variables for each case, with this number occasionally escalating to approximately 120 variables. Moreover, in certain instances, a meticulous review of four million new articles and sources is undertaken to establish a comprehensive and well-rounded understanding of the attack (Ibid.). For instance, in a study conducted by Huamani et al. [18], the GTD served as the primary methodology for predicting the number and location of terrorist attacks, as well as the types of attacks, through the application of ML techniques. In this research, ML algorithms were trained on the GTD datasets to identify optimal methods and generate models that could yield the most accurate predictions (Ibid.). Subsequently, once the results were obtained, the model could be refined and enhanced based on the need for additional features or unmet expectations, thereby rendering it applicable to real-world scenarios. By combining ML techniques and the GTD data, this research revealed a discernible increase in terrorist attacks since 2010, providing precise information about the locations and types of attacks (Ibid.).

To enhance the precision and effectiveness of the GTD, the team of researchers at START has adopted a comprehensive approach that combines both manual and automated strategies [13]. The current methodology involves the initial collection of articles containing information on terrorist attacks or related topics, with potential translations undertaken as needed, thereby narrowing down the vast corpus of online articles to a pertinent subset (Ibid.). Subsequently, ML and language processing techniques are utilised to refine this subset, identifying articles that are most relevant to the topic and of the highest quality. Moreover, manual curation is performed to emphasise unique and relevant information required by the GTD team, ensuring a meticulous and accurate representation of the data. In terms of volume, the GTD researchers review a maximum of 16,000 articles each month for potential inclusion in the database. Furthermore, the coding process is organised into six distinct domains, namely "location, perpetrators, targets, weapons and tactics, casualties and consequences, and general information" [13, p. 11]. In order for an incident or attack to be included in the GTD, it must meet three primary characteristics. Firstly, it must be an intentional act with motives and premeditation. Secondly, the incident must entail some form of threat or violence, which may encompass messages or physical acts directed at humans or properties. Lastly, the database excludes attacks carried out by state or government actors, thereby focusing on incidents perpetrated by sub-national actors [13].

In 2016, there were approximately 1100 terrorist attacks worldwide, resulting in nearly 2400 fatalities [14]. A comprehensive analysis of terrorism in 2019 indicated that Iraq experienced more terrorist attacks than any other country between 2013 and 2017, while Afghanistan witnessed a 16% decline in terrorism-related deaths (Ibid.). However, monthly trends in 2019 revealed that violence attributed to terrorist acts persisted in Iraq, and the previously observed rate of decline was slowing, indicating a cessation of violence reduction (Ibid.). Furthermore, the background report elucidated that although violence had somewhat decreased in Iraq from 2012 to 2019, the influence of terrorist groups had extended to 35 different countries in 2018, increasing from 31 in 2019. Of the new countries added in 2019, Sri Lanka, the Netherlands, and

Mozambique were noteworthy examples (Ibid.). This dataset and background report offer a glimpse of the work undertaken by GTD and emphasise the significance of governments using this information to inform their counter-terrorism strategies.

8.7   Discussion

Terrorist organisations have harnessed the power of the internet, SMPs and other associated technologies to significantly advance their agendas. This technological shift has enabled them to disseminate their propaganda, spread misinformation, and recruit vulnerable individuals at a global scale [17, 38] never witnessed before. In response to this growing threat, LEAs and other governmental entities across the world have enhanced their traditional counterterrorism measures and approaches. Recognising the profound impact of cyberterrorism on national security, government approaches have also included the exploration and deployment of a wide range of technological tools and novel approaches such as ML to address these security threats more effectively [53].

However, concerns regarding the timing and nature of future terrorist activities have continued to persist, particularly since the 9/11 terrorist attacks [11]. These concerns have been further exacerbated by the adaptability of extremist groups in embracing and exploiting technology. While cyberterrorists have not yet demonstrated proficiency in orchestrating diverse and sophisticated cyberattacks aimed at disrupting and compromising national security and critical infrastructure, such as communication systems and vital transportation services, they have nevertheless effectively utilised the internet and SMPs for propagating ideologies, disseminating propaganda, spreading disinformation, and recruiting vulnerable individuals to their cause, as discussed above [17, 38]. Considering this, it is therefore essential to acknowledge that as technology advances, so do the tactics of these individuals and groups, as they adeptly exploit the internet and social media for their malevolent intentions [53]. Therefore, continual vigilance and further advancements in counterterrorism strategies are imperative to combat these evolving threats effectively.

Within this context, national governments bear significant responsibility in countering the phenomenon of cyberterrorism [17]. This responsibility encompasses fostering cooperation not only among various jurisdictions but also between intelligence agencies and law enforcement entities. By establishing a robust legal infrastructure, governments can enhance their capabilities in defending against cyber-attacks. This, in turn, facilitates the effective prosecution of all types of cybercrimes including cyberterrorism. Therefore, to achieve these goals, governments should take proactive steps in financing research and expanding the body of literature concerning the development of countermeasures and technologies capable of tracking and preventing cyber terrorists. Similarly, governments should be tasked with the vital role of educating civilians on identifying cyber terrorist threats and implementing protective measures (Ibid.). Furthermore, considering that an increase in the likelihood of terrorist groups being detected or hacked by LEAs raises the costs associated with committing acts of cyberterrorism (Ibid.), governments should adopt comprehensive counterterrorism strategies. These strategies should cover various dimensions, including psychological, law enforcement, educational, and legal aspects. By doing so, governments can increase the costs and punitive measures associated with terrorists' actions (Ibid.).

8.8 Conclusion

This chapter has delved into the technical aspects of AI in the context of cyber security and cyber terrorism. It has explored the significant role played by ML and DL in detecting extremist content and activities online. By examining these aspects, the chapter highlighted the potential of AI-driven techniques in augmenting cyber security measures and countering cyber terrorism more effectively. The chapter provided a comprehensive examination of various ML techniques applicable to combat cyberterrorism. The exploration of supervised, unsupervised, semi-supervised, and reinforcement learning methods shed light on their effectiveness in addressing cyberterrorism challenges. Moreover, the investigation of essential functions of ML algorithms, such as regression, clustering, and classification, unveiled their utility in identifying patterns and trends in cyberterrorism-related data. The detailed analysis of regression techniques, including linear, polynomial, and ridge regression, further contributed to understanding their implications in the context of cyber security. Furthermore, the discussion on feature extraction and selection methods underscored their significance in mitigating noise and enhancing data processing for cyberterrorism detection.

Similarly, the exploration of the intersection between DL and IDSs, encompassing signature-based and anomaly-based approaches, provided detailed insights into their application in cyber security and cyber terrorism. Additionally, the examination of Single Classifiers, Hybrid Classifiers, and Ensemble Classifiers showcased their relevance and adaptability within the context of cyberterrorism. Finally, the chapter provided an overview of counter-terrorism policing, which contributed to a deeper understanding of the subject matter by analysing the recruitment strategies employed by cyber terrorists. A specific emphasis was placed on delineating the pre-terrorist profiles of cyber terrorists. Additionally, the chapter explored the efforts undertaken by the ISD in countering violent extremism in the online sphere, while offering governmental entities guidance on mitigating and addressing such threats. Furthermore, this examination involved a comprehensive investigation of the GTD, which serves as an accessible open-source repository, encompassing extensive information on international terrorist attacks and incidents worldwide.

Finally, by exploring the various techniques and methods available, this research contributes to the advancement of cyber security and cyberterrorism measures and lays the foundation for future developments in combating extremist content and activities online. As the threat of cyberterrorism continues to evolve, the insights provided in this chapter serve as a valuable resource for researchers, practitioners, and policymakers seeking to strengthen national security while preserving fundamental rights and privacy in the digital age.

References

1. Bandura A (1997) Self-efficacy: the exercise of control. Freeman, New York
2. Bird SJ (2013) Security and privacy: why privacy matters. Sci Eng Ethics 19(3):669–671. https://doi.org/10.1007/s11948-013-9458-z
3. Bishop CM (2006) Pattern recognition and machine learning. Springer. ISBN 978-0-387-31073-2
4. Cai J, Luo J, Wang S, Yang S (2018) Feature selection in machine learning: a new perspective. Neurocomputing 300:70–79
5. Ch R, Gadekallu TR, Abidi MH, Al-Ahmari A (2020) Computational system to classify cyber crime offenses using machine learning. Sustainability 12:4087. https://doi.org/10.3390/su1210 4087
6. Chandrasekar P, Qian K (2016) The impact of data pre-processing on the performance of Naïve Bayes classifier. In: IEEE 40th annual computer software and applications conference, pp 618–619
7. Chen H, Chung W, Xu JJ, Wang G, Qin Y, Chau M (2004) Crime data mining: a general framework and some examples. Computer 37:50–56
8. Cohen K, Johansson F, Kaati L, Mork JC (2014) Detecting linguistic markers for radical violence in social media. Terrorism Polit Violence 26(1):246–256. https://doi.org/10.1080/095 46553.2014.849948
9. Crenshaw M (1992) How terrorists think: what psychology can con-tribute to understanding terrorism. In: Howard L (ed) Terrorism: roots, impact, responses. Praeger, New York, pp 71–80
10. Dash M, Liu H (1997) Feature selection for classification. Intell Data Anal 1:131–156
11. Elovici Y, Kandel A, Last M, Shapira B, Zaafrany O (2004) Using data mining techniques for detecting terror-related activities on the web. J Inf Warfare 3(1):17–29
12. Global Counterterrorism Forum (GCTF) (2017) Zurich-London recommendations on preventing and countering violent extremism and terrorism online. Countering violent extremism (CVE) working group strategic communications initiative. ISD Global. Retrieved from: https://www.isdglobal.org/wp-content/uploads/2021/02/GCTF-Zurich-Lon don-Recommendations-ENG.pdf
13. Global Terrorism Database (GTD) (2021) Codebook: methodology, inclusion criteria, and variables. START. University of Maryland, pp 1–63. Retrieved from: https://www.start.umd. edu/gtd/downloads/Codebook.pdf
14. Global Terrorism Overview (GTO) (2020) Terrorism in 2019: background report. START. Retrieved from (12/04/2022): https://www.start.umd.edu/pubs/START_GTD_GlobalTerror ismOverview2019_July2020.pdf
15. Gupta P, Varshney P, Bhatia MPS (2017) Identifying radical social media posts using machine learning. Technical report. https://doi.org/10.13140/RG.2.2.15311.53926
16. Haq NF, Onik AR, Hridoy MAK, Rafni M, Shah FM, Farid DM (2015) Application of machine learning approaches in intrusion detection system: a survey. IJARAI-Int J Adv Res Artif Intell 4(3):9–18
17. Hua J (2012) How can we deter cyber terrorism? Inf Secur J Glob Perspect 21:102–114
18. Huamani EL, Alicia AM, Roman-Gonzalez A (2020) Machine learning techniques to visualize and predict terrorist attacks worldwide using the global terrorism database. Int J Adv Comput Sci Appl 11(4):526–570
19. Institute for Strategic Dialogue (ISD) (2022) Institute for strategic dialogue. Retrieved from: https://www.isdglobal.org/about/

20. Jabbar MA, Aluvalu R, Reddy SSS (2017) Cluster based ensemble classification for intrusion detection system. In: Proceedings of the 9th international conference on machine learning and computing, pp 253–257

21. Jackson P (2017) If you want to understand anti-fascist movements, you need to know this history. HuffPost. Retrieved 29 Mar 2022: https://www.huffpost.com/entry/anti-fascist-mov ements_b_599b11b8e4b04c532f4348f4

22. Johansson F, Kaati L, Sahlgren M (2016) Detecting linguistic markers of violent extremism in online environments, chap 18. IGI Global, pp 374–390. https://doi.org/10.4018/978-1-5225-0156-5.ch018

23. Karatas G, Demir O, Sahingoz OK (2018) Deep learning in intrusion detection systems. International congress on big data, deep learning and fighting cyber terrorism. IEEE. Retrieved from: IEEE Xplore full-text PDF

24. Khanzode KCA, Sarode RD (2020) Advantages and disadvantages of artificial intelligence and machine learning: a literature review. Int J Libr Inf Sci (IJLIS) 9(1):3

25. Kitrum (2020) How to use machine learning in cybersecurity? Blog. Retrieved from: How to use machine learning in cybersecurity (kitrum.com)

26. Li Y (2017) Deep reinforcement learning: an overview. arXiv preprint arXiv:1701.07274. Accessed 05 Mar 2022: [1701.07274] Deep reinforcement learning: an overview (arxiv.org)

27. Lygre RB, Eid J, Larsson G, Ranstorp M (2011) Terrorism as a process: a critical review of Moghaddam's "Staircase to Terrorism." Pers Soc Psychol Scand J Psychol 52:609–616. https://doi.org/10.1111/j.1467-9450.2011.00918.x

28. Maidamwar PR, Bartere MM, Lokulwar PP (2021) A survey on machine learning approaches for developing intrusion detection system. In: International conference on innovative computing & communication (ICICC), pp 1–8

29. Mandaviya H, Sathwara S (2021) Cyber terrorism-related multimedia detection using deep learning—a survey. In: Gunjan VK, Zurada JM (eds) Proceedings of international conference on recent trends in machine learning, IoT, smart cities and applications, pp 135–143

30. Meloy JR (2011) Approaching and attacking public figures: a contemporary analysis of communications and behaviour. In: Chauvin C (ed) Threatening communications and behaviour: perspectives on the pursuit of public figures. The National Academies Press, Washington, D.C., pp 75–101

31. Meloy JR, Mohandie K, Knoll JL, Hoffmann J (2015) The concept of identification in threat assessment. Behav Sci Law 33(2–3):213–237. https://doi.org/10.1002/bsl.2166. PMID: 25728417

32. Meloy JR, O'Toole ME (2011) The concept of leakage in threat assessment. Behav Sci Law 29(4):513–527. https://doi.org/10.1002/bsl.986. PMID: 21710573

33. Meyer J (2018) Machine learning and tracking terrorists. The Cipher Brief. Retrieved from: Machine learning and tracking terrorists (thecipherbrief.com)

34. Mishra P, Varadharajan V, Tupakula U, Pilli ES (2019) A detailed investigation and analysis of using machine learning techniques for intrusion detection. IEEE Commun Surv Tutorials 21(1):686–728

35. Moghaddam FM (2005) The staircase to terrorism: a psychological exploration. Am Psychol 60(2):161–169

36. Musa US, Chharbra M, Ali A, Kaur M (2020) Intrusion detection system using machine learning techniques: a review. In: International conference on smart electronics and communication (ICOSEC). IEEE, pp 149–155

37. Ngejane CH, Eloff JHP, Sefara TJ, Marivate VN (2021) Digital forensics supported by machine learning for the detection of online sexual predatory chats. Forensic Sci Int Digit Invest 36:2666–2817. https://doi.org/10.1016/j.fsidi.2021.301109

38. Nouh M, Nurse JRC, Goldsmith M (2019) Understanding the radical mind: identifying signals to detect extremist content on Twitter. IEEE, pp 98–103. 978-1-7281-2504-6/19

39. Orsini A, Vecchioni M (2019) How does one become a terrorist and why: theories of radicalization. Department of Political Science. Luiss. Retrieved from: https://tesi.luiss.it/id/eprint/24388

40. Patel S, Sondhi J (2014) A review of intrusion detection technique using various technique of machine learning and feature optimization technique. Int J Comput Appl (0975-8887) 93:43–47
41. Pelzer R (2018) Policing of terrorism using data from social media. Eur J Secur Res. 3:163–179. https://doi.org/10.1007/s41125-018-0029-9
42. Pham NT, Foo E, Suriadi S, Jeffrey H, Lahza HFM (2018) Improving performance of intrusion detection system using ensemble methods and feature selection. In: Proceedings of the Australasian computer science week multiconference, pp 1–6
43. Rish I (2001) An empirical study of the Naïve Bayes classifier. IJCAI workshop on empirical methods in artificial intelligence, vol 3(22), pp 41–46
44. Shah AA, Hayat MS, Awan MD (2015) Analysis of machine learning techniques for intrusion detection system: a review. Int J Comput Appl 119:19–29
45. Simanjuntak DA, Ipung HP, Iim C, Nugroho AS (2010) Text classification techniques used to facilitate cyber terrorism investigation. In: Second international conference on advances in computing, control, and telecommunication technologies. IEEE, pp 198–200. https://doi.org/10.1109/ACT.2010.40
46. Solove DJ (2011) Nothing to hide: the false tradeoff between privacy and security. Yale University Press, London
47. Sonawane TR, Al-Shaikh S, Shinde R, Shaikh S, Sayyad AG (2015) Crime pattern analysis visualization and prediction using data mining. Int J Adv Res Innov Ideas Educ 1:681–686
48. START (2022) Global terrorism database. National Consortium for the Study of Terrorism and Responses to Terrorism. Retrieved from 12 Apr 2022: https://www.start.umd.edu/gtd/about/
49. Sutton RS (1992) Introduction: the challenge of reinforcement learning. Mach Learn 8:225–227
50. Tang J, Alelyani S, Liu H (2014) Feature selection for classification: a review. In: Data classification: algorithms and applications, vol 37, pp 1–25
51. Tausczik YR, Pennebaker JW (2010) The psychological meaning of words: LIWC and computerized text analysis methods. J Lang Soc Psychol 29(1):24–54. https://doi.org/10.1177/026192 7X09351676
52. Tsai CF, Hsu YF, Lin CY, Lin WY (2009) Intrusion detection by machine learning: a review. Expert Syst Appl 36:11994–12000
53. UNICRI & UNCCT. (2021). Algorithms and terrorism: the malicious use of artificial intelligence for terrorist purposes. United Nations Office of Counter-Terrorism (UNOCT). Available at: https://www.un.org/counterterrorism/sites/www.un.org.counterterrorism/files/malici ous-use-of-ai-uncct-unicri-reporthd.pdf. Accessed 28 Sept 2023.
54. Ustebay S, Turgut Z, Aydin MA (2018) Intrusion detection system with recursive feature elimination by using random forest and deep learning classifier. International congress on big data, deep learning and fighting cyber terrorism. IEEE, pp 71–76
55. Van den Hoven J, Lokhorst GJ, Van de Poel I (2012) Engineering and the problem of moral overload. Sci Eng Ethics 18(1):143–155. https://doi.org/10.1007/s11948-011-9277-z
56. Verhelst HM, Stannat AW, Mecacci G (2020) Machine learning against terrorism: how big data collection and analysis influences the privacy-security dilemma. Sci Eng Ethics 26:2975–2984. https://doi.org/10.1007/s11948-020-00254-w
57. Verma C, Verma V (2018) Predictive modeling of terrorist attacks using machine learning. Int J Pure Appl Math 119(15):49–61. http://www.acadpubl.eu/hub/
58. Wu H, Yuan N (2018) An improved TF–IDF algorithm based on word frequency distribution information and category distribution information. In: Proceedings of the 3rd international conference on intelligent information processing, Guilin, Chin, 4–6, pp 211–215
59. Zhang H, Li D (2007) Naïve Bayes text classifier. In: IEEE international conference on granular computing, pp 708–711. https://doi.org/10.1109/GrC.2007.40

Chapter 9
Analysing Ethical, Legal, Technical and Operational Challenges of the Application of Machine Learning in Countering Cyber Terrorism

Abstract This chapter serves as a continuation of Chap. 8, which critically analysed the fundamental Machine Learning (ML) methods underpinning the counterterrorism landscape. Building upon this foundation, the chapter delves deeper into the multifaceted complexities that arise when applying ML techniques within the context of countering cyber terrorism. The primary aim of this chapter is to examine a spectrum of challenges spanning the technical, operational, ethical, and legal dimensions inherent to the application of Artificial Intelligence (AI) and ML in counterterrorism efforts. In addressing ethical and legal considerations, the chapter investigates issues pertaining to individual autonomy, human rights preservation, algorithmic bias, privacy preservation, and the implications of mass surveillance. It also navigates the complex legal landscape, including discussions on hybrid classifiers, the assignment of accountability, and jurisdictional quandaries that emerge in this domain. Furthermore, from a technical and operational standpoint, this chapter delves into the complexities of big data management, the inherent challenges posed by class imbalance, the curse of dimensionality, the identification of spurious correlations, the detection of lone-wolf terrorists, the implications of mass surveillance, and the application of predictive analytics. It also illuminates the critical aspects of transparency and explainability, while shedding light on the potential unintended consequences that can arise from the integration of AI and ML into counterterrorism practices. The chapter contributes significantly to the existing body of knowledge by offering a comprehensive examination of the myriad challenges faced in the application of AI and ML within the context of countering cyber terrorism. Furthermore, by systematically analysing technical, operational, ethical, and legal obstacles, the chapter provides a nuanced understanding of the intricacies involved and lays the groundwork for informed strategies and solutions that ensure both security enhancement and the preservation of fundamental human rights.

Keywords National security · Cyberterrorism · Extremism · Artificial intelligence · Machine learning · Digital forensics · Challenges · Ethics · Legal · Bias · Algorithms · Operational challenges · Counter terrorism policing · Privacy

9.1 Introduction

The internet's ability to serve as a shield against governmental scrutiny and policing renders it an alluring facilitator for radicalisation processes and the execution of terrorist attacks [60]. However, this dual nature presents both advantages and challenges for national intelligence and security agencies. The copious and diverse information and data available online pose one of the most significant challenges faced by these agencies. Estimates suggest that the internet contains approximately 1.8 exabytes of data per day, with about 80% of this data being unstructured and categorised as noisy (Ibid.). Consequently, analysing this vast amount of data manually is unfeasible, thereby necessitating the development of advanced technology such as AI to cope with the increasing volume and sophisticated distribution methods. AI has demonstrated notable success in regulating and addressing terrorist propaganda within cyberspace such as social media platforms (SMPs). For instance, it is reported that a staggering 94% of content attributed to the Islamic State of Iraq was removed with an impressive 99.995% accuracy using AI techniques [51]. Similarly, YouTube asserts that a ML algorithm identifies 98% of videos removed due to terrorist content and propaganda [44]. Thus, the aspiration to develop AI systems capable of autonomously detecting and mitigating cyber threats or identifying and removing terrorist propaganda from the internet is indeed compelling.

Among the principal methods for enhancing and refining AI capabilities are ML algorithms. This process entails machines acquiring knowledge from experience through computational methods [126]. The experiential foundation for computers arises from the databases or datasets, which then culminate in the formulation of learning algorithms that draw insights from the data to construct models [126]. Unlike conventional software, ML systems do not require explicit human instructions for operation, instead, they decipher patterns and internalise implicit rules from the examples embedded within datasets or databases [116]. The potential of AI and ML in curtailing cyber threats is underscored by their capacity both to refine existing and to devise novel cyber security strategies, alongside their aptitude for handling the escalating complexity and volume of data. Their intrinsic ability to uncover concealed insights and intricate patterns without necessitating explicit programming is pivotal, enabling the construction of progressively advanced models to reinforce cyber defence mechanisms [17, 78, 83].

Nonetheless, the realisation of developing AI systems capable of autonomously detecting and mitigating cyber threats or identifying and removing terrorist propaganda from the internet is a complex and challenging task. The fundamental challenge in AI's efficacy lies in its need to be trained for these tasks. Data, which can be sourced from a multitude of contexts, serves as the foundation for AI development [71]. Essentially, data comprises discrete units of information, aggregated and stored within datasets or databases alongside other pertinent information [116]. However, a significant hurdle emerges: while data is pervasive, assembling datasets tailored to specific AI objectives can be a complex undertaking. Therefore, researchers might contend with limitations in access to pertinent datasets [35, 45], which potentially curtails AI effectiveness and scope. Furthermore, it is imperative to approach with caution the impressive statistical data presented by SMPs and government agencies concerning the effective eradication of terrorist content, as elucidated earlier. This caution is especially warranted given that certain platforms utilize automated techniques resembling advanced pattern matching rather than genuine ML methods. This approach entails the comparison of newly generated content against pre-existing flagged and blacklisted materials, frequently resulting in the identification of mere duplicates of content that has previously undergone human moderator review [42].

The use of ML by LEAs and other organisations to detect terrorist content online encounters various other technical challenges, including language translation and efficient examination [60]. The dynamic and noisy nature of language in the constantly evolving online environment poses another obstacle, as ML classifiers and methods struggle to keep up automatically (Ibid.). Furthermore, automated detection and information gathering present challenges due to their limited recall, resulting in certain behaviours and content evading detection (Ibid.). Therefore, to enhance the effectiveness of ML for countering cyberterrorism, it is imperative to address these technical challenges. Furthermore, there are significant ethical and legal challenges arising from the use of ML in counterterrorism. Thus, similar to the need to address technical challenges, understanding and navigating ethical and legal challenges are imperative for the responsible and effective utilisation of AI in addressing cyberterrorism.

This chapter serves as a dedicated examination of the myriad challenges inherent in the application of AI and ML in the context of counterterrorism. Its primary aim lies in conducting a comprehensive analysis of the ethical, legal, technical, and operational complexities that arise when applying ML in counterterrorism practices. By meticulously investigating these challenges, this chapter lays the essential groundwork for the subsequent chapter, Chap. 10, where the focus shifts towards formulating solutions and enhancements aimed at addressing or mitigating the identified challenges. In essence, this chapter provides the vital context and understanding required to bridge the critical link between the analysis of challenges and the pursuit of solutions.

The subsequent sections of this chapter are structured as follows: Sect. 9.2 examines ethical and legal challenges associated with the use of ML in counterterrorism. In doing so, the section investigates issues pertaining to individual autonomy, human rights preservation, algorithmic bias, privacy preservation, and the implications of mass surveillance. It also navigates the complex legal landscape, including discussions on hybrid classifiers, the assignment of accountability, and jurisdictional quandaries that emerge in this domain. Section 9.3 analyses the legal and operational challenges pertaining to the use of AI and ML in counterterrorism. To this end, the section delves into the complexities of big data management, the inherent challenges posed by class imbalance, the curse of dimensionality, the identification of spurious correlations, the detection of lone-wolf terrorists, the implications of mass surveillance, and the application of predictive analytics. It also illuminates the critical aspects of transparency and explainability, while shedding light on the potential unintended consequences that can arise from the integration of AI and ML into counterterrorism practices. Section 9.4 discusses the key findings of this research. Finally, the chapter concludes in Sect. 9.5.

9.2 Ethical and Legal Considerations

The primary concern when assessing the effectiveness of ML in the field of digital policing revolves around ethical and legal considerations. These ethical and legal challenges are paramount due to their potential implications for various key areas, such as human rights and individual privacy. This section delves into the intricate relationship between ethical and legal aspects and the implementation of AI and ML algorithms in the context of counterterrorism. To achieve this objective, it critically examines a wide range of ethical and legal challenges that arise within this context, including ethical challenges, autonomy, human rights issues, various biases (including cultural, racial, confirmation, and selection biases), privacy concerns, and legal issues. As part of this thorough analysis, the section underscores the pivotal importance of aligning with regulations, standards, and ethical principles. This emphasis highlights the complexity of simultaneously safeguarding human rights and privacy in the evolving landscape of counterterrorism technologies. By illuminating the multifaceted nature of these challenges, this section contributes to a more comprehensive understanding of the intricate considerations that underpin the responsible utilisation of AI and ML algorithms in counterterrorism endeavours.

9.2.1 Ethical Challenges

To contextualise the significance of ethics in this research, it is essential to understand its fundamental meaning. Ethics is defined as "a field of study that is concerned with distinguishing right from wrong, and good from bad" [23, p. 21]. Ethicists adopt two main theoretical approaches to justify their moral judgments: consequentialism and deontology [23]. Deontology posits that humans have moral duties, which exist independently of any consequences resulting from their actions, while consequentialists believe that wrong actions lead to negative consequences (Ibid.). The emergence of computer ethics as a researched field in the 1980s marked the first instance of addressing moral responsibility in technology, information development, and public policy (Ibid.). A pivotal milestone in the history of the Internet is the notion of widespread availability of large volumes of data accessible at any time [14]. The rapid growth of this Big Data paradigm has propelled technological advancements at a pace distinct from the development of social technologies, such as governance systems, legal frameworks, and cultural norms (Ibid.). The evolution of these new technologies in the context of countering terrorism online highlights both their potential and their limitations, particularly regarding human adaptability (Ibid.). In contrast to earlier historical periods, wherein it would take several generations to adapt to novel developments and acquire the necessary skills, the present time frame for such adaptability has significantly reduced to approximately 9–15 years (Ibid.). This accelerated rate of change is arguably influenced by the Internet's development, as contemporary challenges demand swift responses and adjustments from the human species (Ibid.).

Within the context of this study, ethics pertains to the moral considerations surrounding data access, collection, storage, and use, with a particular focus on the potential harm it might inflict on the general public (Ibid.). These discussions aim to heighten awareness and provide recommendations to mitigate such harms. It is essential to clarify that ethical challenges in ML do not primarily revolve around the algorithms themselves but rather the manner in which data is employed [32]. Several ethical issues and privacy concerns arise when applying ML in the context of the Internet, prompting discussions on the benefits, potential harm, and the qualifications of those overseeing and utilising such methods [60]. Amidst these considerations, a significant question that this chapter will explore is whether the threat of violent extremism and terrorist behaviour justifies the use of these techniques (Ibid.). A cyberterrorist's exposure of personal information, such as browsing history, on a public platform might lead to social and psychological harms, potentially causing a range of mental health issues [113]. From a consequentialist standpoint, the impact of such an attack is contingent on whether the resulting harm includes material damages, such as unemployment, suicide, or divorce (Ibid.). However, this approach tends to overlook the significance of psychological effects, considering that they might not be readily observable, contributing to a lack of awareness regarding such consequences [12]. Therefore, the primary purpose of ethics should be to ensure that the "development, deployment, and use" of ML technologies adhere to "fundamental rights," "applicable regulations," "core principles," and "values", thus emphasising the importance of "ethical purpose" [25, p. 1].

The intersection of ethics and cyberterrorism represents a relatively nascent yet highly diverse research area. The proliferation of disinformation is deemed unethical, given its infringement on privacy rights, thereby raising concerns regarding the safety of individuals on the Internet [100]. In contemporary society, the impact of such attacks is increasingly pervasive, as they transcend geographical boundaries and manifest within close proximity to individuals (Ibid.). The recent COVID-19 pandemic might have further exacerbated these threats, rendering the task of countering cyberterrorism even more challenging for agencies [2]. The focus of ethical analysis in this context pertains to the algorithms themselves, encompassing data collection practices and subsequent data use [32]. The opacity of ML algorithms, often referred to as "black boxes," complicates the understanding of decision-making processes, necessitating greater transparency and public disclosure of the algorithms' data and methodologies (Ibid.). Furthermore, ML algorithms may inadvertently develop undesired behaviours or decision-making patterns, resulting in unanticipated consequences (Ibid.). Mas a result, understanding the rationale behind specific decisions made by these algorithms can prove challenging, and greater transparency can be achieved through the public disclosure of data and methodologies employed by these algorithms (Ibid.). There exists the possibility for ML algorithms to develop behavioural patterns or decision-making tendencies that might not be necessary or desired, leading to potential issues (Ibid.). It is, therefore, essential to acknowledge the dual nature of ML, including both benefits and limitations. These algorithms have the capability to facilitate predictions and expedite processes, however, some experts contend that these algorithms lack causality, thereby impeding the establishment of a direct cause-and-effect relationship [117]. Thus, drawing definitive conclusions becomes challenging in this context (Ibid.).

The employment of ML-driven mass surveillance harbours the potential for grave misuse and extensive harm to civilian populations. A glaring example of this is evident in the Chinese government's use of ML for predicting potential terrorist activities among the Uyghur Muslim minority in Xinjiang [16]. This application has led to the intensified repression of approximately 13 million civilians in the region. The Chinese government's approach to mass surveillance entails an array of technologies, including license-plate cameras, facial recognition systems, iris scanners, and voice pattern analysers, which collectively subject the Uyghur population to constant scrutiny [67]. Individuals deemed threatening by these algorithms are subjected to interrogation or detention in what the authorities euphemistically term 're-education' camps. Disturbing reports have emerged revealing inhumane practices perpetrated against the ethnic group within these facilities [96]. The Chinese government's application of mass surveillance through AI not only infringes upon the privacy and rights of the Uyghur minority but also raises concerns about the unchecked power wielded by state figures in exploiting these technologies. The case underscores the imperative of ethical considerations, human rights protections, and responsible governance in the use of ML algorithms for surveillance purposes.

In relation to utilising AI to counteract terrorist content on SMPs and other online platforms, this usage also raises profound controversy due to the substantial human rights concerns it engenders, coupled with the failure to address the underlying causes of terrorism. This approach also leaves vulnerable individuals susceptible to terrorist propaganda still exposed to risk [116]. Effectively, this strategy mirrors situational crime prevention online, merely displacing content from one online platform to another without addressing the fundamental issues at the core [29]. This practice of countering terrorism online does not alleviate the root problem itself. One of the most contentious aspects revolves around human rights, particularly the freedom of expression. In the context of UK law, the freedom of expression is outlined as follows: "Everyone has the right to freedom of expression … The exercise of these freedoms … may be subject to such formalities, conditions, restrictions or penalties as are prescribed by law and are necessary in a democratic society." [53]. A notable challenge with AI lies in its struggle to comprehend subcultural nuances, contextual meanings, sarcasm, and subtleties [42]. Consequently, authentic content shared on SMPs could be misconstrued by AI algorithms and subsequently removed. Instances of this nature, where legitimate content is wrongly taken down, are categorised as false positives [115]. Such occurrences directly infringe upon human rights. However, an additional layer of complexity emerges when examining the rise of right-wing terrorism, as highlighted earlier. The boundary between freedom of expression and the propagation of right-wing extremism can be exceedingly delicate. An illustrative example is the case of Donald Trump's ban from Twitter following the Capitol riots. Despite the fact that his account was eventually suspended for inciting violence and contributing to the riot, this action was not taken until several days after the incident [86]. Debates persist to this day concerning whether this action was justifiably taken or not, emphasising the complexity of balancing freedom of expression with the prevention of extremist content dissemination.

AI systems are also susceptible to the challenge of under-blocking, as illustrated by the surge of hate speech from right-wing extremists in 2019 on Facebook. In this instance, the algorithms failed to detect the hate speech due to its intricate and multi-layered cultural context [116]. Conversely, over-blocking has also emerged as an issue. For instance, Facebook previously allowed the presence of terrorist imagery if it was used to express disapproval or condemnation of a group or if it originated from a reputable news source [44]. Similarly, on video platforms such as YouTube, there exists a multitude of 'witness videos' that have played a pivotal role in documenting events such as the Syrian Civil War [104]. However, even with AI implemented on YouTube, numerous videos created by civil society groups and activists aiming to document the atrocities during the Syrian Civil War were inadvertently removed [44]. Indeed, there are perspectives suggesting that despite its inherent inaccuracies, AI remains a viable and effective approach for addressing terrorist content online and on SMPs [51].

9.2.1.1 Autonomy

Human autonomy is recognised as a salient and conspicuous ethical and legal concern in the context of technological advancements, particularly concerning the influence exerted on individuals through technology [95]. In essence, ML algorithms possess the capability to control and manipulate individuals, a phenomenon often referred to as persuasive technology. Persuasive technology aims to shape user behaviour independently by curbing temptations and encouraging the display of desired conduct (Ibid.). The substantial growth in AI has led to increasingly autonomous robots. In the past, algorithms were deterministic and subject to human control before every action, allowing for a clear understanding of the decision-making process. However, concerns have arisen with the advancement of AI, as the decision-making processes of algorithms might become less distinguishable and controllable due to the absence of predetermined rules, potentially leading to inaccuracies in automated decisions (Ibid.). The unpredictability of these autonomous machines and systems raises the likelihood of surprising human counterparts by their rapid learning and independence [37]. The ethical consideration of delegating moral decisions to robots is a matter of significant importance [95]. The central concern revolves around the issue of holding robots or machines accountable for their actions, raising questions about whether such devices should be entrusted with decisions that might involve matters of life and death. The crux of the matter lies in the fact that machines lack the capacity for accountability, prompting a debate about the ethical acceptability of granting them decision-making authority in morally sensitive situations. One argument in favour of autonomous decision-making by machines is that stress plays a crucial role in human decision-making, while machines are immune to the influence of environmental stress and external variables, potentially making them more effective in handling sensitive situations [33]. Nevertheless, this viewpoint could be subject to counterarguments that support human involvement in critical decisions, highlighting the importance of empathy, emotional intelligence, and nuanced understanding, which machines inherently lack.

Additionally, human creators might exert complete control over the systems they design initially. However, once deployed, these systems operate autonomously in diverse environments, leading to the development of autonomy beyond the comprehension of their human creators (Ibid.). Therefore, this unpredictability raises further concerns regarding the potential evolution of autonomous systems in ways that might diverge from human intentions or understanding. The debate about delegating moral judgements to robots highlights the intricate ethical and practical issues that researchers and policymakers must solve. Therefore, in order to ensure ethical deployment and responsible use of autonomous systems, it will be critical to strike a balance between human agency and machine autonomy in decision-making processes. Similarly, as AI continues to evolve, the advanced autonomy exhibited by these machines might challenge human understanding in relation to the degree of automation and rapid adaptability. Therefore, considering the potential ramifications of persuasive technology and increasing autonomy in AI systems, it becomes necessary for researchers, policymakers, and developers to address these challenges. To this end, establishing transparent and ethical frameworks for AI design and implementation will play a pivotal role in protecting human autonomy and ensuring the responsible and fair use of ML technologies.

9.2.2 Human Rights Challenges

ML algorithms can significantly shape human rights dynamics, yielding both constructive and detrimental impacts. As these algorithms become integrated into diverse facets of society, spanning from criminal justice systems to SMPs, their decisions can impact individual liberties and social dynamics. On the affirmative side, ML algorithms carry the potential to bolster the safeguarding and advancement of human rights. For instance, they can be harnessed to identify patterns of discrimination, enable rapid translation of documents for marginalised communities, and even predict potential human rights violations by analysing large datasets. Similarly, ML algorithms could potentially curtail unwarranted encroachments on the privacy of the general populace. This is premised on AI's ability to selectively focus its attention on areas and individuals that exhibit increased vulnerability to associations with terrorism or might present perceived threats to societal order [75]. Yet, the utilisation of ML algorithms simultaneously raises apprehensions about human rights erosion and individual rights violations, prompting contemplation of a society marked by dystopian attributes [77]. This is exemplified by the circumstance in which any internet-connected individual, utilising SMPs, contributes a substantial corpus of public and private data to entities such as Google and Facebook. These platforms, in turn, aggregate and retain this information without individuals fully comprehending the beneficiaries, purposes, or underlying objectives of this data collection process. Consequently, the pervasive collection and analysis of personal data for algorithmic decision-making lay the groundwork for potential privacy violations, enabling surveillance and analysis of actions and preferences without explicit consent.

The impact of ML algorithms on human rights is not limited to their direct consequences but also extends to broader societal implications. Their influence on public discourse, information flow, and even political decision-making can shape the ways in which societies uphold or undermine fundamental rights such as freedom of speech and access to information. Within the context of terrorism, MacDonald [73] posits that there exists a need for establishing a discernible threshold when delineating between content deemed as terrorist in nature and that which qualifies as an exercise of freedom of speech. This presents challenges when online content is incorrectly flagged and categorised as terrorist material, leading to its removal or suppression (Ibid.). This phenomenon bears implications for social media users, as unchecked removal of potentially contentious online content by ML systems could infringe upon individuals' right to freedom of speech, thereby introducing a nuanced ethical quandary [36]. The crux of this concern, however, derives from the inherent variability of the concept of terrorism itself [101]. The lack of a universally accepted definition of terrorism hampers the efficacy of content filtration algorithms. Establishing a consensus on an unambiguous definition of terrorism within the research community could mitigate the inadvertent curtailment of freedom of speech. In this context, the absence of well-defined technological standards for accurately identifying unambiguously terrorist communication creates an environment where constraints on freedom of expression might misalign with the principles of unrestrained speech and religious liberty. Thus, it is vital to acknowledge and anticipate the inherent limitations of ML systems in consistently and accurately detecting terrorist content. The propensity for overcompensation, wherein non-terrorist content is unjustifiably purged from online platforms in an effort to mitigate terrorist content, poses a tangible threat to both freedom of speech and the ethical underpinnings of contemporary digital discourse.

Finally, the potential for algorithmic manipulation, disinformation, and the creation of filter bubbles challenge the integrity of open societies and the diversity of perspectives that underpin human rights values. Despite this, it is imperative to recognise that while ML possesses the propensity to impinge upon human rights, it concurrently plays a pivotal role in fortifying societal resilience against substantial perils, including terrorist attacks. This role is exemplified in its ability to apprise intelligence and LEAs of online terrorist activities.

9.2.3   Challenges of Bias

ML algorithms have demonstrated considerable benefits in various domains, including government, organisations, and society at large [122]. Nonetheless, alongside these benefits, the adoption of these algorithms has also given rise to challenges and repercussions stemming from bias. The decisions made by ML algorithms can be opaque and difficult to challenge, potentially infringing upon the principles of accountability and due process. As a result, there is a risk that these algorithms, if not developed and audited rigorously, might perpetuate existing biases and exacerbate inequality. For instance, biased training data could lead to discriminatory outcomes in criminal justice settings or exacerbate disparities in access to essential services such as healthcare or credit. The majority of AI systems are data-driven and heavily reliant on the quality and integrity of their training data [76]. If this data contains inaccuracies or biases, the algorithms will inevitably assimilate and incorporate them into their decision-making processes, leading to biased outcomes and predictions. Bias can infiltrate the ML process from various sources and through different stages. One potential factor is human intervention, which can disrupt the model's performance and introduce unfair or inaccurate results [117]. Additionally, misuse of the algorithm, such as altering its intended goals and methodologies through human manipulation, can contribute to biased findings (Ibid.). Furthermore, existing research strongly supports the notion that bias in ML often results from cognitive errors made by those involved in the data training process [117]. For instance, engineers or developers of ML algorithms could inadvertently encode their unconscious biases into AI algorithms [125]. Bias could also originate from the underlying societal norms reflected in the training and data sets [90, 117], leading to discrimination and perpetuation of unfair biases by AI systems.

Furthermore, the characteristics of the users themselves can introduce biases into the gathered information considering the fact that user-generated data sources constitute a substantial portion of the data employed in ML [76]. These biases can manifest in various forms, spanning from content production to historical biases (Ibid.). Historical bias within the context of ML refers to pre-existing biases present in the world that may permeate the data collection and utilisation process. On the other hand, content production bias arises from structural and semantic disparities in the data generated by users, particularly evident in language usage on different platforms and accounts, potentially influenced by users' perceptions, age, and gender (Ibid.). Population bias represents another significant form of bias encountered in the ML process, which revolves around demographics, statistical patterns, and user characteristics. It plays a crucial role in data collection and the determination of relevant and significant information. For instance, platform preferences, such as Instagram and Facebook being more popular among women, while Twitter is more commonly used by men, exemplify this form of bias (Ibid.).

Moreover, biases can emerge as a result of the "black-box" nature inherent in algorithm design, where companies frequently withhold the underlying calculations and formulas driving the algorithms [122]. This opacity not only impacts the broader understanding of how these algorithms function but also poses challenges for engineers and designers who are granted access to the formulas. Even with this access, they might encounter challenges in accurately predicting the intricate outcomes and effects that these complex algorithms can produce [122]. Consequently, the human element involved in designing these algorithms becomes a source of biases that inevitably manifest in the predictions they generate [122]. Expanding on the repercussions of the black-box issue, this challenge exacerbates the difficulty of understanding why an AI made a particular mistake [125]. The lack of transparency in algorithmic decision-making complicates the process of diagnosing errors. This leads to the pertinent question of legal responsibility in cases where AI systems make mistakes. Addressing this legal dimension is in itself an intricate problem, with the existing regulatory framework for assigning responsibility being largely underdeveloped and decisions about liability often varying from one instance to another [50].

While biases arising from various sources in the ML process can have wide-ranging implications, a particularly concerning manifestation of bias occurs within algorithmic decision-making, with significant consequences for fields such as counterterrorism. Without diligent examination and integration into the developmental process, ML models can inadvertently foster algorithmic decision-making that exhibits a preference for specific individuals or groups, predicated on protected characteristics. The infusion of human biases, whether overt or latent, invariably culminates in distorted outcomes within the system [77]. Bias manifests when an algorithm consistently endorses one outcome over others [52]. Of particular concern is the prevalence of racial bias when the focus pertains to counterterrorism efforts. Within the societal context, there exists a fallacious narrative asserting that "terrorists are always (brown) Muslims" while concurrently positing that "white people are never terrorists" [30, p. 455]. This narrative is further exacerbated by the disproportionate targeting of young men fitting the aforementioned description [68]. The underpinning of this ingrained bias can be attributed, in part, to the correlation drawn between infamous terrorist incidents and members of the Islamic faith [87].

Furthermore, the media's tendency to sensationalise terrorists and their actions has significantly shaped the public's preconceived notions regarding the outward appearance of a terrorist. Evidently, the portrayal of terrorists and their attacks in the media has contributed to the formulation of stereotypes that influence the perception of a terrorist's visual characteristics. Notably, it has been observed that instances wherein terrorism is driven by Islamic extremism often evoke a greater likelihood of public support for punitive measures [120]. Given this context, the imperative emerges that the individuals responsible for designing ML algorithms must be devoid of any indicators of racial bias. While this might seemingly aid LEAs in tracking non-white individuals associated with Islamic terrorism, it introduces the disconcerting prospect of white-ethnic background terrorists eluding detection [68]. Consequently, the prevailing lack of diversity and inclusivity in the composition of AI system designers constitutes a notable concern. Paradoxically, rather than enhancing the objectivity of judgments, such systems can inadvertently amplify bias and foster discriminatory outcomes. The repercussions of these ML inaccuracies can trigger a cascading sequence of errors, the ramifications of which might persist unnoticed for extended durations. These misjudgements culminate in biased determinations and erroneous allegations of terrorism [64], thereby affording anonymity to numerous individuals who do not conform to the stereotypical image of a terrorist as perceived by society.

9.2.4   Privacy Challenges

The impact of surveillance, particularly in the context of ML, raises significant ethical concerns related to privacy and the ongoing privacy-security debate [49]. While the intricacies of this debate have been discussed previously, this section aims to reiterate the implications of prioritising information security over the privacy of individuals' personal data and information. Two primary privacy issues arise from the Internet: the disclosure of personal information on social media platforms (SMPs) and the monitoring, or surveillance, of users' online activities [23]. Article 8 of the European Convention on Human Rights (ECHR) underscores the right to a private life, free from interference or intervention by public authorities, except when justified by law and relevance to specific situations [60]. The realm of privacy and security issues emerged as a distinct field within computer ethics in the 1980s, concentrating on the ethical responsibilities of computer professionals and users, along with the ethical considerations pertaining to "public policy for information technology development and use" [23, p. 22]. The moral analysis of these ethical concerns seeks to elucidate the facts and strike a balance between the rights and interests involved, while also proposing necessary courses of action (Ibid.). This dynamic underscores the pervasive nature of the privacy-security debate in contemporary society, as various agencies, policy-makers, and researchers grapple with finding a middle ground that ensures a secure society without compromising privacy and human rights [60]. Striking such a balance is paramount to the responsible and ethical deployment of ML and surveillance technologies, safeguarding individual privacy while addressing security concerns for the greater welfare of society.

A study conducted at Cornell University examined the implications of ML on individuals' privacy, with a particular focus on the application of algorithms to identify individuals from blurred images with a high level of accuracy [54]. The researchers sought to assess the performance of artificial neural networks, specifically trained automatic algorithms, in this context. The findings revealed that these ML algorithms achieved an accuracy rate of 71% in identifying individuals from blurred images. To provide a comparative perspective, the researchers conducted a human accuracy test, which yielded a significantly lower rate of 0.19% (Ibid.). The primary objective of this study was not to establish the superiority of ML over human efforts but rather to underscore the potential threat posed by ML techniques to individual privacy through the identification of individuals from obscured images (Ibid.). In practice, pixelated and blurred images find frequent use in media and online platforms, particularly in situations related to radical behaviours and online extremism, where anonymity of faces and location details is desired (Ibid.). The domain of computer science that pertains to security and privacy is referred to as the Moral Importance of Computer Security. This field is concerned with the various mechanisms established to safeguard computer systems from unauthorised exploitation, deletion, and manipulation of information [23]. A fundamental distinction within this domain lies between system security, which pertains to the software protecting against malicious attacks, and information security, which involves safeguarding data residing on disk drives or being transmitted between systems (Ibid.). Information security is associated with ensuring data integrity, confidentiality, and availability.

One aspect closely related to information security is the issue of identity fraud, particularly relevant in cases of cyberterrorism where an individual might attempt to impersonate a member of a terrorist organisation or seek to be detected by the police for terrorism-related activities. Identity fraud, also known as identity theft, involves the unauthorised use or transfer of another person's means of identification with the intent to commit unlawful activities [63]. While instances of identity fraud are not prevalent, it remains a potential threat, necessitating preventive measures. Biometric recognition has emerged as a robust security measure, as it requires the physical presence of the user, reducing the risk of fraud through document falsification, card theft, and password disclosure. However, biometric technology is not immune to vulnerabilities, as biometric systems can be deceived with falsified elements, requiring vigilance from relevant agencies [95]. Therefore, implementing comprehensive security measures that combine biometric recognition with other protective mechanisms becomes imperative in ensuring the integrity of computer systems and safeguarding against identity fraud and related security breaches.

Breaches of computer security or national security pose significant economic repercussions. National security is defined as "the maintenance of the integrity and survival of the nation state and its institutions by taking measures to defend it from threats, particularly threats from the outside" [23, p. 24]. Exploitation and undermining of the system can lead to permanent or temporary damage to the software, resulting in losses across various resources and financial domains. The severity of economic losses can escalate further if valuable information is corrupted [23]. In some cases, the corruption of certain systems can lead to severe and life-threatening indirect consequences, particularly in areas involving medical diagnosis, decision-making, and design professions (Ibid.). Breaches of confidentiality can exacerbate harm and violations when third parties gain unauthorised access to sensitive information and disseminate it (Ibid.). Such actions might infringe upon various rights and interests, including property rights, where information is owned by individuals who have the right to determine its usage and accessibility [47]. Furthermore, the lack of control over personal information online poses significant concerns as data is collected, analysed, and utilised for automated decisions regarding individuals' personality and behaviour [14]. This raises issues with respect to the General Data Protection Regulations (GDPR), as it might violate key principles of the GDPR, such as the collection of only necessary data and ensuring transparency to users regarding the purpose of data collection (Ibid.). The privacy-security debate intertwines with legal considerations, and many arguments are based on false assumptions that the law will consistently safeguard privacy [107]. This is problematic as it may prioritise the government's focus on protecting national security and critical infrastructures, potentially relegating privacy and citizens' rights to a secondary position (Ibid.). As a result, the tension between privacy and security raises complex challenges in determining the appropriate balance in the legal and policy frameworks governing these domains.

9.2.5   Ethical Challenges of Mass Surveillance

A central facet of the ethical challenges arising from the use of ML in counterterrorism practices revolves around the issue of mass surveillance, which necessitates careful examination. Privacy, within the realm of mass surveillance, encompasses the "individual right to exclude others", prohibiting government access to individuals' information [111]. Habermas [46] contributes to this discourse by examining how political power interplays with public activities through the prism of mass surveillance. The perpetuation of public surveillance inevitably restricts public freedoms, encroaching upon the individual's fundamental right to liberty. Consequently, compelling arguments arise, suggesting that the public is no longer adequately safeguarded due to the absence of public agency in determining areas exempt from surveillance [111].

The concept of mass surveillance prompts an examination of its necessity within a democratic society, as there are alternatives that could be considered 'less intrusive' [93]. This practice introduces concerns about potential power imbalances between the government and the public, raising apprehensions regarding the safeguarding of individual rights and the risk of an abuse of power, particularly as mass surveillance continues to evolve [93]. However, it is essential to consider the opposing viewpoint, which argues that the benefits of mass surveillance outweigh the drawbacks. From this alternative perspective, the argument posits that, in the interest of individual liberty, safeguarding the public through the identification of offenders without their knowledge is of paramount importance [102]. Refer to Sect. 9.3.3 for an in-depth examination of the technical challenges associated with mass surveillance.

9.2.6   Legal Challenges

The legal implications of AI and ML technologies introduce intricate challenges that necessitate careful analysis and prompt legal responses. As these technologies continue to evolve, both legal professionals and policymakers must navigate the complexities of addressing liability concerns whilst also being cognizant of the moral implications of their actions. AI and ML techniques represent categories of software that possess the capability to learn from the information they receive, leading to changes in their coding and methods. However, the use of such software raises pertinent legal implications. The rapid evolution of technology has surpassed the pace at which the law can keep up, creating a scenario where legal professionals, including lawyers and judges, encounter AI-related cases without adequate precedents and relevant knowledge [48]. Consequently, they are often required to approach these cases with limited insights and reliance on initial impressions. As AI advances, it exhibits characteristics that render it increasingly akin to human-like capabilities. In the United States, certain corporations and organisations have been granted legal responsibilities and rights, and experts predict that in the near future, machines and robots might also be granted comparable legal rights (Ibid.). The growing capabilities of these AI systems and their potential to replace human decision-making raise concerns regarding liability, necessitating attention from the legal system [94]. Liability issues arise when these technologies make decisions that result in negative consequences for third parties, necessitating legal intervention in such scenarios (Ibid.). However, it is essential to recognise that legislation alone cannot serve "as a substitute" for moral considerations [23, p. 21]. Consequently, agencies must assess not only the legality but also the morality of their actions when deploying AI systems and inputting data. Striking a balance between legal compliance and ethical considerations becomes a pivotal challenge in the responsible use of AI technologies.

9.2.6.1 Hybrid Classifiers

Digital Forensics (DF) is an established discipline encompassing the use of well-established methodologies to analyse and interpret digital evidence extracted from various digital sources. The primary objective of this field is to reconstruct events and facilitate the anticipation of potential behaviours or actions [91]. In essence, it involves the identification and extraction of digital data within the legal context [28]. However, the integration of data collected for ML processes has introduced numerous complexities within the legal domain. Technology's involvement in addressing critical and contentious issues, such as violent extremism and terrorist acts, raises questions concerning the acceptability of predictive outcomes in trial environments and the robustness of their validity to handle such sensitive matters [48]. DF is extensively employed by Law enforcement agencies and governmental bodies to analyse digital data as evidence during digital investigations [18]. Particularly, digital investigations prove invaluable when dealing with anonymous sources of attacks, as they can ascertain the identity of perpetrators by uncovering characteristic clues indicative of gender and age (Ibid., 2019). The conventional process of a DF investigation involves several stages: seizing the pertinent device, acquiring the relevant data, compiling a comprehensive report, and subsequently analysing the gathered information to obtain the requisite insights [103]. Nonetheless, the manual completion of this procedure is time-consuming. Currently, within the existing literature, there exists an extensive array of Digital Forensics process models that have been put forth for consideration.

Numerous Digital Forensics process models have been proposed in the literature, demonstrating their capacity to synergise with ML algorithms and datasets, thereby facilitating more efficient and lucid investigations for the personnel involved. One particular application of this integration involves the analysis of textual content, such as texts and posts, to identify instances of radicalism and extremism [18]. However, a notable challenge in DF lies not in handling anonymity during investigations, but rather in ensuring the preservation of digital evidence for a duration sufficient for its admissibility in a court of law [18]. While the global acceptance of digital evidence signifies a significant milestone in the advancement of DF, several challenges still persist [6]. Some scholars contend that existing frameworks concerning DF do not address the issue of the admissibility of digital evidence from a "holistic perspective" [7, p. 27]. It becomes imperative to meet specific legal requirements for the evidence to be considered admissible. For instance, the collection of evidence at a crime scene necessitates a valid search warrant, failure to adhere to this requirement compromises the reliability of the evidence (Ibid., 2017). Moreover, adherence to prescribed procedures in DF models, tools, and laboratories is crucial for ensuring the evidence's reliability and acceptance in court. Additionally, the assessment of certain evidence necessitates legal authorisation, with due consideration and respect for data protection, human rights, and privacy rights of the involved parties, including both victims and perpetrators (Ibid., 2017). The process of ensuring evidence's admissibility and credibility in court also depends on factors such as relevance, integrity, and authenticity in the DF investigation (Ibid., 2017). These factors are pivotal in establishing the evidential value and accuracy of the presented digital evidence during legal proceedings.

Pre-search and post-seizure warrants constitute essential components within the forensic investigation process, significantly contributing to the legal aspects of digital evidence handling [123]. These warrants serve as crucial documents that ensure due diligence before and after the crime scene examination is conducted. The admissibility of digital evidence in court necessitates its authenticity and persuasiveness to the jurors. Consequently, it has been argued that challenges arise due to the absence of formal procedures governing pre-search and post-seizure warrants (Ibid., 2020). The significance of this stage in the digital investigation process lies in its determination of the admissibility and viability of evidence in a court of law [11]. Compliance with legal standards and requirements is pivotal, any evidence obtained under illegal pretences or through a process that fails to meet legal standards may be deemed inadmissible or subject to scrutiny (Ibid.). This holds true even in cases involving Internet-related crimes, where the conventional concept of a crime scene may not apply directly to digital evidence (Ibid.). Nevertheless, the principles governing evidence collection remain pertinent. In the context of ML investigations, investigators seek to identify various stages in the ML process that might lead to potential faults. These issues can encompass challenges during the training process, or the presence of model malware, wherein the model's performance might exhibit flaws (Ibid.).

Furthermore, a distinct facet known as environment forensics merits consideration, focusing on the assessment of whether the training environment is capable of producing a valid algorithm (Ibid.). The integrity of the algorithm can be compromised if the training environment has been manipulated or does not accurately represent the underlying data, leading to undesired faults and behaviours. Consequently, unfettered access to all elements of the training environment becomes valuable in the context of DF investigations (Ibid.). Another area that significantly contributes to the collection of forensic data is the model authentication process, which assumes particular prominence in the realms of legislation and trials (Ibid.). The objective of this process lies in determining the authenticity of the model, thereby establishing its trustworthiness in handling the pertinent information. Recent advancements in watermarking techniques for ML models offer a means to achieve authenticity.

However, it is worth noting that these watermarks are susceptible to forging and tampering, necessitating the application of additional forensic methodologies to ensure accurate authentication (Ibid.). The phenomenon surrounding DF evidence revolves around the ever-evolving landscape of cyber threats and attacks, rendering the process of evidence gathering challenging and time-consuming. In response, investigators, legal practitioners, and judiciary personnel must adapt and educate themselves on the dynamic nature of these attacks, crime scenes, and the requisite measures to ensure the admissibility and reliability of evidence during trial proceedings [123]. The realisation of this objective, however, proves to be a protracted undertaking due to the continuously evolving landscape of digital crimes. As such, law enforcement agencies and governmental entities are tasked with the responsibility of exercising utmost vigilance in relation to the DF process, ensuring its alignment with the prevailing legal framework (Ibid.).

9.2.6.2 Accountability

ML has garnered significant attention within the legal and medical domains, particularly concerning confidentiality, which presents distinct challenges for both industries and research pursuits [54]. In cases involving legal matters or instances where misconduct adversely affects individuals or groups, a fundamental question arises as to who bears responsibility for detecting radical behavior through ML (Ibid.). This responsibility might lie with the computer operator, the individual responsible for constructing the training dataset, or even the manufacturer of the AI/ML system [66]. The issue of liability attribution requires careful legal consideration, as it gives rise to questions about the intentions behind potential malevolent actions. The concept of accountability concerning machines and AI lacks a universally accepted definition. Nonetheless, current knowledge underscores a close correlation between accountability and verifiability (Ibid.). Despite numerous legal complexities surrounding the notion of accountability, it becomes evident that all stakeholders involved in the production of the ML system bear some degree of responsibility for its outcomes, whether favourable or unfavourable [117]. The ML system, itself, cannot assume responsibility for these consequences, as it lacks the capacity for consideration. Thus, the onus falls upon those engaged in the creation and operation of the system to be held accountable for any ensuing issues (Ibid.). The absence of proper accountability mechanisms could have profound societal implications, emphasising the need for appropriate governance and regulatory measures in the face of advancing ML capabilities.

9.2.6.3 Jurisdictional Challenges

Another formidable challenge that demands thorough deliberation pertains to the policing of online terrorist activity. The internet, including the intricate landscape of the dark web, constitutes a pervasive global phenomenon, shaping the daily lives of an estimated 5 billion individuals worldwide [112]. Nevertheless, a notable observation arises from the fact that even cursory searches on platforms such as Facebook, Twitter and YouTube yield a proliferation of terrorist-related content, indicating a palpable deficiency in online security on a global scale [119]. For criminals, the digital realm and its integrated SMPs represent a convenient instrument to orchestrate their attacks, owing to the expeditious and real-time nature of responses achievable within this domain [108]. This modality of convenience is anticipated to continue bolstering the activities of terrorists, a phenomenon further exacerbated by the continuous emergence of novel online platforms [119].

However, the global ubiquity of the internet engenders a significant predicament. The enforcement of legal ramifications for criminal activities conducted therein is constrained by jurisdictional complexities. This dilemma is particularly pronounced in instances where multiple countries stake claims to the prosecution of an online offender, creating a competitive atmosphere wherein various nations vie for the prerogative to pursue legal action. This scenario is further compounded when the transgressions impact multiple victims situated across diverse geographical locations [22]. Consequently, the effort to combat online terrorism becomes inherently intricate, particularly in regions where access to advanced policing methodologies, such as ML, remains limited. Bangladesh serves as an illustrative example, confronting deficiencies in both technological infrastructure and developmental trajectories, thereby lagging behind in the battle against online terrorism [82]. Notwithstanding these challenges, the unfolding landscape of these unprecedented changes necessitates a comprehensive re-evaluation of approaches governing the reconstruction, reconfiguration, and execution of data processing and policing methodologies [3].

9.3   Technical and Operational Challenges

The challenges surrounding the use of AI and ML in counterterrorism efforts are multifaceted and significant. One crucial aspect is the diversity of terrorist attacks, that manifest with a broad spectrum of motives, planning strategies, and execution methodologies. These variations lead to a complex digital trail left by perpetrators during their preparations [105]. This diversity in attack profiles poses a considerable hurdle for mass surveillance systems, as extracting meaningful insights from individual datasets becomes increasingly complex. The complexity of these variations is often contingent upon the underlying motivations driving their actions—whether they stem from religious beliefs, political agendas, mental health issues, or other factors—as well as the scope of participation and the mode of communication employed, be it offline or encrypted (Ibid.). Consequently, in the context of mass surveillance, extracting relevant insights from individual data sets can prove to be a complex task. To address these complexities, ML algorithms designed for mass surveillance must be trained on a substantial and diverse range of data sets. This approach enhances the potential for identifying connections indicative of potential threats. However, this pursuit introduces a technological and mathematical complexity known as the 'curse of dimensionality', as previously discussed in detail, where the statistical complexity increases as the dataset's dimensions expand [69].

Another significant challenge that arises in the context of applying ML to counterterrorism concerns the paucity and acquisition of sufficient training data for effective monitoring of terrorist activities. This challenge is particularly pronounced when dealing with terrorist groups such as Al-Qaida, where limited accessible online information hampers ML systems' effectiveness [5]. Some have argued that datasets with fewer than 40,000 examples are insufficient for achieving high ML-based decision-making accuracy [57]. The success of ML-based approaches fundamentally hinges on the availability of substantial and regularly updated datasets. Consequently, the acquisition of such a substantial amount of data for ML to proficiently track and combat terrorist organisations such as Al-Qaida often proves to be a formidable task, considering the limited online footprint of these groups. In contrast, larger terrorist organisations, such as the Islamic State, consistently produce extensive online propaganda, providing ample data for ML analysis to inform counterterrorism efforts.

The following subsections will critically analyse the technical challenges that arise when applying ML in the field of counterterrorism. Building on the foundation laid in previous discussions, these critical analyses will illuminate the complexities and intricacies of harnessing ML to address the unique and evolving challenges posed by terrorism in today's digital age.

9.3.1 Big Data

In the context of digital policing, Big Data (BD) holds immense utility as it encompasses vast quantities of digital information through processes such as analysis, extraction, and visualisation [79]. The term BD was coined to describe the use of large volumes of scientific data for visualization purposes. This framework facilitates the prediction of cyber threats in advance, bolstering pre-emptive measures [121]. BD often involves the application of advanced data analytics methods such as "predictive analytics, user behavior analytics", or other sophisticated techniques aimed at extracting value from data [89, p. 1]. Similarly, in counterterrorism, the application of BD proves notably advantageous, particularly in terms of its scale and variety. Leveraging ML techniques, LEAs can effectively analyse and interpret large datasets. By harnessing ML algorithms, they can proactively identify potential threats and allocate their limited resources more efficiently [116]. Furthermore, the predictive capabilities afforded by ML empower LEAs to make informed decisions, prioritise tasks, and deploy specific policing strategies as needed. This strategic approach is instrumental in addressing multifaceted security challenges.

Due to its complexity and sheer volume, BD cannot be efficiently processed using conventional systems or traditional data warehousing methods (Ishwarappa and Anuradha 2015, as cited in [78]). The data generated can take various forms, including structured, semi-structured, or unstructured, originating from diverse sources. Notably, prominent companies such as Google and Facebook generate a substantial demand for BD processing, primarily dealing with vast amounts of unstructured data (Ishwrappa and Anuradha 2015, as cited in [78]). This type of data can pose significant processing challenges due to its content, encompassing "billions of records of millions of peoples information that includes the web social media, images and audio" (Ishwrappa and Anuradha 2015, p. 320, as cited in [78]). However, while BD offers many benefits in digital policing and counterterrorism practices, it simultaneously introduces inevitable challenges, primarily stemming from the scale and sheer magnitude of data it contains [79, 80], a consequence of the rapid expansion of data collection driven by technological advancements. This expansion has outpaced developments in computing, creating a discrepancy between data growth and processing capabilities [3]. Furthermore, the complex and decentralised nature of contemporary security threats presents significant challenges. These "decentralized issues" can afford sophisticated terrorists the opportunity to pose security risks through various means, including cyber threats such as hacking, device interception (e.g., CCTV or IP cameras), and the execution of Denial of Service (DoS) attacks [81].

Aligned with the concept of BD, five distinctive challenges including Volume, Velocity, Variety, Veracity, and Value have been identified [7, 80, 85, 88], as depicted in Fig. 9.1.

The first component among the fundamental V's of BD is "Volume", which serves as the foundation of BD, as it represents the initial scale and quantity of accumulated data. It refers to sheer amount of data available [78, 88] and large volumes of information that an application or system can accommodate [85]. This consideration, however, gives rise to pressing concerns for technological applications, primarily stemming from the imperative for scalable storage solutions and the continual evolution of data analysis techniques [85]. Subsequently, it becomes apparent that the aspect of volume engenders what can be characterised as the 'dark side' of BD [27]. This pertains to the often obscured but significant costs and risks incurred in the processes of data collection, storage, and analysis. Despite these challenges, large corporations recognise substantial advantages in the vast volume of BD, as it allows the acquisition of more comprehensive and diverse datasets on a grander scale. As discussed above, entities such as Facebook and Google leverage this abundance to harness advanced analytics techniques, facilitating the identification of trends and patterns within their datasets [98]. However, it is imperative to acknowledge that the rapid pace at which data is proliferating poses formidable challenges for technology to sustain. Even with concerted efforts such as Moore's Law, originally postulated by Gordon Moore in 1965, which seeks to double the number of transistors on integrated circuits every two years, thereby enhancing processing speed while concurrently maintaining cost-effectiveness [114], technology still grapples with the formidable task of keeping pace with this exponential data growth. Furthermore, it is crucial to

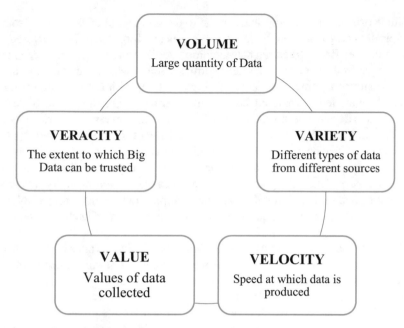

Fig. 9.1 The key components of BD, i.e. the 5Vs [80]

recognise that while significantly large datasets might qualify as BD, this classification remains subjective and adaptable, contingent on the computational capabilities available in the market [43, 79]. Volume presents the most immediate challenge to conventional IT infrastructures.

The second aspect among the 5 V's of BD is "Velocity", which can be best defined as the accelerating pace at which data is generated, processed, stored, and analysed, as well as the speed with which new data is created and circulated [56]. This velocity of data shows a consistent increase with both structured and unstructured data, especially as the world becomes more interconnected and global, with an "increasing frequency of data capture and decision making about 'things' as they move through the world" [88, p. 96]. Notably, decisions made using BD can influence the subsequent data that is collected and analysed, thereby adding an additional dimension to the concept of velocity [88]. Therefore, regulating velocity is of paramount importance, particularly as BD continues to evolve into AI, where analytical systems automatically identify data patterns and leverage them to provide insights [19, 79].

The third dimension of BD, which holds significant relevance in the context of counterterrorism, is Variety. Variety addresses the diversity of data types, which can originate from a broad spectrum of sources, each with its own unique characteristics and values. Data may originate both within and outside an organisation, and the challenge of standardising and managing this diverse data poses significant issues in relation to variety [43]. As discussed, Variety is primarily concerned with the multitude of sources that contribute to the expansive nature of BD. These sources are typically categorised into three distinct segments: Structured, Unstructured, and Semi-structured [99]. Structured data encompasses information that has been meticulously organised and formatted, often residing within a designated 'data warehouse' [99]. In contrast, unstructured data lacks organisation, exhibiting variations in both size and format. Semi-structured data, on the other hand, takes an intermediate position, it retains a degree of informality in its organisation, yet remains understandable [43]. The central challenge concerning data in the context of Variety pertains to its organisation. Raw data is often characterised by its substantial lack of categorisation, owing to its inherent diversity and complexity [85]. This poses significant obstacles in the context of data mining and cleansing processes, as the diversity inherent in unstructured and semi-structured data makes compression and use a formidable task. This, in turn, raises pertinent questions regarding the utility of such diverse data in this particular context.

The fourth element among the 5 Vs is "Veracity", which pertains to the statistical reliability and trustworthiness of data as well as the level of confidence in the data collected [43]. The reliability of data can be impacted by various factors, including its source, processing methods, and the trustworthiness of the infrastructure and facilities involved [34]. Consequently, "the gathered data can be incomplete, disorganised, erroneous, unable to offer any useful insight, or challenging to use" [43, 80, p. 94]. When data lacks completeness, its sheer volume can lead to more confusion than meaningful insights [43]. For instance, in the field of medicine, incomplete data regarding a patient's medication history can pose a life-threatening risk. Therefore, it becomes evident that both the value and veracity of data play pivotal roles in determining the quality and depth of insights that can be extracted from it [43].

The fifth and final component among the 5 Vs of BD is Value, a relatively newer dimension in the realm of BD. Value in this context can be described as the additional worth or advantage that the gathered data contributes to predictive analysis or hypotheses. The value derived from data can fluctuate, contingent upon the nature of the events or processes they represent, encompassing characteristics "such as stochastic, probabilistic, regular or random" (Demchenko et al. 2013, p. 50, as cited in [78]). The extraction of value from BD is imperative as the significance of BD largely hinges on the insights that can be collected from it [43].

As aforementioned, the application of BD in counterterrorism presents a critical challenge: striking the right balance between data depth and accuracy. While increasing the depth of data can improve accuracy, it also requires access to larger and more comprehensive datasets. However, this poses a significant challenge due to the scarcity of recorded terrorist incidents and class imbalances, as discussed previously in detail [65]. Moreover, the urgency to combat terrorism exacerbates the data scarcity issue. Building a sufficiently comprehensive dataset could span years, a less-than-ideal scenario which contradicts the objective of swiftly reducing terrorist attacks. To overcome these obstacles and prevent spurious correlations, as detailed previously, sophisticated solutions are needed to harness the full potential of big data in counterterrorism. Balancing the dataset and addressing spurious correlations are essential components of this endeavour. Introducing additional variables into the dataset can inadvertently lead to the discovery of coincidental patterns, making it challenging to prevent algorithm overfitting [110]. These challenges underscore the significant barriers that affect the classification accuracy of ML algorithms. Misclassifying potential threats or innocent individuals poses a risk, emphasising the need for integrating alternative surveillance methods and human oversight [65]. Even with a high overall accuracy rate, when applied to extensive datasets, the potential for inaccuracies remains a concern, raising ethical considerations. Navigating the complexities of mass surveillance for counterterrorism requires a comprehensive approach that combines multiple strategies to achieve accurate threat assessments.

The use of BD, while practical for collecting information on offenders and their online activities, presents a significantly more expensive policing technique. As data collection increases, the costs escalate due to the need for data manipulation, processing, and storage [4]. Moreover, the continuous advancement of new technologies perpetuates a never-ending cycle of data processing, resulting in higher expenses for those implementing this approach [4]. Consequently, the potential for low-income countries to adopt ML as a policing technique appears unlikely unless more cost-efficient data processing methods emerge. Furthermore, while the intersection of BD and ML in counterterrorism presents a powerful toolset for threat detection and analysis, this synergy also comes with its own set of unique challenges including: class imbalance, curse of dimensionality, and spurious correlations. These hurdles, while common in data-driven fields, take on unique significance in counterterrorism. The following subsections explore each challenge in detail, discussing their implications and strategies for mitigating their impact on the effectiveness of counterterrorism efforts.

9.3.1.1 Class Imbalance

Another significant challenge in the application of ML to counterterrorism pertains to the issue of 'class imbalance' [118]. The issue of class imbalance emerges when ML algorithms are trained to differentiate individuals with a higher probability of engaging in terrorist activities [65, 118]. Class imbalance signifies the uneven distribution of data points between distinct target classes, such as terrorist activities and non-terrorist activities [118]. Within this context, it becomes imperative to maintain a balance within the dataset, ideally achieving an equal representation of positive and negative instances. This balance holds paramount significance as any imbalance can substantially impact the performance of ML models [65, 118]. Consequently, the presence of both positive and negative instances within the dataset is pivotal, as it serves as the foundation for distinguishing between potential terrorists and non-terrorists [118]. While class imbalance might not pose a substantial challenge in smaller datasets, its significance grows exponentially with larger datasets, especially those covering larger populations, such as entire countries or states. In these scenarios, a noticeable prevalence of non-terrorism data in comparison to terrorism-related data becomes evident. This inherent imbalance can introduce biases into ML models, inclining them towards favouring predictions of the majority class (non-terrorism) at the expense of the minority class (terrorism). Such bias has the potential to diminish the models' sensitivity in detecting genuine terrorist threats. In light of these challenges, addressing class imbalance in ML represents a common and pressing concern. Successful resolution[1] necessitates meticulous consideration and the application of strategic approaches to ensure the attainment of accurate and equitable results. Consequently, the recommendations outlined in Chap. 10 can be instrumental in mitigating the issues posed by class imbalance and enhancing the overall effectiveness of ML in countering terrorism-related challenges.

9.3.1.2 Curse of Dimensionality

Another significant challenge in the context of ML is the 'Curse of Dimensionality', that can significantly impact the efficiency and accuracy of models. Coined by Richard Bellman in 1961, this concept refers to the exponential expansion of data volume as the number of features or dimensions grows [15]. The consequences of the Curse of Dimensionality can be far-reaching, posing significant challenges for practitioners. For instance, as datasets grow in size, so does their dimensionality, resulting in challenges such as data sparsity, noise, and overfitting. Moreover, this increase in dimensionality not only hinders the algorithm's ability to extract meaningful insights and patterns but also renders many algorithms computationally infeasible for high-dimensional datasets due to the associated computational complexity [69, 118]. Additionally, this increase in complexity leads to extended processing times, rendering it challenging to train and deploy models efficiently. Furthermore, the complexity of statistical implications increases with the dimensionality of the dataset [118]. This increased complexity can reduce the precision and accuracy of ML algorithms, particularly when dealing with larger datasets. Another consequence of this heightened complexity could be the increased risk of misinterpretation by ML algorithms, emphasising the intricate interplay between dimensionality and the overall performance of algorithms [69, 118].

In the context of online terrorist activity, data diversity, including communication patterns, the number of individuals involved, and motivations, continuously expands the dataset [118], thereby exacerbating the curse of dimensionality. Furthermore, within the framework of digital policing techniques, mass surveillance efforts can further compound this challenge for ML algorithms. While extensive datasets are essential for accurately identifying individuals and enhancing algorithm performance, they inevitably result in higher-dimensional data that must be processed. Consequently, as the dataset dimensionality increases, there might be limitations in the number of recorded terrorist attacks available for training ML models, potentially leading to reduced model accuracy [118]. The aforementioned challenges highlight the significant barriers that impact the classification accuracy of ML algorithms. As the probability of misclassification increases, there is a heightened risk of both

---

[1] See Chap. 10 for a comprehensive set of recommendations.

overestimating and underestimating the threat levels associated with potential terrorists and innocent individuals alike. Furthermore, an environment characterised by extensive monitoring of the population can distort datasets, potentially misleading the authorities. This phenomenon underscores the fact that even a notably high accuracy rate, such as 99.99%, when applied to a dataset encompassing 20 million individuals, would yield at least 2000 instances of inaccurate analysis. While this number is relatively small compared to the overall accuracy level, it still raises ethical concerns intrinsic to such methods. As a result, a comprehensive approach is needed to incorporate additional surveillance strategies and human oversight to navigate the complexities and uncertainties associated with mass surveillance for counterterrorism purposes. Similarly, the multifaceted nature of terrorist activities underscores the necessity for algorithmic models that can handle the complexity presented by diverse data points [105]. Such models aid in identifying meaningful patterns and correlations across various dimensions. This analytical effort plays a pivotal role in optimising the efficacy of mass surveillance, particularly in the context of counterterrorism endeavours.

In summary, the Curse of Dimensionality in ML presents a complex set of challenges, characterised by increased computational complexity, data sparsity, high-dimensional datasets, and an increased risk of overfitting. As demonstrated throughout this section, achieving accurate threat assessments through machine learning algorithms alone is a complex pursuit, laden with potential ethical and practical dilemmas. Therefore, to strategically navigate the Curse of Dimensionality and maximise the benefits of ML in counterterrorism efforts, researchers and practitioners must continuously seek innovative techniques and strategies to mitigate its adverse effects. This will help advance the field of ML across various domains, including counterterrorism. To this end, Chap. 10 explores effective strategies for mitigating the challenges posed by the Curse of Dimensionality and provides a set of comprehensive recommendations and insights to address these intricate issues and further enhance the application of ML in counterterrorism.

9.3.1.3 Spurious Correlations

The occurrence of spurious correlations is another notable challenge in the context of BD [118]. When dealing with vast datasets, the introduction of additional 'characteristics' can lead to the discovery of coincidental patterns [110]. These patterns might be unrelated to the task at hand. For instance, seemingly unrelated details, such as an individual's shopping habits, might exhibit correlations with potential terrorists. This issue persists irrespective of dataset balance, making it challenging to prevent and potentially causing overfitting of algorithms. The challenge arises from the inclusion of numerous features within a dataset, often done to increase the likelihood of identifying meaningful correlations [118]. However, as a consequence of this approach, some discovered patterns might ultimately prove to be inconsequential. Consider, for example, the application of ML in identifying potential terrorists. While ML algorithms can be valuable in recognising patterns associated with terrorist activity, they might also encounter spurious correlations, that, upon closer examination, prove to be irrelevant in capturing an offender's identity or intent. For instance, the algorithm might stumble upon a spurious correlation between a potential terrorist and their shoe size, which has little practical value in the context of counterterrorism [118]. While the potential for discovering meaningless correlations is not entirely preventable, it is essential to acknowledge that some of the patterns and discoveries made in the process might ultimately prove to be of little practical value, rendering them as potentially pointless findings in the broader context of counterterrorism efforts.

In summary, the challenges posed by AI and ML in counterterrorism encompass diverse data sources, the curse of dimensionality, and a delicate balance between data depth and accuracy, compounded by challenges such as data scarcity and class imbalance. Navigating these complexities necessitates comprehensive strategies, algorithmic refinement, and the integration of alternative surveillance methods to ensure effective mass surveillance while addressing ethical considerations. Addressing these issues while preventing spurious correlations is crucial for the successful deployment of machine learning algorithms. Moreover, the potential for misclassification underscores the need for alternative surveillance methods and human oversight in counterterrorism efforts.

9.3.2   Detecting Lone-Wolf Terrorists

Building upon the challenges in the application of ML within counterterrorism, a critical concern involves the detection and prevention of terrorist activities through the analysis of digital footprints [118]. The complexity arises from the unique nature of many terrorist attacks, which present a substantial challenge to ML algorithms. While organized groups are responsible for a significant portion of attacks, there is also a distinct category of lone actors [118]. This shift in the security landscape implies that the digital footprints of potential terrorists can exhibit significant variations, creating isolated data points within training datasets. This evolving security landscape, particularly the rise in lone-wolf terrorism as highlighted by the Global Terrorism Index [55], adds a layer of complexity to identifying potential terrorists based on digital behaviours and patterns. This challenge is magnified by the fact that lone-wolf attackers operate independently and often under the radar, posing a significant threat to national security and society as a whole [8, 62]. In response to this evolving threat, LEAs, security organisations, and security SMPs have resorted to ML as a potent tool in combating lone-wolf terrorism. ML technology offers a promising solution by automating the detection process, enhancing efficiency and accuracy. By analysing vast amounts of data and recognising patterns that may elude human observers, ML systems can identify potential lone-wolf terrorists before they carry out their destructive attacks [58]. This proactive approach not only bolsters security but also enables a more targeted allocation of resources and personnel to effectively mitigate the threat (Ibid.).

However, developing robust ML models for the detection of lone-wolf terrorist activities is complicated by the unique characteristics of these attacks. Lone-wolf attackers often operate in isolation, complicating the identification of consistent digital behaviors and patterns [106, 118]. This complexity has given rise to several critical obstacles that LEAs, security organisations, and researchers must address. The foremost challenge is the scarcity of data specifically related to lone-wolf terrorism. Unlike well-organised terrorist groups, lone-wolf perpetrators leave behind a sparse digital footprint, hampering the training and development of effective ML models [118]. This scarcity of data poses a daunting challenge for training ML models to recognise specific patterns of behaviour, potentially resulting in LEAs receiving notifications about these individuals only when it is too late to prevent an attack. While increasing the size of the training dataset might appear an intuitive approach to enhance algorithm accuracy, practical implementation can be far from straightforward. The crux of the issue lies in the requirement to label the dataset, necessitating the prior identification of potential terrorists—a task that authorities find challenging to execute with complete confidence. A poignant example of this challenge is the Homeland Security's extensive watch list, assembled over the past decade. This list has experienced numerous instances of false positives, leading to individuals being unjustly labelled as dangerous. These incidents underscore the inherent bias that algorithms might inherit from their training datasets. A closer look at the issue through literature analysis reveals a staggering statistic: for every terrorist correctly identified by an algorithm, there is a potential for an alarming 100,000 innocent individuals to be erroneously marked as suspects [41].

Moreover, the inherent unpredictability and variability of lone-wolf behavior and motivations further complicates the application of ML [109]. Similarly, their deviation from traditional terrorist group patterns adds an extra layer of complexity to detection efforts. Furthermore, balancing the need to identify lone-wolf terrorists while minimizing false positives is paramount. Overly sensitive ML models can generate numerous false alarms, straining resources and eroding public confidence. Therefore, striking the optimal trade-off between precision and sensitivity remains a persistent challenge. Another substantial challenge concerns the swift adaptability of lone-wolf terrorists to shifts in security measures and detection techniques [9]. They can rapidly adjust their tactics and strategies to evade ML-based detection systems, rendering previously trained models obsolete. As a result, this necessitates continuous model adaptation and improvement. Moreover, the collection and analysis of data for lone-wolf terrorism detection raise legitimate privacy concerns. Therefore, achieving the right balance between national security imperatives and safeguarding individual privacy rights could give rise to a complex set of ethical and legal dilemmas in the context of ML-driven surveillance [118]. Additionally, ML models can be susceptible to biases within their training data, potentially resulting in unfair profiling or discrimination against specific groups or communities. Hence, prioritising fairness and bias reduction within detection algorithms is essential to maintain public trust and prevent civil liberties violations.

Expanding on the challenges and potential benefits of applying ML for counter-terrorism, facial recognition (FR) technology[2] is another valuable tool that can assist LEAs in their fight against terrorism [92]. Recent advancements in FR technology have enabled the tracking of varying behavioural patterns of potential terrorists, such as identifying suspicious activities such as leaving a suitcase unattended for an extended period or frequent visits to specific locations for photography purposes (Ibid.). However, as with other technological advancements, including ML, FR technology also faces its set of challenges [84]. The inherent disparities among lone terrorists and their attack methods, as previously discussed, still pose a challenge in achieving sufficient training data for FR algorithms. This limitation underscores a significant concern regarding the use of FR and ML in counterterrorism efforts, as inaccurate algorithms can hinder the effectiveness of LEAs in countering terrorist threats. Therefore, while these technologies offer valuable tools for improving counterterrorism efforts, they must be approached with caution, and their limitations must be considered to ensure their effectiveness in addressing the complexities of lone-wolf terrorism [41].

9.3.3 Mass Surveillance and Predictive Analytics

In recent years, traditional policing has undergone a rapid transformation, adapting to the integration of emerging systems and cutting-edge technologies, particularly in the realm of predictive policing and AI [74]. This evolution in law enforcement is fuelled by the expanding scope of police responsibilities, which now extend beyond conventional enforcement duties. As a response to this shift, LEAs have found it imperative to adopt a proactive crime prevention strategy. Consequently, a range of forecasting techniques has been developed to aid in the identification of when and where crimes are most likely to occur ([39], as cited in [78]). Within this period, there have been "two major structural developments": an increase in surveillance technologies and the rise of BD, both of which have aided police practise in both positive and negative ways, as discussed throughout this chapter. Police have effectively used predictive policing as a crime mapping tool to identify where a crime is likely to occur, as well as predictive analytics to identify individuals who are likely to offend ([10], as cited in [78]). As a result, data-driven policing has garnered widespread recognition for its capacity to strengthen police accountability by requiring law enforcement to address discriminatory practices and be held accountable in the process [20, 78]. Furthermore, embracing data-driven decision-making in law enforcement has the potential to enhance the "prediction and pre-emption of behaviours by helping law enforcement deploy resources more efficiently" ([20, p. 982], as cited in [78]). Additionally, the recent advancements in police data management and collection, coupled with the use of forecasting techniques, have empowered law enforcement to better allocate

---

[2] See also Chap. 13 for detailed information about facial recognition technology.

resources with the aim of preventing and ultimately reducing overall levels of crime [39].

However, despite its many potential benefits to LEA's, BDPA presents many ethical and legal challenges such as discrimination and bias readings that demand careful consideration. A primary ethical consideration revolves around algorithmic bias, where machine learning models, when trained on historical data, may perpetuate or worsen existing biases within those datasets. To illustrate, within the realm of criminal justice, predictive analytics systems have faced criticism for disproportionately focusing on minority communities, thus contributing to the perpetuation of systemic inequalities [13]. Moreover, as these technologies continue to evolve and become increasingly integrated into individuals' daily lives, ethical concerns regarding privacy and data anonymity arise [37]. LEAs routinely collect and analyse vast amounts of data, including details such as location, timing, and the nature of criminal activities, prompting some to raise concerns about potential infringements on individuals' privacy and security [38]. Predictive analytics fundamentally relies on extracting insights from individual-level data, which might contain sensitive information. This indiscriminate collection and analysis of personal data can indeed pose a threat to individual privacy and give rise to surveillance-related apprehensions [21, 31]. Therefore, striking a balance between preserving data utility and ensuring privacy protection stands as a significant challenge. Simultaneously, a pivotal issue emerges surrounding consent. Individuals might lack awareness regarding the use of their data or might not be granted the choice to opt out, giving rise to concerns regarding informed consent [31]. Moreover, the opacity of algorithms and the intricate nature of predictive analytics create obstacles for users in understanding the decisions that impact their lives, thereby eroding the principle of autonomy [124].

In light of the comprehensive examination of the challenges associated with BDPA in "Countering Cyberterrorism: The Confluence of Artificial Intelligence, Cyber Forensics and Digital Policing in US and UK National Cybersecurity" [78] book, this section does not intend to delve further into the intricacies of BDPA challenges. Rather, the reader is encouraged to refer to the aforementioned source for a more in-depth examination of these challenges and their implications.

9.3.4 Transparency and Explainability

In the context of ML and its algorithms, the issues of transparency and explainability are paramount in the ongoing debate surrounding the challenges they introduce. Both concepts share similar perspectives, with explainability focusing on the difficulty in interpreting the complex nature of ML algorithms, while transparency pertains to the researcher's obligation to ensure clarity in the decisions made regarding the data [117]. A crucial aspect in this context is the effective communication of the methodology in a manner comprehensible to a wide audience. Explainability can be understood as a continuum that denotes the degree to which humans can comprehend and place trust in ML algorithms (Ibid.). The "black box" approach, as previously mentioned, specifically refers to instances where researchers possess knowledge about the input data fed into the algorithm but struggle to interpret the output data (Ibid.). This perspective significantly influences the discussions on transparency and explainability.

There are certain ML algorithms designed to enhance transparency, but their efficacy depends on the strength of the model, its methodologies, and its outcomes. However, a trade-off must be considered between transparency and potentially sacrificing the accuracy and validity of a more effective model (Ibid.). In some cases, the black box perspective might offer better data quality compared to alternatives, rendering it a successful option (Ibid.). Transparency assumes particular significance when employing AI technologies such as ML, especially when reproducibility is a key objective (Ibid.). Reproducibility, which denotes the ability to replicate research results, is vital for fostering scientific trust and accountability. The capacity to be transparent with data usage and its handling reflects the usability and relevance of such research, and it is integral to predicting the future applications of these models (Ibid.). Conversely, a lack of transparency can pose challenges and limitations to achieving reproducibility (Ibid.). Organisations bear the responsibility of informing individuals about the data they collect, the intended recipients of this data, and the potential implications of data collection on individuals or groups [14]. These practices of transparency are essential in establishing trust and ensuring ethical use of AI technologies in a wide range of applications.

9.3.5   *Unintended Consequences*

As already discussed, AI has emerged as a critical tool in the ongoing battle against cyber threats posed by terrorists. However, this technological advancement brings forth concerns related to unintended consequences and potential paradoxes. One such paradox is the dual nature of AI, where its progress in bolstering cybersecurity might inadvertently provide malicious actors with tools to circumvent defences. In this context, ML techniques, integral to AI, have the potential to be employed by these actors to undermine the very safeguards they were designed to strengthen [116]. This situation presents a significant predicament, especially if AI assumes a central role in future cybersecurity measures. If the defences built upon AI are compromised, cyber terrorists could potentially breach critical infrastructures, thereby inflicting unparalleled damage, potentially leading to catastrophic consequences, including loss of life [72].

On the other hand, as previously discussed, the trajectory of AI development could also yield positive effects, whereby AI's potential is harnessed to mitigate cyber threats, significantly enhance cybersecurity measures, and efficiently handle the increasing volume of data that requires processing [116]. However, the underlying complexity of AI introduces a dilemma known as the 'black box effect', as discussed throughout this chapter. This phenomenon encapsulates the challenge wherein even the designers of the AI might remain oblivious to the intricacies of the data processing undertaken by the algorithm [97]. Consequently, the AI system's decision-making becomes unpredictable, engendering unforeseen outcomes or ramifications [125]. For instance, an AI might inadvertently remove substantial volumes of innocent content or websites without a clear understanding of the rationale. Regrettably, numerous AI systems employing ML algorithms adopt a black box design, thereby rendering them inherently complex to comprehend and govern [97].

This dual-edged nature of AI underscores the need for careful consideration of its role in counterterrorism and cybersecurity. While AI can provide powerful tools for detecting and mitigating cyber threats [83], it must be implemented with a deep understanding of its potential drawbacks and vulnerabilities. Therefore, addressing the paradoxical challenges posed by AI necessitates ongoing research and development in the field of cybersecurity. Furthermore, collaboration between technologists, policymakers, and cybersecurity experts is crucial to strike the right balance between harnessing AI's potential and safeguarding against its unintended consequences. Similarly, efforts to mitigate the black box effect and enhance AI transparency and accountability are imperative [1]. Researchers and engineers are actively exploring methods to make AI decision-making more interpretable and explainable, ensuring that AI systems can be trusted even when dealing with critical security tasks. Additionally, ongoing vigilance is required to stay ahead of malicious actors who might seek to exploit AI vulnerabilities. Cybersecurity professionals must continuously adapt and evolve their strategies to defend against ever-evolving threats. In conclusion, AI's integration into counterterrorism and cybersecurity efforts introduces both promising capabilities and complex challenges. While AI has the potential to significantly bolster defences and enhance threat detection, it also poses risks, particularly when it comes to unintended consequences and the black box effect. A proactive and multidisciplinary approach is essential to harness AI's benefits while safeguarding against its potential misuse. As the technological landscape continues to evolve, addressing these challenges remains critical to ensure the security and resilience of our digital infrastructure.

9.4 Discussion

In the realm of counterterrorism, the integration of ML and deep learning (DL) has undeniably yielded significant advantages. For instance, ML and DL could play a pivotal role in detecting online radicalisation processes and facilitating facial recognition matches with known terrorists [59]. However, the comprehensive assessment of the challenges associated with the use of ML in this chapter has revealed the potential weaknesses of ML that could overshadow its benefits. One such limitation lies in the challenges ML encounters when trying to access sufficient data sets for accurately recognising newly emerging terrorist Organisations online and individuals involved in lone terrorist activities [61]. This dearth of data can hinder the effectiveness of ML systems in certain contexts. Additionally, the use of ML in counterterrorism introduces the risk of infringing upon individuals' human rights, especially concerning freedom of speech online [24]. Similarly, the inherent issue of bias within ML models poses a significant concern. This bias can lead to some terrorists not conforming to societal stereotypes, making it difficult for ML systems to identify them and allowing them to evade detection [70]. As a consequence of these limitations, a significant number of terrorists could remain undetected in the online realm, which presents a formidable problem. These individuals and groups continue to engage in radicalisation, recruitment activities, and use the internet as a means to facilitate discussions and plan future terrorist attacks that are intended to occur offline [26, 40]. Thus, while ML exhibits the capability to identify certain terrorists, its inability to recognise others online could undermine its overall effectiveness in counterterrorism efforts.

Furthermore, the utilisation of AI and ML in the context of SMPs to counter cyber terrorism and terrorist content on social media does offer substantial benefits. However, these technologies also bring forth numerous ethical concerns. These ethical apprehensions encompass potential human rights abuses, particularly regarding freedom of expression, as AI may struggle to comprehend context and interpret posted content accurately [42]. Moreover, the presence of bias in AI poses a worrisome issue, especially when AI models are trained on low-quality data, which can lead to subpar AI performance ([35, 90], as cited in [45]). Another concern pertains to the risk of under-blocking or over-blocking certain types of content, where terrorist propaganda might evade detection or genuine content may be wrongly removed from platforms [44, 116].

Additionally, the development of AI can introduce other challenges, such as the potential for cyber terrorists to utilize AI to bypass cyber defences and access critical systems, thereby endangering lives [72, 116]. The use of black box AI models also presents difficulties, as engineers and researchers may find it challenging to discern the reasons behind AI errors [97].

However, notwithstanding the extensive discussions throughout this chapter on the challenges associated with the utilisation of AI and ML in counterterrorism, there exists a potential for optimism. This optimism stems from the recognition that the integration of AI and ML into counterterrorism endeavours, while presenting complex hurdles, also offers a prospect for addressing these very challenges. In the pursuit of more effective and ethically sound counterterrorism strategies, it becomes imperative to explore innovative solutions. Therefore, in the following chapter, namely Chap. 10, these solutions will be comprehensively examined and articulated, providing a systematic framework to navigate the intricate landscape of technical, operational, ethical, and legal challenges inherent in the application of AI and ML in counterterrorism. To this end, Chap. 10 will seek to facilitate a deeper understanding of the potential of AI and ML not only in enhancing security measures but also in safeguarding fundamental human rights. Through a meticulous exploration of the strategic integration of these technologies, it will endeavour to strike a delicate equilibrium, one that ensures the preservation of privacy and freedom of expression while simultaneously efficaciously countering the evolving threats posed by terrorism. In essence, the upcoming chapter aspires to chart a course that optimizes the benefits derived from AI and ML in the realm of counterterrorism while prudently mitigating the inherent risks, thereby contributing to the establishment of a safer and more secure digital landscape.

9.5 Conclusion

In the realm of digital crime prevention, technology stands out for its adaptability and capacity to acquire diverse capabilities. At the forefront of this technological evolution lie AI and ML. However, the shift towards a greater reliance on these technologies is accompanied by a set of formidable challenges. To this end, in pursuit of a comprehensive understanding of these challenges within the context of counterterrorism efforts, this chapter critically examined a wide range of challenges associated with the increased use of ML in counterterrorism efforts. These challenges span a wide spectrum, encompassing technical, operational, ethical, and legal considerations. Through this in-depth exploration, it has become evident that the enhancement of both ML techniques and counterterrorism strategies involves multifaceted dimensions that require closer attention and refinement [33]. This sentiment, increasingly underscored in the post-pandemic era, resonates deeply among researchers, emphasising the impending integration of autonomous systems into the societal fabric. This trajectory presents a crucial choice: embracing this evolution or risking obsolescence in the face of the evolving sophistication of terrorists and the internet (Ibid.).

One of the central challenges in this domain revolves around the overwhelming proliferation of internet-based information. Specifically, within the context of cyberterrorism, there is a compelling case for shifting towards greater reliance on computerized tools rather than manual human monitoring (Ibid.). This transition underscores the necessity for researchers and counterterrorism entities to chart a strategic path forward. In scenarios where the analysis and aggregation of crucial data for the detection of radical content surpass human capacity, one practical solution is to gradually delegate responsibilities to computational systems. However, this transition, itself, poses other ethical challenges. For example, the potential for labour displacement, although not explicitly discussed in the chapter, can be inferred as an additional challenge, further emphasising the multifaceted nature of the adoption of AI technology.[3] Addressing this concern, along with others, necessitates astute government management to mitigate the possibility of widespread unemployment. Furthermore, the evolving landscape of ML requires meticulous consideration of its ramifications for human rights, striking a delicate equilibrium between technological advancement and the preservation of fundamental human freedoms. The dichotomy between privacy infringement and security enhancement underscores the complexity of the issue at hand, thereby necessitating comprehensive evaluation and prudent decision-making in steering the trajectory of societal evolution.

---

[3] Although this specific challenge might initially appear unrelated to the context of ML in counterterrorism, it shares a common thread of ethical significance.

In conclusion, extensive research into the secure augmentation of technological responsibilities is imperative for optimizing efficacy and outcomes in the context of counterterrorism efforts. The challenges outlined here not only highlight the intricate nature of this endeavour but also emphasise the urgency of addressing them with a balanced and forward-thinking approach. This will be the subject of the study presented in the next chapter, namely Chap. 10.

References

1. Adadi A, Berrada M (2018) Peeking inside the black-box: a survey on explainable artificial intelligence (XAI). IEEE Access 6:52138–52160
2. Ackerman G, Peterson H (2020) Terrorism and COVID-19: actual and potential impacts. Perspect Terror 14(3):60–73
3. Agrawal D, Bernstein P, Bertino E, Davidson S, Dayal U (2011) Challenges and opportunities with big data 2011-1. Purdue University
4. Almeida F, Calistru C (2013) The main challenges and issues of big data management. Int J Res Stud Comput 2(1):11–20
5. Ammar J (2019) Cyber gremlin: social networking, machine learning, and the global war on Al-Qaida—and IS-inspired terrorism. Int J Law Inf Technol 27(3):238–265
6. Antwi-Boasiako A (2018) A model for digital evidence admissibility assessment. In: IFIP international conference on digital forensics. Springer, Cham, pp 23–38
7. Antwi-Boasiako A, Venter H (2017) A model for digital evidence admissibility assessment. In: Peterson G, Shenoi S (eds) Advances in digital forensics XIII. Advances in information and communication technology. Springer, Cham, pp 23–38
8. Appleton C (2014) Lone wolf terrorism in Norway. Int J Hum Rights 18(2):127–142
9. Appleton C (2017) Lone wolf terrorism in Norway. In: Contingencies, resilience and legal constitutionalism. Routledge, pp 19–34
10. Babuta A (2017) Big data and policing an assessment of law enforcement requirements, expectations and priorities. Royal United Services Institute for Defence and Security Studies. Available at: https://static.stage.rusi.institute/201709_rusi_big_data_and_policing_babuta_web.pdf. Accessed 04 Oct 2023
11. Baggili I, Behzadan V (2019) Founding the domain of AI forensics. Computer science: cryptography and security. Available at: https://arxiv.org/abs/1912.06497. Accessed 28 Sept 2023
12. Barbosa J (2020) Cyber humanity in cyber war. In: 15th International conference on cyber warfare and security, pp 20–25
13. Barocas S, Hardt M, Narayanan A (2017) Fairness and machine learning: limitations and opportunities. Available at: https://fairmlbook.org/pdf/fairmlbook.pdf. Accessed 04 Oct 2023
14. Battaglini M (2020) How the main legal and ethical issues in machine learning arose and evolved. Technology and Society. Available at: https://www.transparentinternet.com/technology-and-society/machine-learning-issues/. Accessed 27 Sept 2023
15. Bellman RE (1961) Adaptive control processes: a guided tour. Princeton University Press
16. Bianchi A, Greipl A (2022) States' prevention of terrorism and the rule of law: challenging the 'magic' of artificial intelligence (AI). International Centre for Counter-Terrorism. Available at: https://icct.nl/publication/states-prevention-terrorism-rule-of-law-artificial-intelligence/. Accessed 28 Sept 2023
17. Bishop CM (2016) Pattern recognition and machine learning, 3rd edn. Springer
18. Borj PR, Bours P (2019) Predatory conversation detection. In: IEEE 2019 international conference on cyber security for emerging technologies (CSET), pp 1–6

19. Botelho B, Bigelow SJ (n.d.) Big data. Available at: https://www.techtarget.com/searchdat amanagement/definition/big-data. Accessed: 06 Oct 2023
20. Brayne S (2017) Big data surveillance: the case of policing. Am Sociol Rev 82(5):977–1008
21. Brayne S (2020) Predict and surveil: data, discretion, and the future of policing. Oxford University Press, USA
22. Brenner SW (2006) Cybercrime jurisdiction. Crime Law Soc Chang 46:189–206
23. Brey P (1998) Ethical aspects of information security and privacy. In: Petkovic M, Jonker W (eds) Security, privacy, and trust in modern data management. Springer
24. Brown É (2021) Regulating the spread of online misinformation. In: de Ridder J, Hannon M (eds) The Routledge handbook of political epistemology
25. Burke T, Trazo S (2019) Emerging legal issues in an AI-driven world. Gowling WLG. Available at: https://gowlingwlg.com/en/insights-resources/articles/2019/emerging-legal-issues-in-an-ai-driven-world/?msclkid=560505cdb8c811ecb8492a811b3a4ebe. Accessed 28 Sept 2023
26. Canhoto AI (2021) Leveraging machine learning in the global fight against money laundering and terrorism financing: an affordances perspective. J Bus Res 131:441–452
27. Cappa F, Oriani R, Peruffo E, McCarthy I (2021) Big data for creating and capturing value in the digitalized environment: unpacking the effects of volume, variety, and veracity on firm performance. J Prod Innov Manag 38(1):49–67
28. Casey E (2011) Digital evidence and computer crime: forensic science, computers, and the Internet, 3rd edn. Academic
29. Clarke RV (2016) Situational crime prevention. In: Wortley W, Townsley M (eds) Environmental criminology and crime analysis. Routledge, pp 286–303
30. Corbin C (2017) Fordham law review. Terrorists are always Muslim but never white: at the intersection of critical race theory and propaganda. Fordham Law Rev 86(2):445–485
31. Crawford K, Schultz J (2014) Big data and due process: toward a framework to redress predictive privacy harms. Boston Coll Law Rev 55(1):93–128
32. Delony D (2018) What are some ethical issues regarding machine learning? Techopedia. Available at: https://www.techopedia.com/what-are-some-ethical-issues-regarding-machine-learning/7/33376. Accessed 28 Sept 2023
33. De Visser EJ, Pak R, Shaw TH (2018) From 'automation' to 'autonomy': the importance of trust repair in human-machine interaction. Ergonomics 61(10):1409–1427
34. Demchenko Y, Grosso P, De Laat C, Membrey P (2013) Addressing big data issues in scientific data infrastructure. In 2013 International Conference on Collaboration Technologies and Systems (CTS), pp. 48–55.
35. Duarte N, Llanso E, Loup AC (2018) Mixed messages? The limits of automated social media content analysis. Available at: https://cdt.org/wp-content/uploads/2017/12/FAT-conference-draft-2018.pdf. Accessed 28 Sept 2023
36. Equality and Human Rights Commission (2021) Article 10: freedom of expression. Available at: https://www.equalityhumanrights.com/en/human-rights-act/article-10-freedom-exp ression. Accessed 27 Sept 2023
37. Favaretto M, De Clercq E, Elger BS (2019) Big data and discrimination: perils, promises and solutions. A systematic review. J Big Data 6(1):1–27
38. Ferguson A (2017) Policing predictive policing. Wash Univ Law Rev 94(5):1109–1190
39. Fitzpatrick DJ, Gorr WL, Neill DB (2019) Keeping score: predictive analytics in policing. Annu Rev Criminol 2(1):473–491
40. Gaikwad M, Ahirrao S, Phansalkar S, Kotecha K (2021) Online extremism detection: a systematic literature review with emphasis on datasets, classification techniques, validation methods, and tools. IEEE Access 9:48364–48404
41. Ganor B (2021) Artificial or human: a new era of counterterrorism intelligence? Stud Confl Terror 44(7):605–624
42. Gillespie T (2020) Content moderation, AI, and the question of scale. Big Data Soc 7(2)
43. Gillis AS (2021) 5 V's of big data. TechTarget. Available at: https://www.techtarget.com/sea rchdatamanagement/definition/5-Vs-of-big-data. Accessed 28 Sept 2023

44. Gorwa R, Binns R, Katzenbach C (2020) Algorithmic content moderation: technical and political challenges in the automation of platform governance. Big Data Soc 7(1)
45. Gunton K (2022) The use of artificial intelligence in content moderation in countering violent extremism on social media platforms. In: Montasari R (ed) Artificial intelligence and national security. Springer, Cham, pp 69–79
46. Habermas J (1996) Between facts and norms: contributions to a discourse theory of law and democracy. Polity, Cambridge
47. Halbert D (1999) Intellectual property in the information age: the politics of expanding ownership rights. Praeger
48. Hall B (2017) Top 5 legal issues inherent in AI and machine learning. Internet Law, Internet Lawyer. Traverse Legal. Available at: https://www.traverselegal.com/blog/top-5-legal-issues-inherent-in-ai-and-machine-learning-2/?msclkid=56058f26b8c811ec89136e7a5f52e41d. Accessed 28 Sept 2023
49. Hamilton RH, Davion HK (2021) Legal and ethical challenges for HR in machine learning. Empl Responsib Rights J 34(1):19–39
50. Henz P (2021) Ethical and legal responsibility for artificial intelligence. Discov Artif Intell 1:1–5
51. Home Office, The Rt Hon Amber Rudd (2018) New technology revealed to help fight terrorist content online. UK Government. Available at: https://www.gov.uk/government/news/new-technology-revealed-to-help-fight-terrorist-content-online. Accessed 27 Sept 2023
52. Huang J, Galal G, Etemadi M, Vaidyanathan M (2022) Evaluation and mitigation of racial bias in clinical machine learning models: scoping review. JMIR Med Inform 10(5):e36388
53. Human Rights Act 1998, c. 42. Available at: https://www.legislation.gov.uk/ukpga/1998/42/contents. Accessed 28 Sept 2023
54. Hussain Z (2017) The ABCs of machine learning: privacy and other legal concerns. Law Practices Today. Available at: https://www.lawpracticetoday.org/article/machine-learning-privacy-legal-concerns/#:~:text=%20The%20ABCs%20of%20Machine%20Learning%3A%20Privacy%20and,officer%2C%20has%20said%20that%20%E2%80%9CAI%20will...%20More%20?msclkid=5604c6d8b8c811ec94d09eb8666f6e67. Accessed 28 Sept 2023
55. Institute for Economics & Peace (2019) Global terrorism index 2019: measuring the impact of terrorism. Available at: https://www.visionofhumanity.org/wp-content/uploads/2020/11/GTI-2019-web.pdf. Accessed 27 Sept 2023
56. Ishwarappa, Anuradha (2015) A Brief Introduction on Big Data 5Vs Characteristics and Hadoop Technology. International Conference on Intelligent Computing, Communication and Convergence (ICCC-2015), 48:319–324.
57. Jay P (2017) Transfer learning using Keras. Medium. Available at: https://medium.com/@14prakash/transfer-learning-using-keras-d804b2e04ef8. Accessed 27 Sept 2023
58. Jiang D, Wu J, Ding F, Ide T, Scheffran J, Helman D, Zhang S, Qian Y, Fu J, Chen S, Xie X, Ma T, Hao M, Ge Q (2023) An integrated deep-learning and multi-level framework for understanding the behavior of terrorist groups. Heliyon 9(8)
59. Johnson JM, Khoshgoftaar TM (2019) Survey on deep learning with class imbalance. Journal of Big Data, 6(1):1–54.
60. Johansson F, Kaati L, Sahlgren M (2017) Detecting linguistic markers of violent extremism in online environments. In: Artificial intelligence: concepts, methodologies, tools, and applications, pp 2847–2863
61. Jones SG, Doxsee C, Harrington N, Hwang G, Suber J (2020) The war comes home: the evolution of domestic terrorism in the United States. Center for Strategic and International Studies (CSIS), pp 1–13
62. Kaplan J, Lööw H, Malkki L (eds) (2017) Lone wolf and autonomous cell terrorism. Routledge
63. Koops BJ, Leenes R (2006) Identity theft, identity fraud and/or identity-related crime: definitions matter. Datenschutz und Datensicherheit-DuD 30(9):553–556
64. Kotian I (2020) Basics of machine learning. Kaggle. Available at: https://www.kaggle.com/discussions/general/240136. Accessed 27 Sept 2023

65. Krieg SJ, Smith CW, Chatterjee R, Chawla NV (2022) Predicting terrorist attacks in the United States using localized news data. PLoS ONE 17(6):e0270681
66. Kusters R, Truderung T, Vogt A (2010) Accountability: definition and relationship to verifiability. In: Proceedings of the 17th ACM conference on computer and communications security, pp 526–535
67. Leibold J (2020) Surveillance in China's Xinjiang region: ethnic sorting, coercion, and inducement. J Contemp China 29(121):46–60
68. Levy I, Rozmann N (2022) Differences in attitudes toward terrorists: type of terrorist act, terrorist ethnicity, and observer gender and cultural background. Group Process Intergroup Relat 26(2):476–492
69. L'Heureux A, Grolinger K, Elyamany HF, Capretz MAM (2017) Machine learning with big data: challenges and approaches. IEEE Access 5:7776–7797
70. Liang PP, Wu C, Morency LP, Salakhutdinov R (2021) Towards understanding and mitigating social biases in language models. In: International conference on machine learning, pp 6565–6576
71. Loukides M (2011) What is data science? O'Reilly Media
72. Luiijf E (2014) Definitions of cyber terrorism. In: Akhgar B, Staniforth A, Bosco F (eds) Cyber crime and cyber terrorism investigator's handbook. Syngress, pp 11–17
73. Macdonald S (2018) How tech companies are trying to disrupt terrorist social media activity. Scientific American, The Conversation. Available at: https://www.scientificamerican.com/art icle/how-tech-companies-are-trying-to-disrupt-terrorist-social-media-activity/. Accessed 27 Sept 2023
74. McDaniel JLM, Pease KG (eds) (2021) Predictive policing and artificial intelligence, 1st edn. Routledge
75. McKendrick K (2019) Artificial intelligence prediction and counterterrorism. The Royal Institute of International, Chatham House. Available at: https://www.chathamhouse.org/sites/def ault/files/2019-08-07-AICounterterrorism.pdf. Accessed 27 Sept 2023
76. Mehrabi N, Morstatter F, Saxena N, Lerman K, Galstyan A (2021) A survey on bias and fairness in machine learning. ACM Comput Surv (CSUR) 54(6):1–35
77. Mijatović D (2018) In the era of artificial intelligence: safeguarding human rights. openDemocracy. Available at: https://www.opendemocracy.net/en/digitaliberties/in-era-of-artifi cial-intelligence-safeguarding-human-rights/. Accessed 27 Sept 2023
78. Montasari R (2023) Countering cyberterrorism: The confluence of artificial intelligence, cyber forensics and digital policing in US and UK national cybersecurity. Springer
79. Montasari R (2023) The application of big data predictive analytics and surveillance technologies in the field of policing. Countering cyberterrorism: the confluence of artificial intelligence, cyber forensics and digital policing in US and UK national cybersecurity. Springer, Cham, pp 81–114
80. Montasari R (2023) The potential impacts of the national security uses of big data predictive analytics on human rights. Countering cyberterrorism: the confluence of artificial intelligence, cyber forensics and digital policing in US and UK national cybersecurity. Springer, Cham, pp 115–137
81. Montasari R (2022) Privacy, security and forensics in the Internet of Things (IoT). Springer
82. Montasari R, Jahankhani H (2021) Artificial intelligence in cyber security: impact and implications: security challenges, technical and ethical issues, forensic investigative challenges. Springer
83. Montasari R, Carroll F, Macdonald S, Jahankhani H, Hosseinian-Far A, Daneshkhah A (2021) Application of artificial intelligence and machine learning in producing actionable cyber threat intelligence. In: Montasari R, Jahankhani H, Hill R, Parkinson S (eds) Digital forensic investigation of internet of things (IoT) devices. Springer, Cham, pp 47–64
84. Montasari R (2021) Cyber threats and national security: the use and abuse of artificial intelligence. In: Masys AJ (ed) Handbook of security science. Springer, Cham
85. Najafabadi MM, Villanustre F, Khoshgoftaar TM, Seliya N, Wald R, Muharemagic E (2015) Deep learning applications and challenges in big data analytics. J Big Data 2(1):1–21

86. Ng LHX, Cruickshank IJ, Carley KM (2022) Cross-platform information spread during the January 6th capitol riots. Soc Netw Anal Min 12:133
87. Nowrasteh A (2019) Terrorists by immigration status and nationality: a risk analysis 1975–2017. Cato Institute, p 866. Available at: https://www.cato.org/publications/policy-analysis/terrorists-immigration-status-nationality-risk-analysis-1975-2017. Accessed 27 Sept 2023
88. O'Leary DE (2013) Artificial intelligence and big data. IEEE Intell Syst 28(2):96–99
89. Ongsulee P, Chotchaung V, Bamrungsi E, Rodcheewit T (2018) Big data, predictive analytics and machine learning. In: 2018 16th international conference on ICT and knowledge engineering, pp 1–6. https://doi.org/10.1109/ICTKE.2018.8612393
90. Osoba OA, Welser IV, W (2017) An intelligence in our image: the risks of bias and errors in artificial intelligence. RAND Corporation. Available at: https://www.rand.org/pubs/research_reports/RR1744.html. Accessed 28 Sept 2023
91. Palmer G (2001) A road map for digital forensics research. In: First digital forensic research workshop (DFRWS), 27–30. Utica, New York
92. Porter T (2021) Facial recognition: facing up to terrorism. Available at: https://counterterrorbusiness.com/features/facial-recognition-facing-terrorism#:~:text=It%20is%20then%20able%20to,to%20do%20a%20visual%20comparison. Accessed 27 Sept 2023
93. Privacy International (n.d.) Mass surveillance. Available at: https://privacyinternational.org/learn/mass-surveillance. Accessed 28 Sept 2023
94. Reed C, Kennedy E, Silva S (2016) Responsibility, autonomy and accountability: legal liability for machine learning. Queen Mary School of Law Legal Studies Research Paper, p 243
95. Royakkers L, Timmer J, Kool L, Van Est R (2018) Societal and ethical issues of digitization. Ethics Inf Technol 20:127–142
96. Roche G, Leibold J (2022) State racism and surveillance in Xinjiang (People's Republic of China). Polit Q 93(3):442–450
97. Rudin C, Radin J (2019) Why are we using black box models in AI when we don't need to? A lesson from an explainable AI competition. Harvard Data Sci Rev 1(2)
98. Russom P (2011) Big data analytics. TDWI Res 19(4):1–34
99. Sagiroglu S, Sinanc D (2013) Big data: a review. In: IEEE international conference on collaboration technologies and systems (CTS), pp 42–47
100. Saint-Clare S (2011) Overview and analysis on cyber terrorism. Sch Doct Stud (Eur Union) J 85–98
101. Salaam C, Rawat DB (2022) Terrorism detection and approaches to fairness: a brief literature review. In: IEEE global humanitarian technology conference (GHTC), Santa Clara, CA, USA, pp 277–282
102. Shrestha S (2021) Does surveillance law provide security or threaten privacy? Queen Mary University of London. Available at: https://www.qmul.ac.uk/lac/our-legal-blog/items/does-surveillance-law-provide-security-or-threaten-privacy.html. Accessed 28 Sept 2023
103. Sikos LF (2021) AI in digital forensics: ontology engineering for cybercrime investigations. Wiley Interdiscipl Rev: Forens Sci 3(3):e1394
104. Smit R, Heinrich A, Broersma M (2016) Witnessing in the new memory ecology: memory construction of the Syrian conflict on YouTube. New Media Soc 19(2):289–307
105. Smith BL, Roberts P, Damphousse KR (2017) The terrorists' planning cycle: patterns of pre-incident behavior. In: The handbook of the criminology of terrorism, pp 62–76
106. Smith LG, Wakeford L, Cribbin TF, Barnett J, Hou WK (2020) Detecting psychological change through mobilizing interactions and changes in extremist linguistic style. Comput Hum Behav 108:106298
107. Solove DJ (2011) Nothing to hide: the false tradeoff between privacy and security. Yale University Press, London
108. Soomro TR, Hussain M (2019) Social media-related cybercrimes and techniques for their prevention. Appl Comput Syst 24(1):9–17
109. Spaaij R (2011) Understanding lone wolf terrorism: global patterns, motivations and prevention. Springer

110. Srivastava M, Hashimoto T, Liang P (2020) Robustness to spurious correlations via human annotations. In: Proceedings of the 37th international conference on machine learning. PMLR, pp 9109–9119

111. Stahl T (2016) Indiscriminate mass surveillance and the public sphere. Ethics Inf Technol 18(1):33–39

112. Statista (2022) Number of Internet and social media users worldwide as of July 2022. Available at: https://www.statista.com/statistics/617136/digital-population-worldwide/. Accessed 28 Sept 2023

113. Straub VJ (2021) Beyond kinetic harm and towards a dynamic conceptualization of cyberterrorism. J Inf Warfare 20(3):1–23

114. Tardi C (2022) What is Moore's law and is it still true? Investopedia. Available at: https://www.investopedia.com/terms/m/mooreslaw.asp. Accessed 28 Sept 2023

115. Science, Innovation and Technology Committee (2023) The governance of artificial intelligence: interim report, Ninth Report of Session 2022–23. UK Parliament. Available at: https://publications.parliament.uk/pa/cm5803/cmselect/cmsctech/1769/report.html. Accessed 28 Sept 2023

116. United Nations Interregional Crime and Justice Research Institute (UNICRI), United Nations Counter-Terrorism Centre (UNCCT) (2021) Countering terrorism online with artificial intelligence: an overview for law enforcement and counter-terrorism agencies in South Asia and South-East Asia. Available at: https://www.un.org/counterterrorism/sites/www.un.org.counterterrorism/files/countering-terrorism-online-with-ai-uncct-unicri-report-web.pdf. Accessed 27 Sept 2023

117. UK Statistics Authority (UKSA) (2021) Ethical considerations in the use of Machine Learning for research and statistics. Available at: https://uksa.statisticsauthority.gov.uk/publication/ethical-considerations-in-the-use-of-machine-learning-for-research-and-statistics/. Accessed 28 Sept 2023

118. Verhelst H, Stannat A, Mecacci G (2020) Machine learning against terrorism: how big data collection and analysis influences the privacy-security dilemma. Sci Eng Ethics 26(6):2975–2984

119. Weimann G (2015) Terrorism in cyberspace: the next generation. Woodrow Wilson Centre Press

120. Williamson H, Murphy K (2020) Animus toward Muslims and its association with public support for punitive counter-terrorism policies: did the Christchurch terrorist attack mitigate this association? J Exp Criminol 18:343–363

121. Yamini (2021) Big data analytics in cybersecurity: role and applications. Available at: https://www.analyticssteps.com/blogs/big-data-analytics-cybersecurity-role-and-applications. Accessed 28 Sept 2023

122. Yapo A, Weiss J (2018) Ethical implications of bias in machine learning. In: Proceedings of the 51st Hawaii international conference on system sciences, pp 5365–5372

123. Yeboah-Ofori A, Brown AD (2020) Digital forensics investigation jurisprudence: issues of admissibility of digital evidence. HSOA J Forens Legal Invest Sci 6(1):1–8

124. Yeung K (2017) 'Hypernudge': big data as a mode of regulation by design. Inf Commun Soc 20(1):118–136

125. Yu S, Carroll F (2021) Implications of AI in national security: understanding the security issues and ethical challenges. In: Montasari R, Jahankhani H (eds) Artificial intelligence in cyber security: impact and implications: security challenges, technical and ethical issues, forensic investigative challenges. Springer, pp 157–175. https://doi.org/10.1007/978-3-030-88040-8_6

126. Zhou ZH (2021) Machine learning. Springer

Chapter 10
Addressing Ethical, Legal, Technical, and Operational Challenges in Counterterrorism with Machine Learning: Recommendations and Strategies

Abstract Terrorist attacks, in both traditional and cyber forms, constitute one of the most significant factors contributing to the destabilisation of both national and transnational security. As the preceding chapters in this book have underscored, a comprehensive understanding of the underlying mechanisms behind these attacks is crucial for governments and law enforcement agencies (LEAs) to conduct more effective counterterrorism operations and investigations. This chapter, an integral component of the trilogy comprising Chaps. 8, 9, and 10 in this book, builds upon the foundations laid in its predecessors. While Chap. 8 delved into the technical aspects of using artificial intelligence (AI) and machine learning (ML) to counter cyberterrorism and Chap. 9 focused on the challenges associated with these efforts, the aim of this chapter is twofold. Firstly, it addresses the challenges discussed in Chap. 9 by providing insights, recommendations, and strategic guidance. Secondly, it seeks to illuminate a responsible and effective path forward for the use of ML in bolstering cybersecurity efforts. While each of these three chapters stands independently, readers are encouraged to explore them within the context of one another. These interconnected chapters form a cohesive narrative that progressively clarifies the intricacies of countering cyberterrorism through the lens of ML, providing a comprehensive and holistic understanding of this critical subject matter.

Keywords Cyberterrorism · Counterterrorism · Machine learning · Artificial intelligence · National security · International security · Ethical challenges · Jurisdictional challenges · Bias · Big data · Privacy · Security

10.1 Introduction

Terrorist attacks constitute one of the most significant factors contributing to both national and transnational instability. Therefore, as discussed in the previous chapters of this book, a comprehensive understanding of the underlying mechanisms behind these events is crucial for governments and LEAs to conduct more effective counterterrorism operations and investigations. In particular, in Chap. 8, machine learning (ML), a branch of artificial intelligence (AI), was highlighted as one of the most prominent tools in the fights against terrorism. For instance, ML can be employed to detect suspicious terrorist-related activities [126] or to predict future terrorist attacks. This can be achieved by employing datasets in mass surveillance [21, 126]. This development has been greatly facilitated by the abundance of information available on the internet and social media platforms (SMPs), which has led to the development of numerous algorithms, resulting in more accurate outcomes [88] and consequently enhancing the apprehension of terrorists. Moreover, the implementation of Facial Recognition, a form of AI, has significantly contributed to the high accuracy levels in identifying terrorists, surpassing the capabilities of human recognition. Additionally, Natural Language Processing (NLP), another branch of AI, has enabled the recognition of specific slang words associated with terrorism in online contexts.

Notwithstanding these advancements, a wide range of technical, operational, legal and ethical challenges confront ML in countering terrorism, as discussed in Chap. 9 of the book. For instance, the lack of sufficient online data poses difficulties in effectively tracing terrorists acting alone, as their actions exhibit diverse patterns, and combating newly emerging terrorist organisations online becomes arduous due to the paucity of information associated with groups like Al-Qaida online [4]. These challenges are significant as lone terrorists and emerging groups continue to pose threats to society, exacerbated by the insufficient training and flawed algorithms resulting from limited data sets pertaining to these entities. Another example concerns the potential impingement of freedom of speech due to the application of ML [79] as numerous online items are erroneously labelled as terrorism, leading to their removal or blocking. This situation raises ethical concerns [39]. Furthermore, the issue of bias presents a substantial problem for ML. If not adequately addressed, ML models may perpetuate algorithmic decision-making that favours specific individuals or groups, thereby yielding biased outcomes [84]. Similarly, in the context of countering terrorism, racial bias poses a threat, as societal misperceptions often associate terrorists with ethnic minority groups rather than the white ethnic majority [31]. This bias could result in the oversight of terrorists belonging to the white ethnic majority.

Building upon Chaps. 8 and 9, this chapter shifts its focus away from the technical aspects of AI and ML in countering cyberterrorism and the associated challenges extensively covered in those chapters. Instead, the aim of this chapter is twofold. Firstly, it aims to address the challenges previously discussed in Chap. 9 by providing insights, recommendations, and strategic guidance. Secondly, it seeks to illuminate a responsible and effective path forward for the use of ML in bolstering cybersecurity efforts. It is important to note that while Chaps. 8, 9, and 10 each stand independently, readers are encouraged to explore them within the context of one another. These interconnected chapters form a cohesive narrative that progressively clarifies the intricacies of countering cyberterrorism through the lens of ML, thereby providing a comprehensive and holistic understanding of this critical subject matter. Despite this discussion, Sect. 10.2 of this chapter offers a brief overview of the previously discussed Chaps. 8 and 9.

The subsequent sections of this chapter are structured as follows. Section 10.2 provides a brief overview of Chaps. 8 and 9, as previously discussed. In Sect. 10.3, a comprehensive set of technical and operational recommendations is proffered, aligning with the technical and operational challenges elucidated in Chap. 9. This pertains to the application of ML in counterterrorism endeavours. In a similar vein, Sect. 10.4 offers a diverse array of ethical and legal recommendations that correspond to the ethical and legal challenges involved in the use of ML within counterterrorism initiatives, also expounded upon in Chap. 9. Finally, Sect. 10.5 engages in a discussion of the key findings of the study whilst also concluding the chapter.

10.2 Background to Chapters 8 and 9

Chapter 8 "Machine Learning and Deep Learning Techniques in Countering Cyberterrorism" delved into the technical dimensions of AI concerning cyber security and cyber terrorism. Particularly, the chapter focused on examining the pivotal role that ML and deep learning (DL) played in identifying extremist content and activities on online platforms. Specifically, the chapter provided an in-depth exploration of ML and the associated techniques such as Support Vector Machine (SVM), Decision Trees, and Naïve Bayes. Comprehending the operational mechanics of these techniques is integral to comprehending their utility in detecting radical content online. These algorithms predominantly employ feature selection and classification methodologies to distinguish between radical and non-radical textual content [52]. These classifications enable LEAs to ascertain whether identified radical behaviour poses a threat to national security. In the context of dynamic approaches, the chapter discussed the application of the intrusion detection systems (IDS), exemplifying a mechanism designed to identify attacks and thwart unauthorised system access through continuous activity monitoring [80]. This dynamic approach was juxtaposed with the exploration of DL, inclusive of a thorough examination of distinct classifiers, such as single classifiers encompassing SVMs and Decision Trees.

Turning to the domain of counterterrorism policing, the discourse focused on the formulation of a pre-terrorist profile as a strategic tool to fortify countermeasure strategies. Furthermore, attention was directed towards pertinent organisations such as the IDS and the Global Terrorism Database (GTD), entities that government agencies rely upon to gauge the efficacy of counterterrorism initiatives and analyse the progress of countermeasures. The GTD, administered by Study of Terrorism and Responses to Terrorism (START) (GTD, n.d.), endeavours to identify patterns in global terrorism with the aspiration of eventually formulating predictive models to pre-empt and mitigate terrorist activities. In summary, Chap. 8 offered an expansive panorama of techniques, methodologies, and strategic frameworks employed to detect and counteract radical content, offering a comprehensive assessment of the evolving landscape of counterterrorism efforts. Through an amalgamation of technical methodologies and strategic analyses, the chapter serves to enrich the existing comprehension of the multifaceted domain of counterterrorism operations.

Chapter 9, titled "Ethical, Legal and Technical Challenges in Countering Cyber Terrorism with Machine Learning", delved into a comprehensive examination of the multifaceted challenges that arise when applying machine learning (ML) techniques to combat cyber terrorism. This chapter aimed to gain a deeper understanding of the complexities surrounding the use of ML in this domain, with a particular focus on identifying key technical, operational, ethical, and legal issues. In the context of ethical and legal considerations, the chapter explored a spectrum of challenges encompassing autonomy, bias, privacy, and human rights, as well as legal complexities covering concepts such as hybrid classifiers, accountability, and jurisdictional issues. On the technical and operational front, the chapter investigated challenges related to big data, class imbalance, the curse of dimensionality, spurious correlations, the detection of lone-wolf terrorists, mass surveillance, predictive analytics, transparency, explainability, and other unintended consequences associated with the utilisation of ML in the fight against cyber terrorism. In summary, by delineating key areas of concern, from issues of bias and privacy to the intricacies of jurisdiction and the intricacies of handling massive datasets, this chapter provided a deeper understanding of the multifaceted nature of countering cyber terrorism with ML. The chapter's insights will serve as a valuable resource, guiding future efforts to navigate these challenges and ensure the responsible and effective use of ML technology in the pursuit of enhanced cybersecurity.

10.3   Technical and Operational Recommendations

In light of the comprehensive examination of technical and operational challenges associated with the use of ML within the intricate domain of counter-terrorism, as detailed in Chap. 9, this section aims to provide a robust set of recommendations. These recommendations have been meticulously tailored to address the multifaceted hurdles that were identified in the preceding chapter. This section is divided into five key subsections as follows, each of which addresses a pivotal aspect of this complex landscape. Section 10.3.1 delves into strategies for handling the voluminous data streams inherent to counter-terrorism efforts, proposing innovative methods for efficient data processing and storage. Section 10.3.2 outlines guidelines for ensuring the transparency of ML models, particularly in contexts where the interpretability of decisions is paramount for actionable insights. Section 10.3.3 puts forth strategies to minimise false alarms, a challenge that can have significant operational implications. Section 10.3.4 provides insights into the selection and curation of datasets, crucial for the accuracy and relevance of ML models deployed in counter-terrorism. Finally, Sect. 10.3.5 concludes this section by offering an overview of the recommendations presented in each subsection. This, in turn, provides readers with a comprehensive reference point for navigating the intricate landscape of applying ML techniques to counter-terrorism operations. These recommendations are designed to empower practitioners, policymakers, and researchers in their efforts to harness the potential of

ML while effectively addressing the unique challenges posed by counter-terrorism, ultimately contributing to enhanced security and intelligence outcomes.

10.3.1 Big Data Recommendations

The application of data mining offers a means to analyse the findings derived by ML, enabling the identification and isolation of significant patterns [131]. Within this context, two outlined categories are deemed useful in tackling the problem of big data. The first category involves data processing, where data is manipulated to achieve desired outcomes, while the second category focuses on adapting and modifying existing ML algorithms or developing newer ones [71]. More specifically, the first category includes the concept of 'dimensionality reduction'. By retaining only the most significant variables within the data, the size of data input into algorithms is considerably reduced. This, in turn, results in a reduction of high-dimensional data without substantial information loss [111]. This enables ML algorithms to recognise patterns while mitigating the occurrence of 'spurious correlations', as only relevant data is retained. However, such correlations are not entirely preventable and could still arise despite efforts to mitigate them. Additionally, data cleansing techniques can contribute to data size reduction by eliminating irrelevant and 'noisy' features from audio, image, or video data. Nonetheless, the lack of significant advancements in this area renders it a time-consuming process and less practical for immediate policing needs [71]. The second category involves advancing existing techniques to sustain the functionality of ML. Within this context, leveraging DL methods could offer multiple benefits that might not have been attainable otherwise. For instance, the use of DL enables the extraction and differentiation of relevant data from data containing weak semantics, thus enhancing data relevancy [102].

In relation to addressing issues such as the curse of dimensionality and class imbalance that might lead to inaccuracies, it becomes crucial to ensure accurate and robust model performance in the field of ML. One effective approach that could address these challenges involves employing dimensionality reduction techniques in combination with class imbalance mitigation strategies. This integrated approach, discussed below, enhances the overall quality of ML models and their ability to handle complex datasets.

10.3.1.1 Class Imbalance

With regards to class imbalance, as highlighted in Chap. 9, class imbalance is a critical concern in ML, particularly when dealing with large datasets [63]. Therefore, in tandem with solution for the Curse of Dimensionality, handling class imbalance is paramount for model accuracy. Therefore, the following recommendations are intended to assist both LEAs and researchers in addressing class imbalance more effectively. Integrating the recommended strategies, in turn, can result in more robust and accurate ML models with reduced bias but increased reliability, particularly in scenarios characterised by imbalanced class distributions.

One key approach to address class imbalance is to balance the dataset using over-sampling and undersampling techniques [26]. Oversampling the minority class and undersampling the majority class can assist in creating a more equitable distribution of data points, which is crucial for improving model performance. Additionally, generating synthetic data using methods such as Synthetic Minority Over-sampling Technique (SMOTE) can augment the dataset, further mitigating class imbalance [26]. Another strategy is cost-sensitive learning, where different misclassification costs are assigned to different classes [38, 75]. This encourages the model to prioritise accuracy for the minority class, which again is essential for addressing class imbalance effectively. Ensemble methods, such as Random Forests or AdaBoost, offer another avenue for handling class imbalance. These techniques combine multiple models to enhance predictive performance [100, 115, 132]. By aggregating the predictions of multiple models, they can effectively address the challenges posed by imbalanced datasets. Moreover, class imbalance could be considered as an anomaly detection problem using algorithms such as One-Class SVM or Isolation Forest can be beneficial [73, 99, 104, 127, 137]. These techniques are well-suited for identifying rare instances within imbalanced datasets. In addition, to evaluate the ML models' performance accurately, it is essential to employ appropriate evaluation metrics. Metrics such as precision, recall, F1-score, and AUC-ROC can provide a more comprehensive assessment of model performance in imbalanced datasets, where accuracy can be misleading [33, 109]. Similarly, as emphasised throughout this book, ensuring fairness is crucial in ML models. Therefore, regularly monitoring and mitigating bias in both data and algorithms is essential to avoid perpetuating unfair outcomes [107, 125].

Class imbalance should also involve employing resampling techniques that cater to imbalanced datasets. For instance, oversampling increases the number of instances in the minority class by duplicating or generating synthetic data points [26]. On the other hand, undersampling decreases the number of instances in the majority class by randomly removing data points. Additionally, techniques such as the Synthetic Minority Over-sampling Technique (SMOTE) can generate synthetic instances of the minority class, effectively balancing class distributions [26]. Another valuable approach that could be undertaken to address class imbalance could be cost-sensitive learning [16], which involves the modification of the cost function to impose higher penalties for misclassifying minority class instances [38, 78]. This encourages the model to prioritize correct classification of the minority class, thus addressing the class imbalance issue. Similarly, Ensemble methods such as Random Forests [14, 18], Gradient Boosting [46, 116], or AdaBoost [47] inherently handle class imbalance by combining multiple base learners. These methods can adapt to imbalanced datasets and provide more accurate predictions. Another potential solution could be to consider selecting appropriate evaluation metrics that are sensitive to class imbalance, such as positive predictive value (PPV) plots [101], ROC and Precision-Recall (PRC) plots [58], and Area under the Receiver Operating Characteristic (ROC) curve (AUC) [56]. These metrics offer a more comprehensive understanding of model performance, especially in imbalanced datasets. To ensure reliable model assessment, maintain the same class distribution as the original dataset in each fold of cross-validation, preventing misleading optimistic bias in performance evaluation [17, 25, 36].

However, it is essential to note that, in larger states and countries, imbalanced datasets can result in a reduced representation of non-suspicious individuals, potentially hindering the identification of potential threats, such as terrorists [63]. This imbalance can result in models exhibiting bias towards the majority class, consequently undermining their ability to accurately predict minority class outcomes. Within this context, hybrid approaches that combine innovative techniques with ML classifiers can provide a potential solution to address class imbalance even though these approaches can be time-consuming and expensive [35]. For instance, recent studies, such as [70], have demonstrated the effective use of SMOTE as discussed above to balance training datasets and improve predictive accuracy for real-world scenarios, such as predicting terrorist attack occurrences in the United States [35, 70]. These hybrid techniques demonstrate practicality and efficacy in mitigating class imbalance issues.

To address class imbalance, two other distinct approaches emerge. The first involves modifying the dataset's 'characteristics', which may entail sacrificing comprehensiveness by omitting innocent individuals or introducing potential threat indicators. As previously mentioned, both of these options are demanding and fraught with challenges. Consequently, a middle-ground approach should be sought to tackle the problem effectively. However, this middle-ground approach poses its own set of challenges, as it requires ML algorithms to predict 'potential' terrorists. This concept diverges from the conventional binary classification of positive and negative data. The term 'potential' include individuals with a wide range of beliefs, plans, motivations, and backgrounds [112]. This diversity complicates the task of ML, as it must determine and predict behaviours among individuals who do not precisely fit into predefined categories of terrorists and non-terrorists. Consequently, this diversity could diminish the reliability and predictive power of ML in countering online terrorism.

In summary, class imbalance stands as a significant challenge when applying ML techniques in the context of counterterrorism. Therefore, addressing class imbalance becomes a paramount consideration in the application of ML to counterterrorism efforts. To this end, researchers and practitioners alike must navigate the delicate balance between sensitivity and specificity while remaining acutely aware of the real-world implications of misclassification. Nevertheless, it is crucial to acknowledge that class imbalance also offers opportunities for refining classification algorithms and underscores the pivotal role of data pre-processing, including meticulous data cleansing, before feeding information into ML models [5]. In closing, the proposed solutions outlined in this section are designed to mitigate the challenges posed by class imbalance and enhance the overall effectiveness of ML in addressing terrorism-related issues.

10.3.1.2 Curse of Dimensionality

To assist addressing the challenges associated with the Curse of Dimensionality and therefore enhance the overall effectiveness of ML in countering terrorism-related challenges, the following set of recommendations is provided below:

Regularization methods can be used to prevent overfitting in ML models. Overfitting occurs when a model fits the training data too closely and captures noise or irrelevant details, leading to poor generalization to new, unseen data. Regularization introduces a penalty term into the model's objective function, discouraging it from assigning excessively large weights or coefficients to features. Common regularization techniques include L1 regularization (Lasso), L2 regularization (Ridge) [117, 118], and Elastic Net, among others, that penalize excessive model complexity, thus promoting the selection of relevant features. Although Regularization does not necessarily reduce the dimensionality of the feature space, it instead controls the magnitude of feature coefficients to prevent overfitting.

Another valuable approach is Dimensionality Reduction techniques. Methods such as Principal Component Analysis (PCA) [64] and t-Distributed Stochastic Neighbor Embedding (t-SNE) [124] can effectively transform high-dimensional data into lower-dimensional representations while preserving essential information. In particular, PCA stands out as a powerful tool. PCA can effectively reduce the dimensionality of a dataset while retaining crucial information by identifying the most important features and projecting the data onto a lower-dimensional subspace [65, 130]. This not only mitigates the risk of overfitting caused by high dimensionality but also preserves relevant information that contributes to accurate modelling. Another dimensionality reduction technique that could be considered is feature selection. This technique involves identifying and retaining only the most relevant features while discarding redundant or irrelevant ones [54]. Within this context, utilising techniques such as Mutual Information [15], Recursive Feature Elimination (RFE) [55], or tree-based feature importance [13] allows for the identification and retention of the most informative features, effectively reducing dimensionality.

By reducing the dimensionality of the dataset, feature selection enhances model performance. Furthermore, to tackle the challenge of dimensionality, unsupervised machine learning (UML) techniques present promising solutions [6, 12]. These methods effectively address the complexities inherent in datasets, leading to the creation of enhanced features [6, 12, 113]. This transformation involves converting data from a higher-dimensional space to a more informative, lower-dimensional space, thereby enhancing data quality and facilitating data cleansing [51]. However, it is important to note that, while it is evident that increasing the amount of data from historical incidents, such as terrorist attacks, has the potential to boost accuracy levels [71], practical constraints might impede this approach, as it often necessitates waiting for future incidents to occur. Nevertheless, the suggested dimensionality reduction techniques remain valuable tools, as they enable the representation of data in a lower-dimensional space, preserving essential variance from higher-dimensional environments [130].

Furthermore, Feature Engineering can play a vital role to address challenges pertaining to the Curse of Dimensionality. For instance, by carefully crafting features, practitioners can reduce dimensionality by combining or creating new features that capture underlying data patterns [53]. Similarly, in domains such as Natural Language Processing (NLP), the use of Embeddings becomes indispensable. Techniques such as Word2Vec and Doc2Vec [72, 83] enable the transformation of high-dimensional textual data into lower-dimensional representations. In the same vein, for robustness in high-dimensional data, Ensemble Learning methods, including Random Forests and Gradient Boosting [14, 46] can be effectively employed. These techniques have the advantage of handling feature selection implicitly [14, 46]. Lastly, Data Pre-processing steps such as normalization or standardization [57] can contribute significantly. These processes ensure consistent scales across features, alleviating some of the dimensionality-related challenges [57].

In summary, by integrating the strategies recommended in this section, ML practitioners can navigate the complexities associated with high-dimensional data, constructing models that are not only efficient and accurate but also robust, even within expansive feature spaces. These solutions empower researchers and practitioners alike to extract meaningful insights and predictive power from modern datasets, facilitating the advancement of machine learning in diverse domains.

10.3.2 Transparency and Explainability

The challenge associated with black box models[1] in AI and ML presents a complex situation as discussed in Chap. 9. A recommended approach is to transition away from black box models whenever possible and instead opt for interpretable models, commonly known as white or glass box models [76, 98]. The adoption of interpretable models enables a better understanding of AI decision-making processes compared to black box models. This improved transparency increases public trust in AI systems and allows engineers to identify and rectify mistakes more effectively, ultimately enhancing AI performance [98]. In the context of cyber security, this benefit is particularly advantageous as early identification of security flaws can potentially prevent cyber-attacks and save lives.

Another response to the black box conundrum could include the adoption of an open-source licensing framework for these algorithms. This particular course of action presents a viable resolution by enabling the examination of the underlying source code, thereby promoting transparency and enhancing the comprehension of the decision-making processes involved [34]. The imperative of transparency extends its reach further, intersecting with the foundational principles delineated in the General Data Protection Regulation (GDPR). Within this regulatory framework, the Transparency Principle mandates that organisations elucidate their methods of data collection and the prospective utilization of personal information [9]. Furthermore,

---

[1] Characterised by the opacity of their ML algorithms.

it grants individuals the entitlement to access their own data. However, it is imperative to note that certain provisions, such as those articulated in the Trade Secrets Directive, do allow for instances of non-transparency, permitting access to data only subsequent to automated decision-making procedures [9]. As a result a persuasive recommendation emerges, emphasizing the necessity to anonymize databases, create an ethical framework with clear principles, and craft a code of conduct governing data handling [9].

Moreover, inclusive of a multitude of stakeholders, the imperative of transparency mandates comprehensive communication when automated systems engage in the collection and analysis of data over the Internet. Such communication should serve the purpose of elucidating the nature of the data being collected and the rationale underpinning its utilization [120]. This multifaceted interplay between transparency and stakeholders underscores the utmost importance of cultivating an environment characterized by clarity, accountability, and shared understanding.

In the context of data acquisition and utilization, organisations have a responsibility to provide individuals with a comprehensive understanding of collected data, its intended recipients, and potential implications of data aggregation [9]. When opaque learning algorithms are employed for clandestine data collection, decision-makers face a choice about whether they disclose the motives underlying data collection or not. A prevailing recommendation supports such disclosure to alleviate privacy concerns. USKA (2021) echoes this sentiment by emphasising the need for elucidating factors, including algorithm selection rationale, potential biases, training data, and its applicability in similar studies (Ibid.). The standpoint advanced by [120] adds another crucial dimension to the discourse, advocating for transparency in decision-making processes involving both human and machine elements (Ibid.). This includes rendering these processes readily reviewable, thereby enabling pertinent stakeholders to understand the rationale behind the chosen methodology and algorithm, alongside the possible ethical implications (Ibid.). At the crux of this transparency and explainability paradigm lies the principle of honesty. Encouraging such open communication not only enhances the transparency of algorithms but also augments their real-world applicability.

Overall, the hallmark of transparency and explainability rests upon fostering candid and comprehensive communication. Consequently, by fostering a culture of openness, organisations can engender enhanced understanding, thereby reinforcing the ethical and practical standing of algorithms within broader socio-technological contexts.

10.3.3 False Positive Recommendations

One effective strategy for mitigating false positives involves adjusting the alert threshold [20]. Additional technological advancements that have been proposed in the literature to address this issue. One such technology, as reported by Spathoulas and Katsikas [114], has demonstrated a purported 75% reduction in false positives.

This technology incorporates a three-part post-processing filter, with one critical component known as Neighbouring Related Alerts. This component capitalizes on the observation that terrorist attacks often generate clusters of warnings during their victim-seeking phase, thus providing a more precise and context-aware approach to threat detection. Furthermore, researchers have explored the use of trained datasets specific to virtual screening to minimise the risk of false positives. For instance, Adeshina et al. [2] have developed a general-purpose classifier built on the 'XGBoost' framework, which is an optimised gradient-boosting tree algorithm. This classifier aims to enhance target prediction accuracy, a crucial aspect in countering terrorism effectively [129].

Moreover, prospective actions should be directed towards establishing guidelines that can effectively manage false positives. Similarly, it is essential to consider the refinement of linguistic markers in identifying indicators of threatening and cautionary behaviour in text [62]. Although identifying indicators of threatening and cautionary behaviour in text poses relatively minor challenges, there remains a clear demand for the evolution of techniques for practical implementation (Ibid.). Furthermore, prior to endorsing the integration of machine learning (ML) and big data in counterterrorism, it is imperative to conduct a comprehensive evaluation of a diverse array of algorithms to minimise these false positives and false negatives. Similarly, recognising the limitations and constraints of the applications of ML algorithms is paramount to enhancing algorithm performance and enabling law enforcement to utilize appropriately analysed datasets effectively.

10.3.4 Dataset Recommendations

When utilising ML techniques in counterterrorism solutions, it is essential to consider the characteristics of the datasets employed, as they can significantly impact the accuracy of ML algorithms in addressing questions related to the timing and location of attacks. Despite facing numerous challenges, potential recommendations can be proposed to address the identified issues in counterterrorism efforts. One such recommendation involves addressing the limited availability of datasets, which hinders the accuracy of terrorist identification and the efficacy of monitoring terrorists' online activities. A possible solution to this challenge is to prioritise Natural Language Processing (NLP) techniques within intelligence operations. By doing so, intelligence agencies could promptly identify signs of radicalisation online by recognising specific keywords and activities associated with terrorist ideologies. Nevertheless, the potential lack of data to build high-quality AI poses a significant issue, leading to potential mistakes and ethical concerns.

One approach to tackle this challenge is to encourage the collection of more data. However, resource constraints might hinder data collection efforts, leading to the utilization of poor-quality data and subsequently releasing AI in an unsatisfactory state. Policy makers can play a vital role in developing standards for data quality and incentivising data providers to share data with AI developers [32]. This approach can lead to the development of better-quality AI models, reducing the likelihood of mistakes. Additionally, organisations should be transparent about the performance of their AI algorithms, potentially through legislation or independent research, to provide unbiased and clear information about AI capabilities [49]. Concerning the training dataset, a judicious approach is requisite, whereby residual data is thoroughly expunged post-extraction of pertinent information, lest inadvertent erasures disrupt system functionality or compromise user transparency [120].

Furthermore, to ensure responsible implementation, adversarial debiasing and the adoption of the Cross-Industry Standard Process for Data Mining (CRISP-DM) as a data mining standard can contribute to more ethical and accurate datasets. By adhering to ethical standards, considering legislative safeguards, and evaluating algorithms comprehensively, the effectiveness of ML applications in countering terrorism can be improved, empowering law enforcement agencies to leverage datasets responsibly and effectively. Adversarial debiasing has been recommended as a practice to address issues of over- and underrepresentation in datasets. Therefore, mitigating methods like CRISP-DM, which is widely adopted as the standard procedure for data mining, warrant further investigation to establish more ethical and accurate datasets for preventing terrorist attacks.

In summary, addressing the challenges associated with dataset availability is a crucial step in improving the accuracy and efficacy of machine learning applications in counterterrorism. By focusing on expanding, diversifying, and maintaining datasets, LEAs can enhance their ability to detect and respond to threats in a timely and effective manner, ultimately contributing to a safer society.

10.4   Ethical and Legal Recommendations

Addressing the ethical and legal challenges associated with ML in digital policing and counterterrorism efforts necessitates an evaluation of existing technologies, with a focus on moderating them to align with justifiable practices within the public sphere [93]. Furthermore, certain principles should be considered to adhere to ethical standards. ML, particularly when combined with big data, should be subject to strict monitoring by judicial and legislative branches. Such monitoring helps reduce the risk of breaching individual security and potential biases when targeting potential terrorists. For instance, the EU's AI Act [41] aims to establish a standard regulatory and legal framework for potential AI issues, thereby seeking to avoid the implementation of government-run social scoring systems such as that utilised by the Chinese regime. Similarly, the evolution of ethical and legal implications stemming from ML necessitates an ongoing evaluation of standards, particularly within the ambit of technological development centres [134]. Implementing these recommendations and principles can significantly enhance the ethical and effective use of ML in countering terrorism efforts.

The following subsections offer a wide range of recommendations to address or mitigate human rights, bias, privacy and security, mass surveillance, and legal challenges stemming from the implementation of ML in counterterrorism solutions.

10.4.1 Human Rights Recommendations

As discussed in Chap. 9, the integration of AI and ML in countering online terrorism has raised critical concerns about their potential impact on human rights [121]. Therefore, to effectively address these challenges and strike a balance, a multifaceted approach is necessary. One crucial aspect of this approach is the need for ongoing human rights assessments as a prerequisite for deploying ML systems (Ibid.). These assessments should comprehensively evaluate factors such as need, lawfulness, legitimacy, and proportionality throughout the lifespan of AI-enabled counterterrorism systems (Ibid.). To mitigate potential risks associated with this integration, a proposed strategy to mitigate potential risks involves combining human moderators with AI oversight to discern appropriate content [43]. However, given the vast volume of online content, relying solely on human monitoring is not feasible.

Therefore, implementing moderators within a triage system is essential (Ibid.), a solution that can provide substantial benefits [50]. However, it is crucial to recognise the challenges of moderation work. Placing human moderators solely under machine learning system supervision may not suffice in ensuring effective content moderation while safeguarding human rights [43]. To address this, it is essential to provide human moderators with comprehensive training, sufficient decision-making time, and workplace protections (Ibid.). In addition to these considerations, achieving meaningful oversight requires diverse moderation teams with cultural and linguistic proficiency [8]. These professional moderators should possess cultural knowledge of the platform's location and audience, linguistic expertise, specialized knowledge of relevant laws, and an awareness of user guidelines [96]. However, it is essential to underscore that while incorporating human oversight can mitigate some human rights concerns related to machine learning, it cannot absolve AI systems of their responsibility or accountability [30].

Furthermore, in pursuit of safeguarding fundamental human rights, particularly the freedom of speech, Article 19 [7], a prominent global human rights organisation, has put forth a set of vital recommendations. These recommendations serve as a comprehensive strategy for addressing the potential adverse effects of AI on human rights. One critical recommendation involves "Enhanced Engagement for Mitigating Potential Negative Impact" [7]. To counter concerns surrounding AI's potential infringement upon human rights such as free expression and privacy, comprehensive engagement efforts are imperative. This entails gaining a deeper understanding of the underlying technology, the individuals responsible for its development, and the specific contexts in which it is applied. Another vital aspect is the "Protection of International Human Rights Standards" (Ibid.). Upholding, endorsing, and safeguarding international human rights standards should be a fundamental prerequisite in the deployment, investigation, and advancement of AI technologies. Therefore, it is crucial to define what qualifies as "AI human rights essential systems" and ensure alignment between legislation, regulations, codes of conduct, ethical guidelines, self-regulatory norms, and technological requirements with international human rights standards (Ibid.).

Additionally, ensuring "Transparency and Accountability in Personal Data Usage and Storage" is of paramount importance. Transparency and accountability in the use and storage of personal information must be ensured to establish an environment of trust and confidence for the public. These recommendations, as outlined by Article 19 [7], offer a holistic approach to protecting individuals' rights in the context of AI use. They underscore the significance of upholding ethical principles and international human rights standards in the development and deployment of AI technologies (Ibid.). Considering the far-reaching implications of these practices, it is of pressing importance to strike a balance between safeguarding human rights and advancing technology. The current landscape highlights the urgency of this endeavour, particularly as the future generation, reliant on technological progress, faces the prospect of living in a society devoid of privacy.

In summary, the use of AI and ML in counterterrorism necessitates a comprehensive consideration of human rights concerns. To this end, human rights assessments, continuous reviews, and international collaborations are essential steps in addressing these issues. Similarly, combining human moderation with AI oversight, along with the provision of adequate support and training for human moderators, can help strike a balance between content moderation and human rights protection. Additionally, the adoption of an ethical framework is crucial in countering cyber-enabled terrorism while preserving democratic values. These efforts underscore the critical need to find a delicate equilibrium between technological advancements and the protection of fundamental human rights. The urgency of this endeavour becomes even more apparent when one considers the potential consequences for future generations, who will depend heavily on technological progress but also deserve a society that respects their privacy and freedoms.

10.4.2   Bias Recommendations

As discussed in Chap. 9, one of the primary sources of bias in ML models is biased training data. To mitigate this bias, it is crucial to ensure that the training data used to build models is diverse and representative of the entire spectrum of cyber threats and actors. This goal can be accomplished through a combination of strategies, including employing random sampling techniques, such as those widely recognized for reducing sample bias. Random sampling ensures equal representation of individuals in the training dataset, regardless of their characteristics. Additionally, a more nuanced option is stratified random sampling, which not only selects a sample population representative of the overall population of interest but also ensures the inclusion of all relevant subgroups. This approach prevents the oversight of unde-tected terrorists, particularly those from ethnic backgrounds, such as white ethnicity [90]. Moreover, collecting data from a wide range of sources, including different regions and demographics [37], various threat scenarios, attacker profiles, and attack methods, as well as continually updating datasets to reflect evolving threats, is essen-tial to achieving diversity in the training data. Furthermore, solutions based on robust mathematical models, such as those employing "adversarial debiasing" and "over-sampling" can be of significant value, as they compel ML algorithms to account for both over and underrepresented groups [128].

Furthermore, implementing regular bias audits on ML models used for cyber terrorism detection is essential. Bias audits involve evaluating model predictions across different demographic groups to identify and rectify any disparities. For instance, tools such as IBM's AI Fairness 360 can assist in monitoring and addressing bias in real-time [60]. Another potential solution could involve leveraging CRISP-DM, a widely adopted cross-industry standard procedure for data mining, which has been extensively examined for addressing different ML biases, particularly social bias [124]. This type of bias emerges when available data reflects existing biases in the relevant population, potentially leading to false threat identifications in the context of counterterrorism. To effectively address this issue, data massaging techniques that involves alterations to data formats can be applied to eliminate unnecessary information and remove unwanted factors [119].

Similarly, incorporating algorithmic fairness techniques during model develop-ment can be instrumental in addressing data distribution shift and class imbalance while reducing bias. These techniques, including re-sampling, re-weighting, and adversarial training, aim to balance the model's performance across different groups and minimise disparate impact [27, 106]. This data-centric paradigm for achieving fairness and robustness often involves the use of data reweighting, a technique that encompasses various classical methods. These methods include data resampling [66, 106], leveraging domain-specific expertise [27, 44, 136], estimating weights based on data complexity [69, 74, 81], and utilising class-count information [10, 97]. By integrating these techniques and approaches into model development, ML models can not only mitigate issues related to data distribution shift and class imbalance but also enhance their overall fairness and robustness.

Additionally, considering the paramount importance of transparency in ML models for understanding decision-making processes, it is imperative to employ interpretable models and provide explanations for model predictions. This practice not only enhances accountability but also facilitates the identification and rectifica-tion of concealed biases [77]. It could also allow stakeholders to assess the fairness and reliability of the algorithms used in countering cyber terrorism. Furthermore, ongoing monitoring and auditing of machine learning models are essential to detect and rectify bias. To this end, regular bias assessments should be conducted to identify any disparities in how the algorithm treats different groups. By identifying bias early, adjustments can be made to the model's training data or algorithm to rectify these disparities [103].

In the same vein, developing fairness metrics specific to cyber terrorism detection can help measure and quantify bias. Various fairness metrics have been proposed in the broader literature, such as disparate impact and equal opportunity [82, 92]. While these metrics might not have originated in the specific context of cyber terrorism, their utility can extend to the domain and can be effectively adapted for assessing the fairness of ML models employed in this critical field. Therefore, integrating these fairness metrics into the evaluation process is a crucial step towards the development of more equitable algorithms specifically tailored to cyber terrorism mitigation efforts. By doing so, LEAs can systematically measure and quantify the extent of bias in ML models, allowing for targeted interventions to enhance fairness. Therefore, while fairness metrics might have originated outside the realm of cyber terrorism, their application in this domain is not only feasible but also highly beneficial. By incorporating these metrics and fostering interdisciplinary collaboration, one can make significant strides in creating fairer, more effective, and unbiased ML solutions for countering cyberterrorism.

Similarly, interdisciplinary collaboration between ML experts, ethicists, and domain specialists is essential to address bias effectively. Ethical experts can provide guidance on ethical considerations, while domain experts can help identify potential biases specific to cyber terrorism [23]. This collaborative approach can lead to more robust and unbiased solutions. Furthermore, governments and organisations should establish ethical guidelines and standards [85] for the use of ML in countering cyber terrorism. These guidelines should include provisions for bias mitigation, transparency, and accountability in algorithm development and deployment [19]. Finally, the establishment of diverse and inclusive interdisciplinary teams, encompassing individuals with expertise in ML, cybersecurity, mathematics, data science, cyberterrorism mitigation, law enforcement, psychology, law, social science, and criminology, emerges as a matter of utmost importance. These multidisciplinary teams could demonstrate an increased capacity to effectively identify and rectify biases while concurrently enriching the decision-making process with diverse viewpoints. Therefore, the deliberate promotion of diversity within such teams stands as a pivotal strategy in advancing the creation of solutions distinguished by increased equity and fairness.

Furthermore, a potentially contentious yet apparent resolution lies in shifting the focus away from addressing ML bias, itself, and directing attention towards the societal norms that underlie the creation of unjust structures, subsequently reinforcing biased data (Pen and Hung 2021, as cited in [87]). This perspective underscores the imperative of establishing universally applicable standards. Nonetheless, the inherent societal disparities arising from health and social class distinctions highlight the unequal positioning of individuals within society, rendering the application of identical standards a potential exacerbator of existing inequalities. Hence, there is a pressing need to rectify unjust situations and consider the implementation of distinct standards tailored to the diverse needs of societal segments (Ibid.). Additionally, within the same context, a proposed strategy involves the introduction of a policy schema aimed at alleviating concerns related to "distrust, efficiency, racism, and social equity" (Ibid.). This framework serves as a safeguard for predictive policing and encompasses various facets, including the anticipation of immediate risks and subsequent proactive measures, the identification of socially vulnerable individuals with the provision of necessary assistance, and a commitment to transparency through scrutiny and communication with the broader public (Ibid.).

In summary, addressing biases is a fundamental concern in ML research and development, especially in the context of countering cyberterrorism. As revealed throughout Chap. 9 and this chapter, identifying and mitigating bias in ML algorithms for countering cyberterrorism are a complex yet essential task to ensure equitable and responsible deployment in real-world applications. Therefore, to ensure the responsible deployment of ML models and effective bias mitigation, researchers and practitioners must remain vigilant throughout the phases of data collection, design, and deployment. Furthermore, implementing effective strategies, such as ensuring diverse training data, conducting regular audits, promoting transparency, using fairness metrics, and fostering collaboration with experts, can significantly contribute to reducing bias and enhancing the fairness and effectiveness of ML solutions in the fight against cyberterrorism. It is also important to recognise that addressing bias extends beyond technical challenges; it requires a comprehensive understanding of the social, ethical, and legal dimensions surrounding AI applications. Scholars and practitioners alike must diligently attempt to understand and mitigate biases originating from various sources, including user-generated data, historical biases, content production, population characteristics, and algorithm designs. Such diligence is essential to ensure the responsible and equitable deployment of ML models, and these strategies recommended in this study represent significant steps in this ongoing effort.

This section of the study emphasised the importance of understanding and mitigating biases originating from various sources, including user-generated data, historical trends, content production, population characteristics, and algorithm designs. This holistic perspective is essential for advancing the responsible use of ML in this critical domain. By incorporating the recommended approaches, such as ensuring diverse training data, conducting regular audits, promoting transparency, utilising fairness metrics, and fostering collaboration with domain experts, substantial and meaningful progress can be made in reducing bias and enhancing the fairness and effectiveness of ML solutions in countering cyber terrorism. In this context, it is crucial to recognise that these recommendations represent significant steps in an ongoing effort to ensure the responsible and equitable deployment of ML models while upholding fairness and ethics.

10.4.3 Privacy and Security Recommendations

Addressing the complex challenges associated with preserving the security and privacy of civilians while simultaneously countering cyber terrorism through the application of ML requires careful consideration and the adoption of less invasive approaches. These approaches are not only crucial for the protection of individual rights but are also pivotal in maintaining public trust. Moreover, they play a pivotal role in achieving a delicate equilibrium between the imperatives of security and the fundamental individual liberties.

One of the primary approaches to uphold user privacy entails adhering to stringent data minimisation principles within ML models [61] used for cyber terrorism mitigation. This method involves the collection of the minimum necessary data to achieve desired objectives, a practice in alignment with privacy tenets that concurrently mitigates the risk of data breaches and misuse [1, 67]. To implement this approach effectively, a focus on metadata analysis could be a viable option [68]. Metadata encompasses critical information such as interaction timestamps, duration, and participant addresses in phone and email records, providing valuable insights while preserving the actual content of the datasets [22]. This strategic emphasis on metadata analysis aligns with the principle of data minimisation [83], further enhancing the protection of user privacy within ML-based cyber terrorism mitigation efforts. The introduction of metadata as a means to address ethical challenges related to breaching individual privacy and security has been extensively explored in academic and scientific literature. For instance, [22] conducted research on the use of time-based meta-graphs derived from real-world data on terrorist attacks in Afghanistan and Iraq from 2001 to 2018. Their study purportedly gained successful results in predicting future attacks while still safeguarding individual privacy. This research reinforces the notion that metadata analysis is an ethical and effective approach in counterterrorism efforts. Similarly, data anonymisation and pseudonymisation in the development of cyber terrorism detection algorithms emerges as a viable method. These techniques safeguard the identities of individuals while enabling meaningful data analysis [59].

Augmenting these efforts, privacy-preserving ML algorithms, such as federated learning and homomorphic encryption, play a pivotal role in securing sensitive data during both the training and inference phases. Federated learning enables model training across decentralized devices without the exchange of raw data, while homomorphic encryption permits computations on encrypted data without divulging the underlying information or the need for decryption [1, 11]. This, in turn, safeguards personal information during storage and transmission [48]. In addition to these methods, cryptographic approaches offer further options for preserving privacy in ML. They enable multiple input sources to collaboratively train ML models while ensuring the confidentiality of individuals' personal information in its original form [3]. For instance, cryptographic techniques, such as symmetric key cryptography, can be employed to secure data transmission and storage, tailored to specific security requirements and threats [48]. Adopting such robust data security measures, which incorporate encryption and safeguards for data storage and transmission, also helps mitigate the risk of unauthorised access and data breaches [105].

Furthermore, ensuring transparency throughout the development and deployment of machine learning models is another critical step in addressing privacy challenges. To achieve this, it is paramount to establish clear lines of accountability and implement regular audits of these models to identify and rectify potential privacy violations

[28]. Within this context, enlisting the services of independent third-party auditors can play a significant role in assessing the privacy practices of entities [24, 94] engaged in counter-cyberterrorism efforts. This approach not only helps ensure compliance with privacy regulations but also promotes compliance with industry best practices. Another noteworthy solution for addressing the privacy challenges involves the regular conduct of privacy impact assessments (PIAs) [92]. PIAs serve as a proactive means to evaluate potential privacy risks associated with machine learning-based counterterrorism measures. By identifying and addressing privacy concerns before the deployment of new technologies, PIAs ensure compliance with both legal and ethical standards.

Similarly, it is important to develop and adhere to ethical guidelines for the use of machine learning in counter-terrorism. These guidelines should include provisions for privacy protection, and should be followed rigorously during model development and deployment [45]. Ensuring user privacy, therefore, necessitates the acquisition of informed consent whenever feasible, alongside respecting individuals' choices regarding the use of their data [110]. Moreover, prioritizing user consent and control mechanisms empowers individuals to make informed decisions regarding the utilization of their data in machine learning model training, thereby enhancing their participation in cybersecurity efforts while retaining control over their personal information.

Simultaneously, advocating for and enacting legislation and regulations tailored to address privacy concerns within the context of machine learning for counter-terrorism assumes significance [121]. Such regulations should incorporate stringent penalties for privacy or data protection violations to reinforce their efficacy [42]. Moreover, adherence to the existing privacy regulations and staying apprised of emerging laws in the domains of machine learning and cybersecurity is of paramount importance. Ensuring compliance with established frameworks such as General Data Protection Regulation (GDPR) or regional regulations is pivotal in guaranteeing that machine learning-based counterterrorism measures align with prevailing legal and ethical standards [95]. Complementing these measures, the provision of comprehensive training to law enforcement agencies (LEAs), cybersecurity professionals, and machine learning practitioners on the significance of privacy and the potential risks linked to data collection and analysis is essential. Finally, fostering international cooperation and the sharing of information on best practices for privacy protection in counter-cyberterrorism endeavours can result in more comprehensive and effective solutions.

In summary, to reconcile the imperatives of national security with the preservation of individual privacy rights, this section offered several recommendations that can be implemented within the framework of ML-driven cyber terrorism countermeasures. These recommendations encompass a wide range of strategies, such as incorporating privacy-preserving techniques into ML algorithms, cultivating transparency in the collection and use of data, conducting impact assessments to evaluate potential privacy implications of ML-based systems, prioritising user consent when collecting and utilising personal data, and maintaining strict adherence to relevant regulatory frameworks governing data protection and privacy. Finally, effectively addressing privacy challenges in the context of ML-based cyber terrorism counter-measures requires a comprehensive strategy that combines technical advancements, ethical considerations, and regulatory compliance. By bridging the gap between technological advancements and privacy concerns, such an approach ensures that LEAs and governments harness the benefits of machine learning while safeguarding individual privacy in the fight against cyber terrorism.

10.4.4   Legal Recommendations

To address the challenges associated with the use of AI and ML by government and LEAs in countering terrorism, it is imperative to establish clear legislation that provides a structured framework for AI's operation. Such legislation not only fosters trust among the public but also helps mitigate various risks and concerns. The European Union's commissioning of an AI strategy white paper serves as a positive step towards addressing this issue [40]. With comprehensive legislation in place, the risk of human bias in AI can be reduced, and funding can support larger research teams with access to higher-quality data, minimising the potential for bias in AI development [121]. Moreover, a rational regulatory system established through legislation can clarify legal responsibility, alleviating confusion [135].

Another critical factor to consider when policing cyberspace and countering terrorism concerns the impact of jurisdiction on such operations. Jurisdictional issues can complicate efforts, especially when different countries compete to prosecute offenders. To navigate these challenges, factors such as the nationality of the victim and offender and the extent of damage to state interests and security require consideration [122]. These considerations, while helpful, only partially mitigate the jurisdictional hurdles, given that online terrorism transcends borders, involving multiple nations. Thus, jurisdiction remains a substantial obstacle in combatting online terrorism, requiring a concerted effort to rewrite and mutually agree upon policies among nations. This, in turn, will achieve a balanced law enforcement approach within cyberspace [108]. To address this, conventions such as the "Council of Europe Convention on Cybercrime" have emerged, aiming to establish a universal standard for addressing online crimes. These conventions serve as frameworks for policing the internet, fostering harmonisation between nations in the fight against cybercrime [86]. Such initiatives represent significant strides in dismantling jurisdictional barriers. However, persistent disagreements between countries regarding policing strategies underscore that newly formulated policies will only be fully effective with global cooperation and harmonization in regulating the globally interconnected sphere of the internet [29].

10.5 Discussion and Conclusion

The increasing adoption of AI and ML in digital policing and counterterrorism efforts holds significant promise for strengthening the fight against cyberterrorism. Nevertheless, it is crucial to address concerns associated with the use of these technologies, including the technical, operational, ethical, and legal challenges discussed in Chap. 9. Despite the formidable hurdles AI and ML face, ongoing efforts are being made to tackle these challenges. These efforts exemplify promising developments and strategies aimed at improving their effectiveness in the realms of digital policing and countering cyberterrorism. However, the effectiveness of AI and ML in counterterrorism operations is contingent on achieving global accessibility. This includes eliminating jurisdictional barriers and harnessing enhanced surveillance and image recognition capabilities provided by these technologies to enable global prediction of potential offenders. By fostering widespread understanding and promoting the appropriate use of AI and ML, alongside exploring cost-effective alternatives [134], counterterrorism and digital policing efforts can effectively harness the vast data available on the internet to thwart or mitigate terrorist activities.

To further enhance the impact of AI and ML in the fight against terrorism and cybercrime, it is imperative to promote greater awareness and adoption of ML techniques across diverse jurisdictions and policing domains. This integrated approach not only mitigates technical, operational, ethical and legal challenges but also strengthens the global capacity to counter terrorism online. Furthermore, continued research, international collaboration, and policy harmonisation are crucial to further improve the application of AI and ML in counterterrorism efforts and ensure its responsible and ethical use in law enforcement. Similarly, introducing a globally accessible format for understanding machine learning holds the potential for further developments to combat online terrorism effectively. Furthermore, incorporating more quantitative research to assess the true impact of online terrorism will contribute to a more substantial and well-rounded conclusion regarding the utility of ML in countering terrorism online.

It is important to note that while ML holds the potential to empower counterterrorism organisations and governments in predicting impending terrorist attacks, a cautious acknowledgment prevails. The power of causal inference might be circumscribed, engendering uncertainty in drawing unequivocal conclusions [120]. Therefore, a nuanced and strategic approach, considering both the opportunities and obstacles associated with ML, is indispensable to fortify the realms of cybersecurity and counterterrorism, thereby navigating the intricacies of an evolving digital landscape.

In conclusion, this chapter has offered a comprehensive set of solutions and recommendations to address the intricate technical, operational, ethical, and legal challenges that were examined in the preceding chapter, namely Chap. 9. By synthesising these challenges into actionable insights, the chapter has undertaken to equip policymakers, practitioners, and researchers with a roadmap for harnessing the power of ML in countering cyberterrorism effectively. These solutions underscore the importance of international collaboration, responsible implementation, and continual research to ensure the responsible and ethical use of ML in safeguarding digital landscapes. As the society advance into an increasingly interconnected and digital future, these recommendations help to illuminate the path toward a more secure and resilient cyber environment.

References

1. Abadi M, Chu A, Goodfellow I, McMahan HB, Mironov I, Talwar K, Zhang L (2016) Deep learning with differential privacy. In: Proceedings of the 23rd ACM conference on computer and communications security, pp 308–318
2. Adeshina YO, Deeds EJ, Karanicolas J (2020) Machine learning classification can reduce false positives in structure-based virtual screening. Proc Natl Acad Sci 117(31):18477–18488
3. Al-Rubaie M, Chang JM (2019) Privacy-preserving machine learning: threats and solutions. IEEE Secur Priv 17(2):49–58
4. Ammar J (2019) Cyber Gremlin: social networking, machine learning, and the global war on Al-Qaida–and IS-inspired terrorism. Int J Law Inf Technol 27(3):238–265
5. Anonymous. (2017). Imbalanced Data: how to handle imbalanced classification problems. Analytics Vidhya. Available at: https://www.analyticsvidhya.com/blog/2017/03/imbalanced-data-classification/. Accessed 28 Sept 2023
6. Arel I, Rose DC, Karnowski TP (2010) Deep machine learning—a new frontier in artificial intelligence research [research Frontier]. IEEE Comput Intell Mag 5(4):13–18
7. Article 19 (2018) Privacy and freedom of expression in the age of artificial intelligence. Privacy International. https://www.article19.org/wp-content/uploads/2018/04/Privacy-and-Freedom-of-Expression-In-the-Age-of-Artificial-Intelligence-1.pdf. Accessed 29 Sept 2023
8. Barrett PM (2020) Who moderates the social media giants? A call to end outsourcing. Center Bus Human Rights. https://bhr.stern.nyu.edu/tech-content-moderation-june-2020?_ga=2.150 003060.37337998.1684949533-1469338793.1684422925. Accessed 30 Sept 2023
9. Battaglini M (2020) How the main legal and ethical issues in machine learning arose and evolved. Technol Soc. https://www.transparentinternet.com/technology-and-society/machine-learning-issues/ (Accessed: 27/09/2023).
10. Bauder, R. A., Khoshgoftaar, T. M. & Hasanin, T. (2018). An Empirical Study on Class Rarity in Big Data. *17th IEEE International Conference on Machine Learning and Applications (ICMLA)*, Orlando, FL, USA, 785–790.
11. Bellavista P, Foschini L, Mora A (2021) Decentralised learning in federated deployment environments: a system-level survey. ACM Comput Surv (CSUR) 54(1):1–38
12. Bengio Y, Courville A, Vincent P (2013) Representation learning: a review and new perspectives. IEEE Trans Pattern Anal Mach Intell 35(8):1798–1828
13. Breiman L, Friedman J, Olshen R, Stone C (1984) Classification and regression trees (Wadsworth statistics/probability). Routledge
14. Breiman L (2001) Random forests. Mach Learn 45:5–32
15. Brown G (2009) A new perspective for information theoretic feature selection. In: Proceedings of the 12th international conference on artificial intelligence and statistics, pp 49–56
16. Brownlee J (2020) Cost-sensitive learning for imbalanced classification. Mach Learn Mastery. https://machinelearningmastery.com/cost-sensitive-learning-for-imbalanced-classification/. Accessed 30 Sept 2023
17. Brownlee J (2020) A gentle introduction to k-fold cross-validation. Mach Learn Mastery. https://machinelearningmastery.com/k-fold-cross-validation/. Accessed 30 Sept 2023
18. Brownlee J (2021) Bagging and random forest for imbalanced classification. Mach Learn Mastery. https://machinelearningmastery.com/bagging-and-random-forest-for-imbalanced-classification/#:~:text=Another%20approach%20to%20make%20random,on%20misclassif ying%20the%20minority%20class. Accessed 30 Sept 2023

19. Buolamwini J, Gebru T (2018) Gender shades: intersectional accuracy disparities in commercial gender classification. In: Conference on fairness, accountability and transparency, pp 77–91

20. Burgio DA (2020) Reduction of false positives in intrusion detection based on extreme learning machine with situation awareness. Doctoral dissertation, Nova Southeastern University

21. Camacho-Collados M, Liberatore F (2015) A decision support system for predictive police patrolling. Decis Support Syst 75:25–37

22. Campedelli GM, Bartulovic M, Carley KM. (2021). Learning future terrorist targets through temporal meta-graphs. Scientific Reports, 11(1), 8533–8533. https://doi.org/10.1038/s41598-021-87709-7

23. Campolo A, Sanfilippo MR, Whittaker M, Crawford K (2017) AI Now 2017 report. AI Now Institute. https://ainowinstitute.org/publication/ai-now-2017-report-2. Accessed 30 Sept 2023

24. Cavoukian, A. (2012). Operationalizing Privacy by Design: A Guide to Implementing Strong Privacy Practices. Available at: https://gpsbydesigncentre.com/wpcontent/uploads/2021/08/Doc-5-Operationalizing-pbd-guide.pdf.

25. Cawley GC, Talbot NL (2010) On over-fitting in model selection and subsequent selection bias in performance evaluation. J Mach Learn Res 11:2079–2107

26. Chawla NV, Bowyer KW, Hall LO, Kegelmeyer WP (2002) SMOTE: synthetic minority over-sampling technique. J Artif Intell Res 16:321–357

27. Choudhury P, Starr E, Agarwal R (2020) Machine learning and human capital complementarities: experimental evidence on bias mitigation. Strateg Manag J 41(8):1381–1411

28. Caplan R, Donovan J, Hanson L, Matthews J (2018) Algorithmic accountability: a primer. Data Soc. https://datasociety.net/library/algorithmic-accountability-a-primer/. Accessed 02 Oct 2023

29. Clough J (2014) A world of difference: the Budapest convention on cybercrime and the challenges of harmonisation. Monash Univ Law Rev 40(3):698–736

30. Clyde A (2021) Human-in-the-loop systems are no panacea for AI accountability. Tech Policy Press. https://techpolicy.press/human-in-the-loop-systems-are-no-panacea-for-ai-accountability/#:~:text=Technology%20and%20Democracy-,Human%2Din%2Dthe%2DLoop%20Systems%20Are,No%20Panacea%20for%20AI%20Accountability&text=Recent%20research%20and%20reporting%20by,based%20on%20past%20biased%20data. Accessed 30 Sept 2023

31. Corbin C (2017) Fordham law review. Terrorists are always muslim but never white: at the intersection of critical race theory and propaganda. Fordham Law Rev 86(2):445–485

32. Cowls J, Tsamados A, Taddeo M, Floridi L (2021) The AI gambit: leveraging artificial intelligence to combat climate change—opportunities, challenges, and recommendations. AI Soc 38:283–307

33. Davis J, Goadrich M (2006) The relationship between Precision-Recall and ROC curves. Proceedings of the 23rd International Conference on Machine Learning, 233–240.

34. Delony D (2018) What are some ethical issues regarding machine learning? Techopedia. https://www.techopedia.com/what-are-some-ethical-issues-regarding-machine-learning/7/33376. Accessed 28 Sept 2023

35. Desuky AS, Hussain S (2021) An improved hybrid approach for handling class imbalance problem. Arab J Sci Eng 46:3853–3864

36. Domingos P (2012) A few useful things to know about machine learning. Commun ACM 55(10):78–87

37. Dwork C, Hardt M, Pitassi T, Reingold O, Zemel R (2012) Fairness through awareness. In: Proceedings of the 3rd innovations in theoretical computer science conference, pp 214–226

38. Elkan C (2001) The foundations of cost-sensitive learning. In: Proceedings of the seventeenth international joint conference on artificial intelligence (IJCAI'01) 17(1):973–978

39. Equality and Human Rights Commission (2021) Article 10: freedom of expression. https://www.equalityhumanrights.com/en/human-rights-act/article-10-freedom-expression. Accessed 27 Sept 2023

40. European Commission (2020) Artificial intelligence—a European approach to excellence and trust (Whitepaper). https://commission.europa.eu/system/files/2020-02/commission-white-paper-artificial-intelligence-feb2020_en.pdf. Accessed 29 Sept 2023

41. European Commission (2021) Proposal for a regulation of the European Parliament and of the council: laying down harmonised rules on artificial intelligence (Artificial Intelligence Act) and amending certain union legislative acts, COM(2021) 206 final. https://eur-lex.europa.eu/legal-content/EN/TXT/?uri=CELEX:52021PC0206. Accessed 06 Nov 2023

42. European Parliament (2020) The impact of the general data protection regulation (GDPR) on artificial intelligence. In: EPRS | European Parliamentary Research Service, Scientific Foresight Unit (STOA). https://www.europarl.europa.eu/RegData/etudes/STUD/2020/641530/EPRS_STU(2020)641530_EN.pdf. Accessed 02 Oct 2023

43. Federal Trade Commission (FTC) (2022) Combatting online harms through innovation. Technical Report to Congress. https://www.ftc.gov/reports/combatting-online-harms-through-innovation. Accessed 30 Sept 2023

44. Fern A, Yoon S, Givan R (2006) Approximate policy iteration with a policy language bias: solving relational Markov decision processes. J Artif Intell Res 25:75–118

45. Floridi L (2018) Soft ethics, the governance of the digital and the general data protection regulation. Philos Trans Roy Soc A Math Phys Eng Sci 376(2133):20180081

46. Friedman JH (2001) Greedy function approximation: a gradient boosting machine. Ann Stat 29(5):1189–1232

47. Freund Y, Schapire R (1999) A short introduction to boosting. J Jap Soc Artif Intell 14(5):771–780

48. Gaire R, Ghosh RK, Kim J, Krumpholz A, Ranjan R, Shyamasundar RK, Nepal S (2019) Crowdsensing and privacy in smart city applications. In: Smart cities cybersecurity and privacy. Elsevier, pp 57–73

49. Gillespie T (2020) Content moderation, AI, and the question of scale. Big Data Soc 7(2)

50. Gillespie T (2018) Custodians of the internet: platforms, content moderation, and the hidden decisions that shape social media. Yale University Press

51. Gudivada V, Apon A, Ding J (2017) Data quality considerations for big data and machine learning: going beyond data cleaning and transformations. Int J Adv Softw 10(1):1–20

52. Gupta P, Varshney P, Bhatia MPS (2017) Identifying radical social media posts using machine learning. GitHub, California

53. Guyon I (2008) Practical feature selection: from correlation to causality. In: mining massive data sets for security, advances in data mining, search, social networks and text mining, and their applications to security, 27–43.

54. Guyon I, Elisseeff A (2003). An Introduction to Variable and Feature Selection. Journal of Machine Learning Research, 3, 1157–1182.

55. Guyon I, Weston J, Barnhill S, Vapnik V (2002) Gene selection for cancer classification using support vector machines. Mach Learn 46:389–422

56. Hanley JA, McNeil BJ (1982) The meaning and use of the area under a receiver operating characteristic (ROC) curve. Radiology 143(1):29–36

57. Hastie T, Tibshirani R, Wainwright, M (2009) Statistical learning with sparsity the lasso and generalizations. CRC Press.

58. He H, Garcia EA (2009) Learning from imbalanced data. IEEE Trans Knowl Data Eng 21(9):1263–1284

59. Hintze M, El Emam K (2018) Comparing the benefits of pseudonymisation and anonymisation under the GDPR. J Data Protect Privacy 2(2):145–158

60. IBM Developer Staff (2018) AI Fairness 360. IBM. https://www.ibm.com/opensource/open/projects/ai-fairness-360/. Accessed 30 Sept 2023

61. ICO (2023) How should we assess security and data minimisation in AI? Information Commissioner's Office. https://ico.org.uk/media/for-organisations/uk-gdpr-guidance-and-resources/artificial-intelligence/guidance-on-ai-and-data-protection-2-0.pdf. Accessed 2 Oct 2023

62. Johansson F, Kaati L, Sahlgren M (2016) Detecting linguistic markers of violent extremism in online environments (Chapter 18). IGI Global. 374–390. https://doi.org/10.4018/978-1-5225-0156-5.ch018

63. Johnson JM, Khoshgoftaar TM (2019) Survey on deep learning with class imbalance. J Big Data 6(1):1–54
64. Jolliffe IT (2002). Principal Component Analysis. Springer.
65. Jolliffe IT, Cadima J (2016) Principal component analysis: a review and recent developments. Philos Trans Roy Soc A Math Phys Eng Sci 374(2065)
66. Kahn H, Marshall AW (1953) Methods of reducing sample size in Monte Carlo computations. J Oper Res Soc Am 1(5):263–278
67. Kairouz P, McMahan HB, Avent B, Bellet A, Bennis M, Bhagoji AN et al (2021) Advances and open problems in federated learning. Found Trends Mach Learn 14(1–2):1–210
68. Kift P, Nissenbaum H (2016) Metadata in context-an ontological and normative analysis of the NSA's bulk telephony metadata collection program. I/S J Law Policy Inf Soc 13(2):333–372
69. Kremer J, Stensbo-Smidt K, Gieseke F, Pedersen KS, Igel C (2017) Big universe, big data: machine learning and image analysis for astronomy. IEEE Intell Syst 32(2):16–22
70. Krieg SJ, Smith CW, Chatterjee R, Chawla NV (2022). Predicting terrorist attacks in the United States using localized news data. PloS One, 17(6), e0270681.
71. L'Heureux A, Grolinger K, Elyamany HF, Capretz MAM (2017) Machine learning with big data: challenges and approaches. IEEE Access 5:7776–7797. https://doi.org/10.1109/ACCESS.2017.2696365
72. Le Q, Mikolov T (2014) Distributed representations of sentences and documents. In: Proceedings of the 31st international conference on machine learning, 1188–1196. PMLR.
73. Lee J, Lee YC, Kim JT (2020) Fault detection based on one-class deep learning for manufacturing applications limited to an imbalanced database. Journal of Manufacturing Systems, 57, 357–366.
74. Lin TY, Goyal P, Girshick R, He K, Dollár P (2017) Focal loss for dense object detection. In: Proceedings of the IEEE international conference on computer vision (ICCV), pp 2980–2988
75. Ling CX, Sheng VS (2008) Cost-sensitive learning and the class imbalance problem. Encyclopedia of Machine Learning, 231–235.
76. Linardatos P, Papastefanopoulos V, Kotsiantis S (2020) Explainable AI: a review of machine learning interpretability methods. Entropy 23(1):18
77. Lipton ZC (2018) The mythos of model interpretability: in machine learning, the concept of interpretability is both important and slippery. Queue 16(3):31–57
78. López V, Fernández A, Moreno-Torres JG, Herrera F (2012) Analysis of preprocessing vs. cost-sensitive learning for imbalanced classification. Open problems on intrinsic data characteristics. Expert Syst Appl 39(7):6585–6608
79. Macdonald S (2018) How tech companies are trying to disrupt terrorist social media activity. Scientific American, The Conversation. https://www.scientificamerican.com/article/how-tech-companies-are-trying-to-disrupt-terrorist-social-media-activity/. Accessed 27 Sept 2023
80. Maidamwar PR, Bartere MM, Lokulwar PP (2021) A survey on machine learning approaches for developing intrusion detection system. In: Proceedings of the international conference on innovative computing & communication (ICICC), pp 1–8
81. Malisiewicz T, Gupta A, Efros AA (2011) Ensemble of exemplar-SVMs for object detection and beyond. In: IEEE international conference on computer vision, pp 89–96
82. Mehrabi N, Morstatter F, Saxena N, Lerman K, Galstyan A (2021) A survey on bias and fairness in machine learning. ACM Comput Surv 54(6):1–35
83. Mikolov T, Chen K, Corrado G, Dean, J (2013). Efficient estimation of word representations in vector space. arXiv preprint arXiv:1301.3781
84. Mijatovik, D. (2018). In the era of artificial intelligence: safeguarding human rights. Open Democracy. https://www.opendemocracy.net/en/digitaliberties/in-era-of-artificial-intelligence-safeguarding-human-rights/
85. Mittelstadt BD, Allo P, Taddeo M, Wachter S, Floridi L (2016) The ethics of algorithms: mapping the debate. Big Data Soc 3(2):1–21
86. Montasari R, Carroll F, Mitchell I, Hara S, Bolton-King R (eds) (2022) Privacy, security and forensics in the internet of things (IoT). Springer

87. Montasari R (2023) The application of big data predictive analytics and surveillance technologies in the field of policing. In countering cyberterrorism: the confluence of artificial intelligence, cyber forensics and digital policing in US and UK National Cybersecurity, 81–114. Cham: Springer International Publishing

88. Morrell J (2021) Does more data equal better analytics? Datameer. https://www.datameer. com/blog/does-more-data-equal-better-analytics/. Accessed 29 Sept 2023

89. Mourby M, Mackey E, Elliot M, Gowans H, Wallace SE, Bell J et al (2018) Are 'pseudonymised' data always personal data? Implications of the GDPR for administrative data research in the UK. Comput Law Secur Rev 34(2):222–233

90. Newburn T (2017) Criminology, 3rd edn. Routledge

91. Oetzel MC, Spiekermann S (2014) A systematic methodology for privacy impact assessments: a design science approach. Eur J Inf Syst 23(2):126–150

92. Pagano TP, Loureiro RB, Lisboa FV, Peixoto RM, Guimarães GA, Cruz GO et al (2023) Bias and unfairness in machine learning models: a systematic review on datasets, tools, fairness metrics, and identification and mitigation methods. Big Data Cogn Comput 7(1):1–31

93. Privacy International (2022) Mass surveillance. https://privacyinternational.org/learn/mass-surveillance. Accessed 28 Sept 2023

94. Raji ID, Smart A, White RN, Mitchell M, Gebru T, Hutchinson B, et al (2020) Closing the AI accountability gap: defining an end-to-end framework for internal algorithmic auditing. In: Proceedings of the 2020 conference on fairness, accountability, and transparency, pp 33–44

95. Regulation 2016/679. Regulation (EU) 2016/679 of the European Parliament and of the Council of 27 April 2016 on the protection of natural persons with regard to the processing of personal data and on the free movement of such data, and repealing Directive 95/46/EC (General Data Protection Regulation) (Text with EEA relevance). https://eur-lex.europa.eu/legal-content/EN/TXT/?uri=CELEX:32016R0679. Accessed 02 Oct 2023

96. Roberts ST (2019) Behind the screen. Yale University Press

97. Roh Y, Heo G, Whang, SE (2019) A survey on data collection for machine learning: a big data - AI integration perspective. IEEE transactions on knowledge and data engineering, 33(4):1328–1347

98. Rudin C, Radin J (2019) Why are we using black box models in AI when we don't need to? A lesson from an explainable AI competition. Harvard Data Sci Rev 1(2)

99. Ruff L, Vandermeulen R, Goernitz N, Deecke L, Siddiqui SA, Binder A, Kloft M (2018) Deep one-class classification. Proceedings of machine learning research (PMLR), 4393–4402

100. Sagi O, Rokach L (2018) Ensemble learning: A survey. Wiley Interdisciplinary Reviews: Data Mining and Knowledge Discovery, 8(4):e1249

101. Saito T, Rehmsmeier M (2015) The precision-recall plot is more informative than the ROC plot when evaluating binary classifiers on imbalanced datasets. PLoS ONE 10(3):e0118432

102. Saleem TJ, Chishti MA (2021) Deep learning for the internet of things: potential benefits and use-cases. Digital Commun Netw 7(4):526–542

103. Saleiro P, Kuester B, Hinkson L, London J, Stevens A, Anisfeld A et al (2018) Aequitas: a bias and fairness audit toolkit. https://arxiv.org/abs/1811.05577. Accessed 30 Sept 2023

104. Schölkopf B, Platt JC, Shawe-Taylor J, Smola AJ, Williamson RC (2001) Estimating the support of a high-dimensional distribution. Neural Computation, 13(7):1443–1471

105. Schneier B (2007) Applied cryptography: protocols, algorithms, and source code in C, 2nd edn. Wiley

106. Seiffert C, Khoshgoftaar TM, Van Hulse J, Napolitano A (2008) Resampling or reweighting: a comparison of boosting implementations. In: 20th IEEE international conference on tools with artificial intelligence, pp 445–451

107. Selbst AD (2017) Disparate impact in big data policing. Georgia Law Review, 2(1):109–195

108. Simou S, Kalloniatis C, Gritzalis S, Mouratidis H (2017) A survey on cloud forensic challenges and solutions. Secur Commun Netw 9(18):6285–6314

109. Sofaer HR, Hoeting JA Jarnevich CS (2019) The area under the precision-recall curve as a performance metric for rare binary events. Methods in Ecology and Evolution, 10(4):565–577

110. Solove DJ (2006) A taxonomy of privacy. Univ Pa Law Rev 154(3):477–560

111. Sorzano COS, Vargas J, Montano AP (2014) A survey of dimensionality reduction techniques. arXiv preprint arXiv:1403.2877
112. Spaaij R (2011) Understanding lone wolf terrorism: global patterns, motivations and prevention. Springer Science & Business Media
113. Spanoudes P, Nguyen T (2017) Deep learning in customer churn prediction: unsupervised feature learning on abstract company independent feature vectors, pp 1–22. arXiv:1703.03869
114. Spathoulas GP, Katsikas SK (2013) Enhancing IDS performance through comprehensive alert post-processing. Comput Secur 37:176–196
115. Sun J, Li H, Fujita H, Fu B, Ai W (2020) Class-imbalanced dynamic financial distress prediction based on Adaboost-SVM ensemble combined with SMOTE and time weighting. Information Fusion, 54:128–144
116. Tanha J, Abdi Y, Samadi N, Razzaghi N, Asadpour M (2020) Boosting methods for multi-class imbalanced data classification: an experimental review. J Big Data 7:1–47
117. Tibshirani R (1996). Regression shrinkage and selection via the lasso. Journal of the Royal Statistical Society Series B: Statistical Methodology, 58(1):267–288
118. Tikhonov AN (1963) Solution of incorrectly formulated problems and the regularization method. Soviet Mathematics, 4:1035–1038
119. Tong H, Bell D, Tabei K, Siegel MM (1999) Automated data massaging, interpretation, and e-mailing modules for high throughput open access mass spectrometry. Journal of the American Society for Mass Spectrometry, 10(11):1174–1187. https://doi.org/10.1016/S1044-0305(99)00090-2
120. UK Statistics Authority (UKSA) (2021) Ethical considerations in the use of Machine Learning for research and statistics. Available at: https://uksa.statisticsauthority.gov.uk/publication/ethical-considerations-in-the-use-of-machine-learning-for-research-and-statistics/. Accessed 28 Sept 2023
121. United Nations Interregional Crime and Justice Research Institute (UNICRI) & United Nations Counter-Terrorism Centre (UNCCT) (2021) Countering terrorism online with artificial intelligence: an overview for law enforcement and counter-terrorism agencies in South Asia and South-East Asia. https://www.un.org/counterterrorism/sites/www.un.org.counterterrorism/files/countering-terrorism-online-with-ai-uncct-unicri-report-web.pdf. Accessed 27 Sept 2023
122. United Nations Office on Drugs and Crime (2019) Sovereignty and jurisdiction. https://www.unodc.org/e4j/en/cybercrime/module-7/key-issues/sovereignty-and-jurisdiction.html. Accessed 29 Sept 2023
123. van Giffen B, Herhausen D, Fahse T (2022) Overcoming the pitfalls and perils of algorithms: a classification of machine learning biases and mitigation methods. J Bus Res 144:93–106
124. Van der Maaten L, Hinton G (2008) Visualizing data using t-SNE. Journal of Machine Learning Research, 9(11):2579–2605
125. Veale M, Binns R (2017) Fairer machine learning in the real world: Mitigating discrimination without collecting sensitive data. Big Data & Society, 4(2):1–17
126. Verhelst H, Stannat A, Mecacci G (2020) Machine learning against terrorism: how big data collection and analysis influences the privacy-security dilemma. Sci Eng Ethics 26(6):2975–2984
127. Villa-Pérez ME, Alvarez-Carmona MA, Loyola-Gonzalez O, Medina-Pérez MA, Velazco-Rossell JC, Choo KKR (2021) Semi-supervised anomaly detection algorithms: A comparative summary and future research directions. Knowledge-Based Systems, 218:106878
128. Vokinger KN, Feuerriegel S, Kesselheim AS (2021) Mitigating bias in machine learning for medicine. Commun Med 1(1):25
129. Wang MX, Huang D, Wang G, Li DQ (2020) SS-XGBoost: a machine learning framework for predicting Newmark sliding displacements of slopes. J Geotech Geoenviron Eng 146(9):04020074
130. Wang X, Sloan IH (2007) Brownian bridge and principal component analysis: towards removing the curse of dimensionality. IMA J Numer Anal 27(4):631–654

131. Wu X, Zhu X, Wu GQ, Ding W (2014) Data mining with big data. IEEE Trans Knowl Data Eng 26(1):97–107
132. Yang X, Lo D, Xia X, Sun J (2017) TLEL: A two-layer ensemble learning approach for just-in-time defect prediction. Information and Software Technology, 87:206–220
133. Yapo A, Weiss J (2018) Ethical implications of bias in machine learning. In: Proceedings of the 51st Hawaii international conference on system sciences, pp 5365–5372
134. Younas M (2019) Research challenges of big data. SOCA 13:105–107
135. Yu S, Carroll F (2021) Implications of AI in National Security: Understanding the Security Issues and Ethical Challenges. In R. Montasari & H. Jahankhani (Eds.), Artificial intelligence in cyber security: Impact and implications: Security challenges, technical and ethical issues, forensic investigative challenges (pp. 157–175). Springer Nature. https://doi.org/10.1007/978-3-030-88040-8_6
136. Zadrozny B (2004) Learning and evaluating classifiers under sample selection bias. In: ICML '04: proceedings of the twenty-first international conference on machine learning, p 114
137. Zoričák M, Gnip P, Drotár P, Gazda V (2020) Bankruptcy prediction for small- and medium-sized companies using severely imbalanced datasets. Econ Model 84:165–176

Part IV
Artificial Intelligence and National and International Security

Chapter 11
The Dual Role of Artificial Intelligence in Online Disinformation: A Critical Analysis

Abstract This chapter critically analyses the multifaceted role of artificial intelligence (AI) in the realm of online disinformation. Specifically, it delves into the potential of AI techniques to create highly realistic disinformation and efficiently disseminate it across a vast audience on social media platforms (SMPs). Moreover, the study explores the use of AI as a means to combat this deleterious phenomenon, while simultaneously examining the associated challenges and ethical quandaries. To address these complex issues, the chapter proposes a comprehensive set of recommendations aimed at mitigating these concerns. The chapter concludes that ethical considerations must be an integral part of the development and implementation of AI tools, given the inadvertent negative consequences that might arise from their use. This includes addressing the inherent limitations that give rise to ethical concerns and counterproductive outcomes. The chapter contributes to the academic discourse by addressing the multifaceted challenges posed by AI-generated disinformation.

Keywords Artificial intelligence · Machine learning · Online disinformation · AI · ML · Deepfakes · Recommendation algorithms · Social bots

11.1 Introduction

False, inaccurate, and misleading information deliberately designed, presented, and promoted to manipulate individuals and undermine democratic processes constitutes a longstanding phenomenon [3, 10]. However, in recent decades, the pervasiveness of disinformation has surged significantly, coinciding with the emergence of novel technologies, notably AI [2, 3, 11, 16, 21, 23, 28, 43]. AI can be defined as the capacity of a computer system to perform tasks akin to human intelligence, encompassing learning and decision-making abilities [20, 41]. Machine Learning (ML), a subfield of AI, entails training computer systems on extensive datasets to recognise undefined patterns and discern valuable information without explicit programming [23, 41, 44]. This AI capability has become highly appealing to malicious actors as it empowers them to fabricate and disseminate highly realistic disinformation, which

© The Author(s), under exclusive license to Springer Nature Switzerland AG 2024 229
R. Montasari, *Cyberspace, Cyberterrorism and the International Security in the Fourth Industrial Revolution*, Advanced Sciences and Technologies for Security Applications, https://doi.org/10.1007/978-3-031-50454-9_11

rapidly reaches a vast audience [2, 3, 16, 21, 24, 43]. Conversely, AI technologies have also been recognised as potent tools to counter disinformation, enhancing the efficiency and accuracy of detection and removal approaches [3, 21, 28].

This chapter critically examines the dual nature of AI, emphasising its role in both propagating and countering online disinformation, while also examining the ethical implications associated with its application in this context. To this end, the chapter explores the mechanisms of deep learning (DL), with a particular focus on the Generative Adversarial Network (GAN) framework, which plays a pivotal role in creating authentic yet deceptive content. While existing research often focuses on visual and audio deepfakes, this chapter accentuates the significant potential of AI, exemplified by GPT-2, in generating textual content that simulates genuine news articles. This capability effectively blurs the distinction between human- and AI-generated content [16, 23, 24, 44]. The primary contribution of this chapter lies in illuminating the manner in which AI-generated disinformation can induce confusion and foster mistrust in users by convincingly mimicking human-generated texts [16, 23, 24]. Furthermore, it explores the convergence of diverse content formats, where the amalgamation of media enhances the perceived credibility of synthesised information [2, 37]. By elucidating these complex dynamics, the chapter provides valuable insights into the evolving strategies of disinformation campaigns and underscores the urgency of addressing the associated challenges in an increasingly AI-driven information landscape.

The subsequent sections of this chapter are structured as follows: Sect. 11.2 delves into the role of AI in the generation of disinformation, while Sect. 11.3 examines the role of AI in its dissemination. In Sect. 11.4, the role of AI in countering disinformation is examined, with a critical analysis of the associated challenges. Moving forward, Sect. 11.5 presents recommendations that can be implemented to mitigate these challenges and address the ethical concerns surrounding the use of AI. Finally, Sect. 11.6 presents a discussion and conclusion.

11.2 The Role of AI in the Production of Disinformation

In light of the rapid advancements within the domain of AI, a burgeoning concern revolves around the use of these techniques for the automatic generation of disinformation through the manipulation and autogeneration of fake content across various formats [1–3, 16, 21, 24, 37, 43]. This category of content is commonly referred to as Deepfakes, primarily denoting audio and visual content. In this context, deepfakes are distinguished as sophisticated digitally manipulated audio and visual content portraying non-existent entities or entities in which facial or bodily features have been manipulated to simulate a different identity [2, 3, 16, 21, 33, 43, 44, 50]. The creation of deepfakes hinges upon the principles of DL [50], a subset of ML that capitalises on artificial neural networks to glean insights from substantial datasets [20, 43, 44]. Specifically, the GAN framework comprises two interconnected networks: the generator, responsible for content creation, and the discriminator, tasked with discerning the authenticity of input images [15, 44, 50]. Research on the subject of deepfakes elucidates the realism of such content and its potential manifold effects [25, 36]. These effects consist of alterations, for instance, in user attitudes towards political figures [9] and the erosion of trust in news sources [42].

In spite of the predominant emphasis on images and videos in scholarly investigations, it is noteworthy that AI technologies, including GPT-2, hold the potential to generate textual content that closely emulates genuine news articles. Such generated content can be proliferated en masse, thereby inundating the information ecosystem with a surge of material centred around a specific subject [16, 23, 24, 44]. Empirical studies in this domain reveal that while users can often distinguish between AI and human-generated texts, AI tools still have the capability to induce a sense of confusion and cultivate mistrust effectively [16, 23, 24]. Given the rapid trajectory of technological advancement, it is becoming increasingly feasible for individuals with limited technical prowess to craft plausible disinformation [43]. The authenticity of such disinformation can be notably augmented through the amalgamation of various formats [2, 37]. For instance, [4] posited an assumption wherein:

> (…) one may generate controversial text for a speech supposedly given by a political leader, create a 'deep fake' video (…) of the leader standing in the UN General Assembly delivering the speech (trained on the large amount of footage from such speeches), and then reinforce the impersonation through the mass generation of news articles allegedly reporting on the speech. (p. 4)

In summary, the rapid advancement in AI technology has raised concerns about its potential use in generating disinformation, particularly through deepfakes, which involve digitally manipulated audio and visual content. Deepfakes are created using DL, specifically the GAN framework, which consists of a content generator and an authenticity discriminator. Research on deepfakes highlights their realistic nature and their impact on user attitudes and trust in news sources. While most research focuses on image and video deepfakes, it is important to note that AI technologies such as GPT-2 can also produce highly convincing textual content resembling genuine news articles. Although users can often distinguish between AI-generated and human-generated texts, AI tools still have the capacity to sow confusion and mistrust. Finally, as technology advances, even individuals with limited technical skills can create convincing disinformation, further amplified by combining different content formats.

11.3 The Role of AI in the Dissemination of Disinformation

In addition to its potential role in generating disinformation, AI techniques can also be harnessed for the rapid dissemination of disinformation to targeted audiences [3, 9, 27, 43, 44]. Social media companies have strategically employed targeted advertising and content delivery to specific users based on their online activities, thereby increasing user engagement and platform usage, and consequently, enhancing profits derived from advertising companies [1–3, 8, 24, 27, 31, 33, 35, 43]. This practice, however, poses certain challenges, as research indicates that users tend to engage more with content that aligns with their existing beliefs and emotional stimuli, even if the information is untrue [27, 35, 43, 45, 51]. Consequently, SMPs are likely to reinforce this behaviour by recommending more disinformation to users who have already interacted with such content [8], leading to the potential entrapment of individuals within disinformation filter bubbles [40]. These filter bubbles represent distinct information universes that surround individuals with ideas that resonate with their pre-existing beliefs, including false notions and manipulated perceptions of reality, thereby diminishing informational diversity [1, 3, 21, 31, 35, 37, 40, 45].

Furthermore, the issue is compounded by the use of social bots by malicious actors. Social bots are user accounts equipped with algorithms that autonomously post and share content on various online platforms [2, 3, 12, 14, 18, 21, 34, 35, 45]. These bots play a significant role in disseminating and amplifying disinformation by targeting vulnerable individuals and promoting trending topics [1–3, 12, 14, 16, 18, 21, 23, 33–35, 37, 45]. This phenomenon could contribute to an echo chamber effect, where individuals predominantly encounter information that aligns with their pre-existing beliefs, while opposing viewpoints are minimised. This, in turn, reinforces their pre-dispositions under the illusion that they represent the prevailing perspective [1, 14, 35, 37, 43, 45].

The impact of using technological advancements to generate and disseminate disinformation remains a subject of ongoing investigation, and the existing literature tends to converge on the notion that it exerts negative effects on both individual users and broader societies [3, 8, 16, 18, 24, 43, 50]. Notably, exposure to highly realistic disinformation, coupled with the phenomenon of filter bubbles and echo chambers, can significantly influence users' opinions, engender confusion and uncertainty, and impede their capacity to make well-informed decisions [1, 3, 21, 31, 42, 43]. The ramifications of online disinformation extend beyond individual experiences, posing threats to broader societal structures. Therefore, trust in media, news outlets, and institutional frameworks might be compromised due to the proliferation of disinformation [1–3, 12, 16, 18, 21, 24, 34, 35, 43–45, 50, 51].

In summary, AI techniques not only are capable of generating disinformation but can also rapidly disseminate it to specific audiences. Social media companies leverage targeted advertising and content delivery based on user activities, potentially reinforcing pre-existing beliefs and emotional responses, even if the information is false. This practice might trap individuals in disinformation filter bubbles, diminishing informational diversity. Additionally, malicious actors further compound the issue by using social bots to amplify and spread disinformation, potentially creating echo chambers that reinforce existing beliefs. Finally, the impact of using technology for disinformation appears to be largely negative, affecting individual opinions, causing confusion, and hindering informed decision-making. It also poses broader societal threats by eroding trust in media, news outlets, and institutional frameworks.

11.4 The Role of AI in Countering Disinformation and Associated Challenges

Due to the potential adverse effects of online disinformation on both users and democratic processes, the imperative to counter its dissemination becomes paramount. However, the sheer volume of information propagated across online platforms renders manual detection unfeasible [3, 6, 11, 21, 39, 43, 44, 50]. Consequently, the adoption of AI has emerged as an increasingly favoured approach to tackle online disinformation, owing to its capability for the automatic analysis and processing of vast datasets, thereby enhancing the efficiency and speed of disinformation detection [3, 6, 11, 13, 15, 21, 28, 39, 43, 44, 50]. Of particular note, SMPs have widely employed such AI systems to diminish the visibility and engagement with disinformation, pre-empt its dissemination, and remove it from their platforms [3, 11, 21, 27, 28, 51].

Nonetheless, as discussed in the subsequent sections, the detection and elimination of disinformation content present formidable challenges. Consequently, SMPs have prioritised the identification and removal of fake accounts, which are responsible for a significant portion of the disinformation circulating online [51]. As asserted by Meta [29], their AI tools successfully identify 99.30% of the fake accounts, leading to their removal before users report them. Despite their programming to emulate human online behavior [2, 3, 12, 14, 18, 21, 34, 35, 45], these fake accounts often exhibit discernible characteristics and behavioural patterns, such as an abnormal surge in activity during the initial dissemination of disinformation. Thus, leveraging AI techniques, SMPs have succeeded in automatically identifying these patterns and proactively removing fake accounts [1, 6, 12, 21, 33, 51].

The current literature prominently discusses the use of the same tools employed for generating fake content to also identify manipulated and autogenerated content. These tools become well-suited for detection purposes due to their familiarity with the data patterns commonly associated with autogenerated content [2, 15, 16, 23, 38, 44, 50]. Empirical investigations have substantiated the effectiveness of these AI tools in detecting indicators of manipulation, such as irregularities in eye features [7], colour components [26], noise patterns [22], and word predictability [38, 48]. Despite the ongoing evolution of these techniques, autogenerated content still exhibits artificial fingerprints, intrinsic to the generating tool, which prove challenging to erase [2, 15, 50]. Studies in this domain have demonstrated that tools designed to focus on the identification of such artifacts can effectively differentiate genuine content from autogenerated content [46, 49]. Consequently, the literature highlights the successful employment of AI tools in combating the spread of disinformation by effectively detecting autogenerated content and the bots responsible for spreading disinformation.

Despite AI's potential benefits, its use in countering online disinformation also brings limitations, challenges, and unintended consequences [5, 8, 11, 21, 39, 51]. AI technologies are still in developmental stages and currently lack the capability to effectively detect more intricate instances of disinformation [3]. As discussed previously, disinformation is deliberately designed to appear credible, thereby deceiving, influencing, and manipulating individuals [3, 10], making it inherently challenging to differentiate between fake and genuine information. This task is further complicated within the present information ecosystem, where information conflicting with an individual's perspective is readily labelled as "fake news" [35]. Moreover, the existence of more complex forms of disinformation poses a substantial challenge for AI technologies [3]. Disinformation can be generated and disseminated in diverse formats, often adopting a multimodal nature. This, in turn, renders it more credible and poses a challenge for automatic detection, considering that most systems assume a predefined generation model [21, 27, 37, 39].

Additionally, as discussed throughout this book, AI systems might exhibit biases [5, 21, 47]. The development of these technologies involves human contributors with inherent limitations in knowledge and personal values, which might inadvertently influence the system's functioning. Moreover, the data used to train these algorithms can be incomplete, unrepresentative, or infused with human bias [3, 21, 27, 30, 47]. For instance, if an algorithm is trained on data limited to a specific context or language, it will likely fail to identify disinformation related to other topics or events or presented in another language [37]. Furthermore, AI systems could erroneously infer causality or correlation [3, 21, 27, 30, 47]. This limitation is intertwined with the complexity and opacity of AI systems. As these algorithms are self-learning in nature, they might reach a level of sophistication where even their developers cannot fully comprehend and elucidate their decision-making processes [8, 21, 47]. Consequently, if these systems make erroneous decisions, such as misclassifying and removing legitimate content, their complexity will hinder the ability to explain the rationale behind such decisions [47].

As a result, these challenges can significantly impact the efficiency and accuracy of AI systems, potentially leading to errors in their outcomes. These errors pose critical ethical concerns due to their potential adverse impacts on users and societies [21, 47]. Specifically, these issues can give rise to occurrences of false negatives and false positives [3, 8, 21, 27]. False negatives manifest as the erroneous classification of disinformation as accurate information, permitting harmful content to evade detection and circulate unchecked within online spaces [3, 27, 28, 39, 43]. Consequently, such content has the potential to engender confusion and distrust among users, indirectly influencing elections and undermining democratic processes since individuals require access to high-quality information for making informed decisions [1, 3, 21, 31, 42, 43]. Conversely, false positives pertain to the misidentification of accurate content as disinformation. Consequently, legitimate information disseminated online, such as news reports, might face removal due to misclassification. Such missteps have the potential to pose a threat to open discourse, culminating in violations of civil liberties and instances of censorship, thereby posing a challenge to democratic principles [3, 27, 28, 39, 43].

In summary, the need to counter online disinformation is crucial due to its potential adverse effects on users and democratic processes. However, the sheer volume of online content makes manual detection unfeasible, leading to the adoption of AI for more efficient and speedy detection. AI systems are being used by SMPs to diminish disinformation's visibility and pre-empt its dissemination. Despite the benefits of AI, its use in countering disinformation also presents challenges. For instance, AI technologies might struggle with intricate forms of disinformation, and biases in the data and algorithms can influence their functioning. As a result, these challenges can lead to errors, including false positives and false negatives, with the potential to impact user trust and democratic processes.

11.5 Recommendations

This section explores various approaches and recommendations aimed at addressing the challenges posed by AI systems in the context of countering the dissemination of disinformation online. Disinformation constitutes a multifaceted and dynamic phenomenon, involving diverse actors, objectives, formats, and creation and dissemination methods [1]. Consequently, the development and implementation of comprehensive multimodal detection frameworks are necessary. These holistic multimodal detection frameworks should consist of text and visual features while analysing content, organisation, emotions, and manipulation levels [1, 2, 37]. This holistic approach could facilitate a deeper understanding of disinformation, ultimately enhancing the effectiveness and accuracy of AI models. Additionally, considering the rapid pace of innovation in disinformation production and dissemination, countermeasure tools must be consistently updated to maintain efficacy [3, 50]. In this regard, it is suggested that SMPs allow access to their repository of collected autogenerated media to utilise as training data for AI systems, ensuring the continual enhancement of detection programs [19].

Recognising that more intricate instances of disinformation are often detected through human perception, a consensus in the literature favours a combination of AI systems and human moderation [21, 28, 51]. The collaborative approach entails the automatic reporting of content or accounts by AI to human moderators, who then analyse the material based on the platforms' guidelines, rendering decisions on whether it constitutes disinformation or not [21, 51]. Moreover, the involvement of experts, particularly in cases with potential implications for civil liberties, is recommended to assess algorithm design, decision-making processes, and outcomes [28]. Therefore, to mitigate the challenge of algorithmic bias, enhanced accountability and transparency should be essential considerations [3, 8, 21, 27, 28, 35, 43]. For instance, the proposed Algorithmic Accountability Act in the United States aims to mandate companies to audit their AI systems for bias and discrimination, thereby subjecting data and model generation processes to increased scrutiny [21, 32].

Additionally, stakeholders should invest in ethical tools that facilitate the understanding and mitigation of biases within these systems [5]. Furthermore, an approach advocated in the literature involves reconfiguring the economic model of social media, which currently relies on engagement and profit [1, 11, 35, 43, 45]. As previously mentioned, disinformation thrives on SMPs due to its ability to generate high levels of user engagement, leading recommendation algorithms to promote such content more prominently. To counteract this phenomenon, it is imperative for these platforms to implement design changes that foster interaction with diverse perspectives [1, 11, 17, 27, 35, 43, 45]. However, stakeholders must provide incentives for these changes, as alterations to platform designs can lead to reduced engagement and subsequently impact profitability.

Similarly, to effectively combat online disinformation and its propagation through AI techniques, enhanced cooperation among a diverse range of actors, including academia, SMPs, and governments on a global scale, is essential [17, 21, 27, 35, 43, 45, 51]. Collaborative efforts should involve the sharing of best practices and the development of robust tools to establish enduring and sustainable solutions [21]. When content removal measures are applied on SMPs, companies must adhere to local laws, necessitating a common understanding between governments and social media entities regarding the scope of content moderation [51]. Furthermore, platforms should engage in mutual collaboration and information-sharing regarding removed content to curtail its dissemination on other platforms [17]. It is crucial to acknowledge that disinformation represents more than just a technological predicament, thereby demanding solutions that transcend technical measures [1]. Consequently, stakeholders are advised to invest in media literacy and digital awareness initiatives to enhance critical thinking and bolster resilience against online disinformation and manipulated content among users and democratic institutions [16, 17, 21, 23, 27, 28, 43, 44].

In summary, this section explored approaches and recommendations for addressing the challenges posed by AI systems in countering online disinformation. Given the multifaceted nature of disinformation, it is essential to develop comprehensive multimodal detection frameworks that consider various content aspects. These frameworks should be regularly updated to keep pace with evolving disinformation tactics. It was also revealed that combining AI systems with human moderation was favoured to detect more complex instances of disinformation. Additionally, experts' involvement was recommended to assess algorithms and ensure transparency and accountability. Similarly, ethical tools to mitigate biases and reconfiguring the economic model of SMPs are suggested as ways to combat disinformation more effectively. In the same vein, cooperation among academia, SMPs, and governments on a global scale was considered crucial. Moreover, sharing best practices, developing robust tools, and aligning with local laws were deemed essential steps. Finally, media literacy and digital awareness initiatives were also recommended to enhance critical thinking and resilience against disinformation among users and institutions.

11.6 Discussion and Conclusion

The phenomenon of disinformation, while not novel, has evolved significantly due to the advent of novel technologies, enabling malicious actors to enhance both efficiency and scope of its dissemination [2, 3, 11, 16, 21, 23, 28, 43]. In parallel with developments in the field of AI, the capability to automatically generate remarkably authentic disinformation is poised to become readily accessible and user-friendly [2, 3, 9, 16, 21, 24, 43]. Moreover, a considerable portion of mainstream SMPs integrates recommendation algorithms, which have been subject to critique due to their propensity to reinforce pre-existing beliefs and incentivise emotive and provocative content [8, 43]. The susceptibility of these algorithms to manipulation by malicious agents is evident, facilitated by the deployment of social bots to fabricate trending topics and direct disinformation toward vulnerable recipients, thereby augmenting its efficacy [2, 3, 12, 14, 18, 21, 34, 35, 45].

Consequently, an escalation in the dissemination of increasingly credible disinformation is anticipated within the information ecosystem. This will potentially lead individuals into the confines of filter bubbles and echo chambers, where a persistent exposure to disinformation prevails, consequently diminishing access to divergent and credible content [1, 3, 14, 21, 31, 35, 37, 40, 43, 45]. The ramifications of this phenomenon can be detrimental to both users and society at large. As a result, it often leads to confusion, eroded trust in news and institutions, and a weakening of democratic discourse. This is due to the fact that the availability of high-quality, diverse information is essential for fostering open discussions and achieving consensus [1, 3, 21, 31, 42, 43].

Given these implications, the prompt detection of disinformation and the associated disseminating accounts assumes significance in diminishing its audience and curtailing potential risks [30, 37]. However, the proliferation of online information coupled with the refinement of content manipulation technologies renders the detection and differentiation between genuine and false information increasingly complex [2].

In view of AI systems' capacity for automated analysis and processing of voluminous datasets, these technologies hold potential to enhance the efficiency and speed of online disinformation detection [3, 6, 11, 13, 15, 21, 28, 39, 43, 44, 50]. Hence, this chapter posits AI as a dual-edged instrument, capable of not only intensifying the efficacy and scope of disinformation but also augmenting the potency of countermeasures [3, 21]. However, notwithstanding the recognition of AI-powered systems as a cost-effective means to counteract online disinformation, these systems are presently in a phase of ongoing development, accompanied by an array of challenges and limitations that evoke ethical apprehensions [21, 28]. The impetus behind these measures is to mitigate the propagation of online disinformation, owing to its potential adverse impact on users and the democratic fabric [1, 3, 21, 31, 42, 43]. Nevertheless, it is noteworthy that the application of these tools might also exert repercussions upon users and societal domains [21, 28], thus potentially engendering a paradoxical outcome. Furthermore, the juxtaposition of fundamental values, including the freedom of speech and the right to access high-quality information [3], is underscored when the deployment of AI technologies is directed at addressing this issue. In addition, the complex task of striking equilibrium between these values is exacerbated by the inherent tension, as mitigating the risks affiliated with one value can inadvertently amplify those associated with the other.

In light of the negative and potentially counterproductive ramifications that the use of AI technologies for combating online disinformation could yield, the imperative emerges for the development and implementation of AI systems to be underpinned by ethical considerations [5, 16, 21, 28, 39]. The absence of such ethical grounding could potentially culminate in inadequately designed technologies that pose potential risks to users and societies [5, 16, 21, 28, 39]. Additionally, the efficacy of AI-driven measures in counteracting online disinformation and its associated impacts is contingent upon the mitigation of the challenges and limitations inherent to these approaches. As a consequence, further research is indispensable to identify the most effective strategies for mitigating the limitations intrinsic to AI technologies. Future studies should also pivot towards an in-depth analysis of the effects of autogenerated content on both users and society at large. Furthermore, given the successful application of bots in targeting disinformation to specific individuals, a potentially equivalent use case could involve the dissemination of legitimate and diverse information to users deemed more susceptible to disinformation. However, this notion warrants additional investigation that focuses on the efficacy and potential implications of such an approach.

References

1. Akers J, Bansal G, Cadamuro G, Chen C, Chen Q, Lin L, Mulcaire P, Nandakumar R, Rockett M, Simko L, Toman J, Wu T, Zeng E, Zorn B, Roesner F (2019) Technology-enabled disinformation: summary, lessons, and recommendations. http://arxiv.org/abs/1812.09383
2. Benzie A, Montasari R (2022) Artificial intelligence and the spread of mis-and disinformation. In: Montasari R (ed) Artificial intelligence and national security. Springer International Publishing, pp 1–18
3. Bontridder N, Poullet Y (2021) The role of artificial intelligence in disinformation. Data Policy 3. https://doi.org/10.1017/dap.2021.20
4. Bullock J, Luengo-Oroz M (2019) Automated speech generation from UN general assembly statements: mapping risks in AI generated texts. http://arxiv.org/abs/1906.01946
5. Chowdhury R, Mulani N (2018) Auditing algorithms for bias. Harvard Business Review. Retrieved 26 Apr 2023, from https://hbr.org/2018/10/auditing-algorithms-for-bias
6. Crowell C (2017) Our approach to bots and misinformation. Twitter Blog. Retrieved 26 Apr 2023, from https://blog.twitter.com/en_us/topics/company/2017/Our-Approach-Bots-Mis information
7. Demir I, Ciftci UA (2021) Where do deep fakes look? Synthetic face detection via gaze tracking. In: ACM symposium on eye tracking research and applications 1–11. https://doi.org/10.1145/3448017.3457387
8. Dieu O, Montasari R (2022) How states' recourse to artificial intelligence for national security purposes threatens our most fundamental rights. In: Montasari R (ed) Artificial intelligence and national security. Springer International Publishing, Cham, pp 19–45
9. Dobber T, Metoui N, Trilling D, Helberger N, de Vreese C (2021) Do (microtargeted) deepfakes have real effects on political attitudes? Int J Press Polit 26(1):69–91. https://doi.org/10.1177/1940161220944364
10. European Commission (2018) A multi-dimensional approach to disinformation: report of the independent high level group on fake news and online disinformation. Publications Office of the European Union

11. Fard AE, Lingeswaran S (2020) Misinformation battle revisited: counter strategies from clinics to artificial intelligence. Companion Proc Web Conf 2020:510–519. https://doi.org/10.1145/3366424.3384373
12. Ferrara E, Varol O, Davis C, Menczer F, Flammini A (2016) The rise of social bots. Commun ACM 59(7):96–104
13. Frissen V, Lakemeyer G (2018) Ethics and artificial intelligence. Bruegel. Retrieved 26 Apr 2023, from https://www.bruegel.org/blog-post/ethics-and-artificial-intelligence
14. Gorodnichenko Y, Pham T, Talavera O (2021) Social media, sentiment and public opinions: Evidence from #Brexit and #USElection. Eur Econ Rev 136:103772
15. Gragnaniello D, Cozzolino D, Marra F, Poggi G, Verdoliva L (2021) Are GAN generated images easy to detect? A critical analysis of the state-of-the-art. In: 2021 IEEE international conference on multimedia and expo (ICME), 1–6
16. Helmus TC (2022) Artificial intelligence, deepfakes, and disinformation: a primer. RAND Corporation. https://doi.org/10.7249/PEA1043-1
17. Helmus TC, Kepe M (2021) A compendium of recommendations for countering Russian and other state-sponsored propaganda. RAND Corporation. https://doi.org/10.7249/RR-A894-1
18. Howard PN, Woolley S, Calo R (2018) Algorithms, bots, and political communication in the US 2016 election: the challenge of automated political communication for election law and administration. J Inform Tech Polit 15(2):81–93. https://doi.org/10.1080/19331681.2018.1448735
19. Hwang T (2020) Deepfakes: a grounded threat assessment. Centre for Security and Emerging Technologies, Georgetown University. Retrieved 26 Apr 2023, from https://cset.georgetown.edu/publication/deepfakes-a-grounded-threat-assessment/
20. Kavlakoglu E (2020). AI vs. machine learning vs. deep learning vs. neural networks: what's the difference? IBM Cloud. Retrieved 26 Apr 2023, from https://www.ibm.com/cloud/blog/ai-vs-machine-learning-vs-deep-learning-vs-neural-networks
21. Kertysova K (2018) Artificial intelligence and disinformation: how AI changes the way disinformation is produced, disseminated, and can be countered. Secur Human Rights 29(1–4):55–81
22. Koopman M, Macarulla Rodriguez A, Geradts Z (2018) Detection of deepfake video manipulation. The 20th Irish machine vision and image processing conference, 133–136
23. Kreps S (2020) The role of technology in online misinformation. Foreign Policy
24. Kreps S, McCain RM, Brundage M (2022) All the news that's fit to fabricate: AI-generated text as a tool of media misinformation. J Exper Polit Sci 9(1):104–117. https://doi.org/10.1017/XPS.2020.37
25. Lee J, Shin SY (2022) Something that they never said: multimodal disinformation and source vividness in understanding the power of AI-enabled deepfake news. Media Psychol 25(4):531–546. https://doi.org/10.1080/15213269.2021.2007489
26. Li H, Li B, Tan S, Huang J (2020) Identification of deep network generated images using disparities in color components. Signal Process 174:107616. https://doi.org/10.1016/j.sigpro.2020.107616
27. Llansó E, van Hoboken J, Leerssen P, Harambam J (2020) Artificial intelligence, content moderation, and freedom of expression. Transatlantic working group
28. Marsden C, Meyer T (2019) Regulating disinformation with artificial intelligence: effects of disinformation initiatives on freedom of expression and media pluralism. European Parliament
29. Meta (n.d.) Community standards enforcement report. Retrieved 26 Apr 2023, from https://transparency.fb.com/data/community-standards-enforcement/
30. Ogundokun RO, Arowolo MO, Misra S, Oladipo ID (2022) Early detection of fake news from social media networks using computational intelligence approaches. In: Lahby M, Pathan A-SK, Maleh Y, Yafooz W (eds) Combating fake news with computational intelligence techniques, vol 1001. Springer, pp 71–89
31. Pariser E (2011) The filter bubble: what the internet is hiding from you. Penguin Press
32. Robertson A (2019) A new bill would force companies to check their algorithms for bias. The Verge. Retrieved 26 Apr 2023, from https://www.theverge.com/2019/4/10/18304960/congress-algorithmic-accountability-act-wyden-clarke-booker-bill-introduced-house-senate

33. Rosenbach E, Mansted K (2018) Can democracy survive in the information age? Belfer Center for Science and International Affairs, 30
34. Santini RM, Salles D, Tucci G (2021) When machine behavior targets future voters: the use of social bots to test narratives for political campaigns in Brazil. Int J Commun 15:1220–1243
35. Schirch L (2021) The techtonic shift: how social media works. In: Schirch L (ed) Social media impacts on conflict and democracy: the techtonic shift. Routledge
36. Shin SY, Lee J (2022) The effect of deepfake video on news credibility and corrective influence of cost-based knowledge about deepfakes. Digit J 10(3):412–432. https://doi.org/10.1080/216 70811.2022.2026797
37. Shu K, Bhattacharjee A, Alatawi F, Nazer TH, Ding K, Karami M, Liu H (2020) Combating disinformation in a social media age. WIREs Data Min Knowl Discovery 10(6):e1385. https:// doi.org/10.1002/widm.1385
38. Strobelt H, Gehrmann S (n.d.) Catching a unicorn with GLTR: a tool to detect automatically generated text. Retrieved 26 Apr 2023, from http://gltr.io/
39. Thomas-Evans M (2022) A critical analysis into the beneficial and malicious utilisations of artificial intelligence. In: Montasari R (ed) Artificial intelligence and national security. Springer International Publishing, pp 81–99
40. Tomlein M, Pecher B, Simko J, Srba I, Moro R, Stefancova E, Kompan M, Hrckova A, Podrouzek J, Bielikova M (2021) An audit of misinformation filter bubbles on YouTube: bubble bursting and recent behavior changes. In: Fifteenth ACM conference on recommender systems, 1–11. https://doi.org/10.1145/3460231.3474241
41. Tyagi N (2020) 6 major branches of artificial intelligence (AI). Analytics steps. Retrieved 26 Apr 2023, from https://www.analyticssteps.com/blogs/6-major-branches-artificial-intellige nce-ai
42. Vaccari C, Chadwick A (2020) Deepfakes and disinformation: exploring the impact of synthetic political video on deception, uncertainty, and trust in news. Social Media Soc 6(1):2056305120903408. https://doi.org/10.1177/2056305120903408
43. Waldemarsson C (2020) Disinformation, deepfakes & democracy: the European response to election interference in the digital age. The Alliance of Democracies Foundation
44. Walorska AM (2020) Deepfakes & disinformation. Friedrich Naumann Foundation for Freedom. Retrieved 26 Apr 2023, from https://shop.freiheit.org/#!/Publikation/897
45. Wardle C, Derakhshan H (2017) Information disorder: toward an interdisciplinary framework for research and policy making. Council of Europe
46. Yu N, Davis L, Fritz M (2019) Attributing fake images to GANs: learning and analyzing GAN fingerprints. In: 2019 IEEE/CVF international conference on computer vision, 7555–7565. https://doi.org/10.1109/ICCV.2019.00765
47. Yu S, Carroll F (2021) Implications of AI in national security: understanding the security issues and ethical challenges. In: Montasari, Jahankhani H (eds) Artificial Intelligence in cyber-security: impact and implications: security challenges, technical and ethical issues, forensic investigative challenges. Springer International Publishing, pp 157–175
48. Zellers R, Holtzman A, Rashkin H, Bisk Y, Farhadi A, Roesner F, Choi Y (2019) Defending against neural fake news. advances in neural information processing systems, 32
49. Zhang X, Karaman S, Chang S-F (2019) Detecting and simulating artifacts in GAN fake images. In: 2019 IEEE international workshop on information forensics and security, 1–6. https://doi. org/10.1109/WIFS47025.2019.9035107
50. Zobaed S, Rabby F, Hossain MI, Hossain E, Hasan MS, Karim A, Hasib KM (2021) Deep-fakes: detecting forged and synthetic media content using machine learning. In: Montasari R, Jahankhani H (eds) Artificial intelligence in cyber security: impact and implications: security challenges, technical and ethical issues, forensic investigative challenges. Springer
51. Zuckerberg M (2018) A blueprint for content governance and enforcement. Facebook. Retrieved 26 Apr 2023, from https://m.facebook.com/nt/screen/?params=%7B%22note_ id%22%3A751449002072082%7D&path=%2Fnotes%2Fnote%2F&refsrc=http%3A%2F% 2Fwww.google.com%2F&_rdr.

Chapter 12
Responding to Deepfake Challenges in the United Kingdom: Legal and Technical Insights with Recommendations

Abstract The rapid progression of Deep Machine Learning (DML) and Artificial Intelligence (AI) technologies over the past decade has ushered in a new era of digital innovation. Alongside these developments emerges a complex landscape fraught with both technical and legal intricacies. This chapter delves into the multifaceted phenomenon of deepfakes, exploring the challenges they pose at the intersection of law and technology. To this end, the chapter examines the intricate interplay between legal and technical challenges inherent to deepfake technology. A particular focus is placed on the inherent biases within deepfake detectors, illuminating their implications. Furthermore, the evolution towards 3D-GAN (Three-Dimensional Generative Adversarial Network) technology is analysed, uncovering the potential challenges it might present. Within the context of the United Kingdom (UK), this chapter highlights the notable absence of comprehensive legislation specifically addressing deepfakes, with no imminent policy changes on the horizon. Furthermore, an in-depth analysis of the ethical and legal complexities surrounding deepfake pornography is undertaken. Additionally, the chapter delves into the far-reaching implications of disinformation and the exacerbating role that deepfake technology can play in amplifying its impact. The chapter concludes by advocating for a collaborative effort that combines legislative reforms and enhanced digital literacy to effectively mitigate the potential threats posed by deepfake technology.

Keywords Artificial intelligence · Deep machine learning · Deepfakes · Disinformation · Machine learning · Online legislation

12.1 Introduction

Within the rapidly evolving realm of technological innovation, the emergence of deepfakes presents a landscape characterised by uncharted territory and profound uncertainty. The terms 'unregulated' and 'uncertain' aptly encapsulate the essence of this research, for deepfakes, as a technological phenomenon, defy conventional legal and regulatory boundaries. The remarkable pace at which this technology evolves has outpaced legislative frameworks, rendering effective regulation an intricate challenge. As deepfake technology continues to advance, it brings with it a host of concerns that span from broader societal implications to the intricate challenges faced by digital forensic analysts and legal entities. To this end, this chapter seeks to delve into the multifaceted complexities surrounding deepfakes, offering a comprehensive understanding of the challenges arising from this evolving technological landscape. It explores both technical and legal aspects while also illuminating the broader societal and ethical implications that are pertinent in this digital era.

The subsequent sections of this chapter are structured as follows. Section 12.2 provides contextual information to set the study within its broader context. Section 12.3 forms the central foundation of this research. It critically examines technical and legal challenges associated with deepfakes. This section illuminates the limitations of existing detection technologies, revealing how they struggle to keep pace with the rapidly evolving landscape of deepfake creations. Moreover, it analyses the current legal framework surrounding deepfake technology, offering insights into the prospects of impending legislation. This examination includes considerations of the possible form such legislation might take and its implications for various stakeholders. Within the same section, deepfake pornography emerges as a critical area of focus. To this end, the section investigates the legal provisions available to protect individuals ensnared in the web of image-based abuse, seeking to ascertain whether existing laws offer respite for victims or serve as interim safeguards against the pernicious effects of deepfake pornography. Finally, this section explores the phenomenon of disinformation and highlights how deepfake technology can exacerbate the already perilous landscape of misinformation, raising questions about the dissemination of false narratives and their societal impact. In Sect. 12.4, a discussion of findings is presented. Finally, in Sect. 12.5, the chapter concludes by summarising key points and findings, encapsulating the intricate landscape of deepfake technology and its far-reaching implications in today's digitally driven world.

12.2 Background

In the year 2023, it is an infrequent occurrence for a criminal event to transpire without the involvement of digital components, rendering the necessity for digital forensics an integral component of the legal process. This rapid integration of digital forensics into the legal framework has presented a formidable challenge for both law enforcement agencies and specialists in the field of digital forensics. The emergence of "deepfake" technology can be traced back to 2014, when the concept of generative adversarial networks (GANs) was initially introduced, albeit without its subsequent nomenclature [14]. As this technology progressed towards a deep learning paradigm, resulting in significant improvements in quality, deepfakes emerged as a prominent technological phenomenon, gaining widespread recognition on social media platforms, most notably Reddit.

There exist four prominent categories of deepfakes, encompassing deepfake pornography, deepfakes within political campaigns, deepfakes tailored for commercial purposes, and creative iterations of deepfake applications [21]. The formidable realism achieved by deepfakes stems from intricate digital manipulations. Several contributing factors play a pivotal role in enhancing the quality of these deceptive digital creations. Among these factors, the quality and quantity of data collected pertaining to the individual targeted for the deepfake transformation hold paramount importance. The accumulation of diverse data sources, such as images and videos captured from various angles and under different lighting conditions, correlates directly with the resultant quality of the deepfake. Subsequently, an autoencoder or GAN is employed to consolidate and assimilate this data, enabling the system to acquire a profound understanding of the underlying probability distribution that governs the generation of these inputs [2]. The GAN comprises two neural networks: a generator and a discriminator, each assuming distinct and pivotal roles in the deepfake creation process. The generator and discriminator engage in a competitive interplay, with the generator responsible for producing digital manipulations while the discriminator undertakes the challenging task of identifying flaws and inconsistencies within the generated content through iterative trial and error procedures [32]. This adversarial dynamic between the two components drives the refinement of authenticity, striving to achieve the utmost hyper-realism attainable, predicated on the dataset provided to the network. The parameters governing the GAN can be fine-tuned and adjusted to further enhance the quality of the output. Consequently, the system becomes poised to generate highly persuasive content that is exceedingly difficult to distinguish as counterfeit, whether in the form of images or videos. A notable advancement in the realm of deepfake technology, as of 2022, is the introduction of the 3D-GAN. This innovative approach leverages geometric principles to enable the machine to generate reconstructions based on a single image of an individual. By harnessing geometry, this advancement offers the potential to establish consistency across various viewing angles, addressing a longstanding challenge in current deepfake technology where imperfections are often discernible [3].

12.3 Challenges

12.3.1 Technical Challenges

Deepfakes, in their technical underpinnings, generally fall into one of two distinct categories: face re-enactment and face swapping. In the former, the manipulation entails distorting the facial expressions and speech of an individual, creating the illusion that they are engaged in actions or uttering statements they did not actually perform. The latter, face swapping, emerges as the more prominent technique, particularly in contexts involving more sinister applications such as deepfake pornography. In this technique, two videos are merged, effecting the seamless blending of an individual's facial features onto the body of another person [16]. One pressing concern associated with these creations is the pervasive absence of consent from one or both parties involved. As the demand for deepfake videos continues to grow, the development and refinement of deepfake detection mechanisms have become an ongoing imperative to keep pace with the escalating sophistication of this technology. Regrettably, the lack of standardisation in the deployment of deepfake detectors in real-world scenarios engenders numerous challenges for digital forensic analysts grappling with this evolving landscape.

Detection approaches commonly pivot toward deep learning networks, as opposed to conventional machine learning (ML) models, when tasked with identifying manipulated media [4]. However, the specific type of detector employed plays a pivotal role in determining which manipulations within a video it can effectively recognise. This discrepancy underscores one of the primary technical challenges confronting analysts in this domain, namely the preprocessing of videos. The use of preprocessing techniques introduces alterations to the inherent characteristics of videos, thereby affecting detection outcomes. As these deceptive videos continue to evolve and incorporate more intricate nuances and subtleties, the pre-processing phase becomes increasingly time-consuming and resource-intensive, involving the arduous tasks of collecting, cleansing, and meticulously labelling vast amounts of media data. Furthermore, the calibration of detectors to minimise false positives poses a resource-intensive challenge in itself. Achieving optimal detector performance necessitates the provision of ample, unbiased data to facilitate the generation of the most accurate results. This calibration process, while critical for enhancing detection precision, requires a substantial investment of resources and expertise to be able to address the complex deepfake technologies.

Bias can infiltrate detection models through various channels, encompassing training data bias, feature bias, testing data bias, and social bias. The significance of training data in configuring DML models cannot be overstated, as even slight adjustments to this foundational element can significantly skew the outcomes in favour of certain biases. In situations where meticulously curated and tested training data is not readily available, the utilisation of the triplet loss function becomes a viable alternative. The triplet loss function operates by training neural networks to discern object representations by initially exposing them to similar objects and subsequently contrasting them with dissimilar objects. This method has demonstrated enhanced efficiency, particularly in scenarios where training data is scarce. Notably, the triplet loss function has exhibited success in diverse contexts, spanning intra-datasets and cross-datasets, signifying a promising stride towards enhancing the accuracy of detection mechanisms [1].

As technology such as deepfake detection becomes increasingly prevalent, the ramifications of bias are poised to extend far beyond the realm of mere technical concerns. Indeed, they have the potential to have a profound impact on pivotal, life-changing decisions within legal processes in the future. The imperative task of identifying and mitigating bias within detection systems necessitates a comprehensive and meticulous analysis of each specific implementation. This scrutiny is essential to ascertain that biases have not inadvertently crept into the system during the pre-processing phase, where data alterations and manipulations might inadvertently introduce or reinforce biases. The consequences of failing to address bias within these systems could reverberate through legal proceedings, impacting individuals' rights and outcomes in ways that demand the utmost attention and vigilance from both technology developers and legal practitioners.

Currently, the field of deepfake detection primarily operates in a reactive manner, with detectors relying on the identification of inconsistencies within the creation of videos and subsequently exploiting these flaws. An overarching concern in this context is the accessibility of video manipulation tools to unskilled individuals, resulting in the proliferation of deepfake content. While lower-quality deepfakes often exhibit signs of manipulation, such as inconsistencies in eye colour or earlobe shape, developers continually strive to rectify these imperfections within widely used applications. Moreover, scalability poses a formidable challenge for current deepfake detectors. The computational intensity required for analysing substantial volumes of media remains a significant hurdle, particularly when the need arises for real-time identification and removal of deepfake content from online platforms. The technology has yet to fully catch up with the demands of such rapid and large-scale content analysis. In addition to computational constraints, deepfake detectors must possess a high degree of adaptability to effectively contend with new forms of deceptive content. This adaptability necessitates labour-intensive updates and refinements as the content creators employ increasingly sophisticated techniques, making it increasingly challenging to detect such manipulations accurately. Thus, the ongoing evolution of deepfake technology underscores the need for detectors to continually evolve and improve to maintain their efficacy.

12.3.2 Legal Challenges

In addition to the customary legal complexities and challenges that often accompany cybercrimes, such as navigating geographical intricacies and addressing legislative disparities that can impede extradition treaties, deepfakes present their own set of distinctive and pressing issues. Despite calls for specific legislation targeting deepfake-related offences dating back to their inception, there is currently no singular comprehensive legal framework safeguarding victims of such actions within the United Kingdom, the European Union, or the majority of stated in the United States (U.S.) [15]. Notably, in January 2023, China emerged as a trailblazer by enacting legislation that explicitly prohibits the production of deepfakes without user consent [10]. Renowned for their stringent policies concerning cyber threats and content regulation, these laws stand out as the most comprehensive transnational measures presently in existence. The primary objective of these regulations is to curtail the dissemination of disinformation and protect individuals from the circulation of non-consensual deepfake videos. Furthermore, any consensual content generated through the use of AI must bear a watermark to notify viewers of its AI-based origins [8].

These developments mark significant steps toward addressing the multifaceted challenges posed by deepfake technology and the need for legal frameworks to govern its usage.

In June 2021, a bill titled the "Unsolicited Explicit Images and Deepfake Pornography Bill" was introduced in the House of Commons; however, it was subsequently withdrawn in March 2022. Current initiatives pertaining to deepfakes in the U.K. predominantly centre around the issue of deepfake pornography, encompassing both image and video-based content. Specifically, the Online Safety Bill, currently under consideration, aims to criminalise the sharing of pornographic deepfakes in England and Wales [22]. This proposed amendment to the Online Safety Bill seeks to "broaden the scope of current intimate image offences" [23]. Notably, the legislative focus in the U.K. appears to be primarily victim-oriented, concentrating on the pornographic dimension of deepfakes, while potentially overlooking the broader implications related to disinformation campaigns and humiliation attempts. The proposed UK legislation predominantly addresses the primary harm associated with deepfake videos, which is the absence of consent, but this is narrowly defined within a pornographic context and does not cover political manipulation or non-consensual humiliation efforts.

Indeed, the perils associated with the influence of deepfakes extend well beyond the realm of deepfake pornography. While this form of exploitation is unquestionably damaging and warrants substantial attention, it is crucial to recognise that there exist broader and equally significant risks, notably the capacity to disseminate disinformation that has yet to receive adequate acknowledgment. In the context of the U.K., the prevailing legislative framework primarily focuses on prosecuting cases involving the creation of deepfakes rather than the content they produce. Consequently, creators of deepfake content can potentially face legal action under defamation or copyright laws [31]. This approach underscores the legal system's efforts to address the core issues associated with deepfakes, emphasising the need for accountability and consequences for those who produce deceptive and potentially harmful content. However, it also highlights the evolving nature of the legal response to deepfakes and the ongoing exploration of regulatory approaches to mitigate their multifaceted threats.

Beyond specific legislation directly targeting deepfakes, several other existing legal frameworks can potentially apply to address malicious deepfake activities. These include Fraud [11], Defamation [6, 7, 9]. For instance, the Fraud Act of 2006 can be invoked when deepfakes are used for fraudulent purposes, such as deceiving individuals or organisations for financial gain. Similarly, the Defamation Act of 2013 can come into play when deepfake content is employed to tarnish an individual's reputation through false statements or malicious representations. In the same vein, the Data Protection Act of 2018 could be relevant if personal data is used in the creation of deepfakes, particularly if such data is misused without consent. Additionally, the Computer Misuse Act of 1990 can be invoked when deepfakes are involved in cybercrimes or illicit activities. In this context, it can be employed to cover offences related to unauthorised access, hacking, and other malicious actions involving computer systems. The applicability of these laws depends on the specific circumstances and nature of the deepfake incident. It is worth emphasising that the legal landscape surrounding deepfakes is still evolving, and the interpretation and application of existing laws in the context of deepfakes could vary from case to case. Legal professionals and lawmakers are continually adapting and refining these frameworks to address the unique challenges posed by deepfake technology and its potential for harm.

The case in 2019 involving a British energy company falling victim to a fraudulent transfer of $243,000 serves as a stark illustration of how DL and AI can be exploited for criminal purposes [29]. In this instance, criminals employed DL to mimic the voice of the CEO, using it to place a phone call to the CEO of the parent company. The impersonator instructed the CEO to urgently transfer funds to a Hungarian supplier. The funds were swiftly transferred to Mexico, and investigators were unable to recover them. As of now, the perpetrators behind this offence remain at large. If this case were to proceed to court, fraud would likely constitute the primary offence. However, it is essential to consider the use of AI in such circumstances as an extension of the crime itself. While orchestrating fraud on such a significant scale, particularly targeting high-value individuals or organisations, is inherently challenging, advancements in AI and DL significantly facilitated this criminal act. Therefore, it is plausible to assert that the success of this fraudulent scheme would have been improbable without the use of AI technology. Consequently, the integration of AI into fraudulent activities raises complex legal and ethical questions, underscoring the necessity for legal systems to adapt to the evolving landscape of technologically facilitated crimes.

12.3.3 The Deepfake Defence

The emergence of deepfake technology within AI presents profound credibility challenges for digital forensic analysts. Currently, there is no singular tool or methodology that can provide a foolproof, 100% accurate determination of whether a video or image is a deepfake creation or an authentic representation. This has posed a considerable challenge to digital forensic analysts, who have a longstanding history of employing image authentication techniques within courtrooms [19]. However, the proliferation of sophisticated deepfake technology has added a layer of complexity and ambiguity to this process. Thus, the ready availability of such advanced technology has significant implications for the trustworthiness of evidence presented within courtrooms. It erodes the confidence in the authenticity of image or video-based evidence, casting doubt on its credibility. As a result, the legal system and digital forensic community face an uphill battle in adapting to the evolving landscape of deepfake technology. This underscores the pressing need for innovative approaches and tools to address this challenge and restore trust in digital evidence.

The emergence of the 'deepfake defence' represents a complex addition to the legal system, wherein videos that are formally admitted as genuine evidence are subsequently claimed to be false or manipulated [31]. This defence strategy would necessitate digital forensic analysts to undertake the formidable task of definitively proving that the presented evidence has not been tampered with—a challenge that is exceedingly difficult to surmount given the capabilities of today's technology. Although the deepfake defence has not yet made a substantial appearance in court-rooms, the increasing popularity of AI-based deepfake technology suggests that it might become a significant source of complexity and contention within the legal process on a global scale. Video evidence has traditionally served as a valuable tool in strengthening eye-witness testimonies and corroborating facts within courtrooms. However, the proliferation of deepfakes introduces the disconcerting possibility that video evidence, instead of bolstering credibility, could undermine it, casting doubt on the authenticity of recorded events and exacerbating the challenges faced by legal systems worldwide. As a result, the rise of the deepfake defence underlines the urgent need for legal experts and digital forensic professionals to proactively address this evolving landscape and develop robust strategies for preserving the integrity of evidence in an era of advanced AI technology.

Furthermore, as the capabilities of deepfake technology continue to advance, it is theorised that the authentication of media might increasingly require the involve-ment of eyewitnesses to validate its authenticity. Depending solely on digital forensic analysts to verify media is becoming increasingly difficult and less reliable, particu-larly in the face of highly convincing deepfakes [19]. While the legal system has long grappled with issues related to unreliable eyewitnesses and camera tricks, the scale, complexity, and realism of deepfakes amplify this problem considerably. A substan-tial study conducted in the UK underscore the significance of unreliable witness testimony as a primary cause of miscarriages of justice over the past five decades [30]. It is therefore crucial to recognise that relying heavily on eyewitness testimonies over digital media could potentially be counterproductive and represent a regressive step within the courtroom. Similarly, it is important to acknowledge that balancing the reliance on both digital media evidence and eyewitness testimony while ensuring their credibility will be a formidable challenge for legal systems. Therefore, the evolving landscape of technology necessitates innovative and adaptable approaches to evidence authentication to preserve the integrity of the legal process and minimise the potential for miscarriages of justice.

12.3.4 The Complications of Deepfake Pornography

The criminalisation of revenge porn within the U.K. in 2015 has brought to light a broader range of offences involving explicit imagery that have been occurring, driven by the misuse of advancing technology for malicious purposes. Several image-based sexual offenses have emerged, including actions such as 'cyberflashing', which involves sending unsolicited explicit images or videos, as well as various forms of online harassment. Importantly, the perpetrators of these offences are not limited to disgruntled ex-partners, as is often the case in revenge porn situations, but can include individuals ranging from strangers to colleagues. Additionally, for those who may not possess advanced technical skills, there is the possibility of paying a small fee to someone who can create these images or videos on their behalf. However, the democratisation of deepfake technology has made the creation process accessible to a broader audience, including those with limited technical capabilities. This accessibility has expanded the potential for the misuse of technology, presenting a complex challenge for law enforcement and legal systems in addressing these evolving forms of digital misconduct.

The term 'deepfake pornography' has its origins in an anonymous user on the popular online platform Reddit and is a combination of 'DL' and 'fake' [32]. Initially, this technology was predominantly employed using the faces of celebrities with the primary objective of creating images and videos that appear as realistic as possible. It is important to note that, despite the severe consequences it can have on victims, deepfake pornography is not currently protected under revenge porn laws. The proliferation of deepfake pornography can be directly linked to the accessibility of the technology and the ease with which individuals can download the necessary programs and pornographic content for free [18]. Addressing the issue of deepfake pornography presents a complex challenge, and solutions are currently more reactive than preventative in nature. Thus, this evolving landscape of technology and its potential for misuse accentuate the need for ongoing efforts to adapt legal frameworks and develop proactive measures to mitigate the harmful impact of deepfake pornography.

A promising risk reduction solution for addressing the issue of deepfake pornography involves the use of video and image perceptual hashing. This method, employed elsewhere for internet regulation, could prove effective in combating and mitigating the harm caused by deepfake pornography. The process of video and image hashing involves creating unique identifiers or 'hashes' for digital media based on their perceptual characteristics. Even if the content is altered or distorted to evade detection and subsequently reuploaded to the same or different platforms, perceptual hashing can still identify it [17]. Consequently, implementing perceptual hashing for the detection and removal of deepfake pornography has the potential to enhance the effectiveness of content moderation efforts. By employing this technology, online platforms and regulators can more efficiently and accurately identify and address instances of deepfake pornography, thereby reducing its proliferation and minimising the harm caused to victims. However, it is essential to continually refine and adapt these technologies to keep pace with the evolving tactics employed by those creating and disseminating deepfake content.

Many of the major social media platforms (SMPs) currently engage in collaborative efforts with the Global Internet Forum to Counter Terrorism (GIFCT) to combat the proliferation of terrorist material online. This collaboration involves the sharing of hashes among platforms. By sharing hashes, these platforms can expedite the removal of terrorist content, thereby minimising the time it takes for such material to spread and potentially cause harm [13]. While this collaborative approach has been effective in countering terrorist content, the application of similar technology and collaboration to combat deepfake pornography is still in development and has not been implemented on the same large-scale collaborative level. Nonetheless, the potential for such collaborative efforts to mitigate the spread and harm associated with deepfake pornography is promising, and ongoing developments in this field could lead to more robust measures in the future.

In addition, campaigns such as the "TakeItDown" Project, led by the National Center for Missing and Exploited Children (NCMEC), employ hash-based methods to combat child pornography. Notably, this project collaborates directly with one of the largest global pornography networks, Pornhub, to identify and remove illicit content [24]. However, a noteworthy distinction arises when addressing adult victims, particularly those over the age of 18 who have fallen victim to revenge porn or deep-faked media. Projects such as StopNCII which operate in a similar fashion using hash-based techniques aim to combat the proliferation of such content [28]. Interestingly, platforms that engage in collaborations with StopNCII tend to be primarily SMPs, rather than those specifically focused on sharing pornography or explicit adult content. As of April 2023, platforms such as Facebook, Instagram, TikTok, Bumble, Reddit, and OnlyFans have partnered with StopNCII in efforts to combat deepfake pornography and revenge porn. This discrepancy underscores a significant challenge that there appears to be a lack of incentive or collaboration from pornographic platforms to actively address and remove deepfake pornography. The contrasting responses from SMPs and adult content-sharing platforms highlight the complex and evolving landscape of content moderation and the differing approaches taken by various online platforms in addressing this critical issue.

Indeed, the incentive for platforms to remove child pornography is clear, driven by both ethical considerations and the illegality of hosting such content. However, when addressing deepfake pornography, the motivations to take proactive measures might not align as strongly, and the response from platforms might be less robust. Therefore, to address this gap and encourage platforms to adopt more proactive policies and measures against deepfake pornography, there might be a need for external pressure, whether from the public or through legislative actions. Public awareness campaigns, advocacy efforts, and legal frameworks that hold platforms accountable for hosting or disseminating deepfake pornography could play a pivotal role in driving change. The recognition of the potential harm caused by deepfake pornography and the need for platforms to take responsibility in combatting it is a crucial step in mitigating its negative impact on individuals and society as a whole.

12.3.5 Disinformation

The findings of a 2022 survey reveal that a significant 38% of individuals residing in the UK rely on SMPs as their primary source of news [33]. This increasing trend raises concerns, particularly in light of the prevalence of fake news and disinformation easily disseminated online. The emergence of deepfakes poses a potential exacerbation of this issue, as it has the capacity to enhance the credibility of such false information, particularly among those who heavily depend on social media for their news consumption. SMPs offer a two-way mode of interaction and communication, making them particularly well-suited for the rapid dissemination of breaking news or "sharing information of immediate importance" [34, p. 287]. However, this very attribute also renders them susceptible to the spread of disinformation and deepfakes, which can be propagated swiftly, potentially reaching a vast audience before corrective measures can be taken. Consequently, the convergence of deepfakes and the reliance on social media for news underscores the need for increased vigilance, critical media literacy, and robust fact-checking mechanisms to mitigate the risks associated with the spread of fake news and disinformation in the digital age.

The rapid spread of information on the internet, facilitated by SMPs, has made it easier than ever for content to go viral. This ease of dissemination presents a significant challenge, especially in relation to deepfakes. Users often share content without verifying its source or accuracy, contributing to the rapid spread of disinformation and deepfake content. The question of liability for the spread of convincing disinformation is a complex, thereby raising considerations about the responsibility of both SMPs and individual users. Platforms play a pivotal role in enabling the dissemination of content, and they are increasingly under scrutiny for their roles in regulating and moderating content. Users, on the other hand, bear a degree of responsibility for their actions in sharing content, even if they do so unintentionally or without full awareness of its potential harm. The emergence of political deepfakes, in particular, poses a significant challenge for journalism and public trust. These manipulated videos undermine one of the fundamental principles of journalism, which is to report on actual events and facts. Emphasising this, [5] state that they "violate the main principle of journalism—it's impossible to show what doesn't exist" (p. 1851). Repairing the damage to public trust caused by the proliferation of deepfakes in journalism will be a daunting task, and it underscores the urgency of developing effective countermeasures, regulations, and media literacy initiatives to address the challenges posed by this technology.

The use of deepfakes for political interference has indeed emerged as a concerning trend. An illustrative example is the deepfake featuring Ukrainian President Zelensky, in which he appeared to urge Ukrainians to surrender. This particular deepfake was of relatively low quality and lacked the believability required for it to become a viral means of disinformation. However, it is suspected that due to the Russian accent featured in the video, this was part of a broader Russian disinformation campaign with potential geopolitical implications. This incident serves as a real-world example of how deepfakes can be utilised for political interference and is considered one of the first instances of such usage [26]. While, to date, there have not been examples of deepfakes causing destabilisation at a national level in any countries, it is unwise to dismiss the potential for deepfakes to be weaponised in future political contexts or conflicts. Therefore, the increasing sophistication of deepfake technology and its potential for creating convincing and misleading content underscore the need for heightened vigilance, international cooperation, and efforts to develop effective strategies for countering deepfake-based disinformation campaigns that might seek to manipulate public perception and influence political outcomes.

The danger posed by political deepfakes extends beyond individual deceptive videos to the potential for foreign actors to flood communication channels with numerous such deepfake content. This strategy, akin to a disinformation barrage, has the potential to overwhelm audiences and create a significant resource challenge for government bodies tasked with countering such campaigns. This tactic is not new to the realm of disinformation. State and non-state actors have previously employed flooding techniques to disseminate false or misleading information on a large scale. The introduction of deepfake media into this environment adds another layer of complexity and potency to such campaigns. The convincing nature of deepfakes, coupled with their potential to generate a high volume of deceptive content, makes them a powerful tool for those seeking to manipulate public perception and influence political discourse. States and governments that lack the necessary resources to effectively remove and respond to deepfake-based disinformation campaigns might find themselves particularly vulnerable to these tactics. As such, developing strategies to detect, mitigate, and combat the potential flood of political deepfakes is imperative to safeguarding the integrity of communication channels and ensuring the accuracy of information in the face of evolving disinformation threats.

Former US President Donald Trump has encountered deepfake content during his time on Twitter (now rebranded to X), often using the platform to share videos of various kinds. These videos have included those aimed at mocking political opponents, such as a retweeted deepfake of Joe Biden "twitching his eyebrows and lolling his tongue" [12], as well as deepfakes depicting himself, such as Trump's image superimposed into the film Independence Day [25]. The argument made in defence of Trump in these cases is that, despite being deepfakes, these videos are readily identifiable as fake and, therefore, not intended to deceive. This situation raises a significant challenge for legislative bodies and regulators when considering the regulation of deepfakes. Namely, it prompts questions about whether the harm caused by deepfakes and the intentions behind their creation should be factors considered in regulatory efforts. The distinction between deepfakes that are clearly intended to deceive and those that are used for satirical or humorous purposes presents a complex issue in terms of policy and regulation. Consequently, striking a balance between addressing the harmful impact of deceptive deepfakes while allowing for creative expression and political satire is a challenge that lawmakers and policymakers will need to grapple with as they seek to address the evolving landscape of deepfake technology and its potential implications for public discourse and trust.

It is also important to consider the issue of consent which is indeed a crucial consideration when individuals allow the use of their likeness in deepfake media. While this might reduce the harm for the person featured in the content, it does not necessarily eliminate the potential harm for those who view and believe the deepfake. In the case of Trump's retweet of a deepfake featuring opposition candidate Biden, it could be argued that this is a clear instance of spreading disinformation—a form of information that is knowingly false and intended to cause harm, in this case, to the reputation of Biden. However, because the deepfake was not entirely convincing or realistic, a defence could be mounted that audiences could readily identify the video as disinformation, thereby minimising the harmful effects on the audience. On the other hand, when individuals share content that features themselves in deepfake media, the issue of consent comes into play. Even if the deepfake is intended for satirical or humorous purposes, there might still be concerns about whether the individuals portrayed have given their informed consent for such use. Ensuring that consent is obtained and that individuals have agency over how their likeness is used in deepfake content is a significant ethical consideration in this context. Thus, balancing the freedom of expression and creativity in the digital age with the potential for harm and deception remains a complex challenge, and policymakers will need to navigate these issues carefully as they develop regulations and guidelines to address deepfakes and their impact on society.

Currently SMP Facebook, now under the umbrella of Meta, has been one of the businesses at the forefront of taking accountability for deepfakes on its platforms. In January 2020, Facebook introduced a new policy in response to the need to combat disinformation, particularly in the context of US elections. This policy focused on addressing AI-generated and manipulated videos that have the potential to misinform viewers by depicting individuals saying words that they did not actually say [20]. While the primary intention behind this policy was to combat disinformation and misleading content in the context of elections and political discourse, it also extended to deepfake pornography. Facebook's proactive stance on deepfake content could indicate the platform's recognition of the potential harm and deception associated with such media. By implementing policies to address deepfake content and disinformation, Facebook is setting a precedent for other SMPs and online communities to follow suit in tackling the challenges posed by deepfakes and related content. These efforts underscore the importance of proactive measures in safeguarding the integrity of online information and protecting users from the potentially harmful effects of manipulated media.

12.4 Discussion

12.4.1 Social Media

The coronavirus pandemic has prompted a significant shift toward a more virtual way of life for individuals around the world. This transformation has led to a substantial increase in individuals' usage of social media and altered their online behavior compared to the pre-pandemic era [27]. In light of these changes, the importance of regulating online content has become more important than ever before. Additionally, education and digital literacy have taken on greater significance as tools to navigate and mitigate the potential harms associated with technologies such as deepfakes. A balanced approach that combines both regulation and education is likely to be the most effective defence against potential deepfake harms. While legislation can set boundaries and provide legal consequences for malicious uses of deepfake technology, digital literacy and education empower users to critically assess the media they consume and discern between authentic and manipulated content. When individuals are informed, digitally literate, and trained to think critically about the media they encounter, they are better equipped to recognise deepfakes and mitigate their harmful effects. Ultimately, the harms associated with deepfake technology hinge on the element of realism. By enhancing digital literacy and promoting critical thinking skills among users, the potential for manipulation and deception can be reduced, thereby weakening the malicious effects of deepfakes and fostering a more resilient online community.

12.4.2   Political Damage

The use of deepfakes in political contexts raises significant concerns, primarily due to their potential for manipulation and harm. While it is acknowledged that deep-fakes are not always employed for sinister purposes, a substantial portion of them is intended to manipulate audiences or humiliate individuals. These manipulations range from creating malicious pornography involving individuals to spreading disin-formation through political figures. This accessibility to deepfake technology, even for those with limited technical expertise, underscores its potential as a dangerous tool. Although political elections have not been directly impacted by deepfakes to date, it is crucial to recognise that they pose a real threat for future political processes. The development of technologies such as 3D-GAN has made it possible to create hyper-realistic videos that are incredibly convincing. These videos have the potential to cause severe reputational damage, which might prove extremely challenging to rectify. The combination of deepfakes' accessibility, their potential for hyper-realistic deception, and their use in political contexts highlights the importance of addressing this issue proactively. Therefore, it is important for policymakers and digital plat-forms to remain vigilant and to be able to develop robust strategies to detect, counter, and raise awareness about the risks associated with deepfake technology to safeguard the integrity of political processes and protect individuals from harm.

12.4.3   Legislation

The absence of specific legislation addressing deepfakes in the UK can be attributed to the relatively new and rapidly evolving nature of this technology. Creating effec-tive laws to govern deepfakes is challenging because the technology is continuously adapting and evolving. Legislation that attempts to cover all aspects of deepfake creation, distribution, and impact can rapidly become outdated as the technology advances. While legislative efforts are essential for addressing the challenges posed by deepfakes, they cannot be the sole solution. As previously stated, advocating for digital literacy is a crucial component of the defence against deepfakes. Increasing education and awareness among the public enables individuals to make informed decisions about the content they encounter online and equips them with the critical thinking skills needed to identify disinformation and manipulated media. However, digital literacy is a short-term solution and cannot be relied upon exclusively to combat malicious deepfakes. Thus, a comprehensive approach to addressing this issue should include a combination of legislative measures, technological advancements in deepfake detection, public awareness campaigns, and collaboration between government, industry, and civil society to effectively mitigate the risks posed by deepfake technology.

12.5 Conclusion

Deepfakes have the potential to disrupt democracies, strain international relations, and cause harm on various levels. In the legal arena, deepfakes can complicate court proceedings, challenging the authenticity of evidence and eyewitness testimony. Furthermore, they can facilitate and exacerbate cybercrimes, such as fraud, by adding new dimensions of deception and manipulation. On a personal level, deepfake pornography and other forms of malicious deepfake content can inflict significant emotional and reputational damage on individuals. Additionally, the spread of highly convincing disinformation through deepfakes can undermine trust in media, government institutions, and public figures. Addressing these threats requires a multifaceted approach, including legal and regulatory measures, advancements in deepfake detection technology, public education, awareness campaigns, and digital literacy initiatives.

The cautious approach of states in implementing specific legislative guidelines for deepfake content could be due to the rapidly evolving nature of this technology. Clearly defining what should be prohibited and what can be allowed within the realm of deepfakes is a complex task that requires careful consideration. Therefore, when crafting legislation related to deepfakes, a two-fold focus on harm reduction is essential. Firstly, it should consider the consent of all parties involved in the creation and distribution of deepfake content. Ensuring that individuals have given informed consent for the use of their likeness in such content is crucial for ethical and legal reasons. Secondly, legislation should assess the potential impact on audiences and whether the disinformation disseminated through deepfakes could have harmful effects on society at large. This includes considering the broader implications for public trust, misinformation, and societal stability. Additionally, it is important to recognise that while deepfake detectors are a valuable tool in identifying and mitigating malicious deepfakes, they are not a comprehensive long-term solution. The 'cat-and-mouse' game between creators and detectors underscores the need for a multifaceted approach as discussed above.

Furthermore, safeguarding against the threats posed by deepfakes relies on a collaborative effort involving legislative implementations, online platforms hosting user-generated content, and individual awareness and responsibility. Individuals play a vital role in this ecosystem by developing digital literacy skills that allow them to discern between authentic and manipulated content. Being able to critically evaluate the source and fact-check information found online is crucial in navigating the digital landscape, especially as deepfake technology becomes more prevalent. AI and DL can and should be viewed as positive tools that enhance knowledge and offer numerous benefits to society. When harnessed responsibly, these technologies can drive innovation, improve efficiency, and advance various fields. Therefore, in this new age of digital culture, it is imperative to include education on disinformation in the online domain. Equipping users with the knowledge and skills to identify and combat disinformation empowers them to coexist with emerging technologies such as deepfakes and make informed decisions about the media they consume. This educational approach fosters a more resilient and informed digital society, where individuals can navigate the digital landscape with confidence and discernment.

Finally, as deepfake technology continues to evolve, it is crucial for states and stakeholders to adapt and develop robust strategies to mitigate its negative impacts, to stay ahead of the curve, and to protect individuals, institutions, and the integrity of information from the potential harms associated with deepfakes.

References

1. Bondi L, Daniele Cannas E, Bestagini P, Tubaro S (2020) Training strategies and data augmentations in CNN-based deepfake video detection. In: 2020 IEEE international workshop on information forensics and security (WIFS). https://doi.org/10.1109/wifs49906.2020.9360901
2. Brain IGG, Goodfellow I, Brain G, Profile GBV, de Montréal JP-AU, Pouget-Abadie J, de Montréal U, de Profile UMV, de Montréal MMU, Mirza M, de Montréal BXU, Xu B, de Montréal DW-FU, Warde-Farley D, de Montréal SOU, Ozair S, de Montréal ACU, Courville A, de Montréal YBU, Metrics OMVA (2020) Generative adversarial networks. Commun ACM. Retrieved 20 Apr 2023, from https://doi.org/10.1145/3422622
3. Chan ER, Lin CZ, Chan MA, Nagano K, Pan B, de Mello S, Gallo O, Guibas L, Tremblay J, Khamis S, Karras T, Wetzstein G (2022) Efficient geometry-aware 3D generative adversarial networks. In: 2022 IEEE/CVF conference on computer vision and pattern recognition (CVPR). https://doi.org/10.1109/cvpr52688.2022.01565
4. Charitidis P, Kordopatis-Zilos G, Papadopoulos S, Kompatsiaris I (2020) Investigating the impact of pre-processing and prediction aggregation on the deepfake detection task. Retrieved 19 Apr 2023, from https://arxiv.org/abs/2006.07084
5. Chudinov AP, Koshkarova NN, Ruzhentseva NB (2019) Linguistic interpretation of Russian political agenda through fake, deepfake, post-truth. J Siberian Federal Univ Human Soc Sci: 1840–1853. https://doi.org/10.17516/1997-1370-0492
6. Computer Misuse Act 1990, c. 18. Available at https://www.legislation.gov.uk/ukpga/1990/18/contents. Accessed 12 Sept 2023
7. Data Protection Act 2018, c. 12. Available at https://www.legislation.gov.uk/ukpga/2018/12/contents/enacted. Accessed 12 Sept 2023
8. Daws R (2023) China's deepfake laws come into effect today. AI News. Retrieved 21 Apr 2023, from https://www.artificialintelligence-news.com/2023/01/10/chinas-deepfake-laws-come-into-effect-today/
9. Defamation Act 2013, c. 26. Available at https://www.legislation.gov.uk/ukpga/2013/26/contents/enacted. Accessed 12 Sept 2023
10. Fitri A (2023) China has just implemented one of the world's strictest laws on deepfakes. Tech Monitor. Retrieved 21 Apr 2023, from https://techmonitor.ai/technology/emerging-technology/china-is-about-to-pass-the-worlds-most-comprehensive-law-on-deepfakes
11. Fraud Act 2006, c. 35. Available at https://www.legislation.gov.uk/ukpga/2006/35/contents. Accessed 12 Sept 2023

12. Frum D (2020) The very real threat of trump's deepfake. The Atlantic. Retrieved 30 Apr 2023, from https://www.theatlantic.com/ideas/archive/2020/04/trumps-first-deepfake/610750/
13. Global Internet Forum to Counter Terrorism (GIFCT) (2023) What is the hash-sharing database? GIFCT. Retrieved 30 Apr 2023, from https://gifct.org/?faqs=what-is-the-hash-sharing-database
14. Goodfellow IJ, Pouget-Abadie J, Mirza M, Warde-Farley D, Ozair S, Courville A, Bengio Y (2020) Generative adversarial networks. Commun ACM 63(11):139–144
15. Grady P (2023) EU proposals will fail to curb nonconsensual deepfake porn. Center for Data Innovation. Retrieved 2 May 2023, from https://datainnovation.org/2023/01/eu-proposals-will-fail-to-curb-nonconsensual-deepfake-porn/#:~:text=There%20are%20no%20specific%20laws,be%20reported%20and%20taken%20down.
16. Le B, Tariq S, Abuadbba A, Moore K, Woo S (2023) Why do deepfake detectors fail? Retrieved 19 Apr 2023, from https://arxiv.org/abs/2302.13156
17. Ma Q, Xing L (2021) Perceptual hashing method for video content authentication with maximized robustness. EURASIP J Image Video Process 2021(1). https://doi.org/10.1186/s13640-021-00577-z
18. Mania K (2022) Legal protection of revenge and deepfake porn victims in the European Union: findings from a comparative legal study. Trauma, Violence Abuse: 152483802211437. https://doi.org/10.1177/15248380221143772
19. Maras M-H, Alexandrou A (2018) Determining authenticity of video evidence in the age of artificial intelligence and in the wake of deepfake videos. Int J Evid Proof 23(3):255–262. https://doi.org/10.1177/1365712718807226
20. McCabe D, Alba D (2020) Facebook says it will ban 'deepfakes'. The New York Times. Retrieved 30 Apr 2023, from https://www.nytimes.com/2020/01/07/technology/facebook-says-it-will-ban-deepfakes.html
21. Meskys E, Liaudanskas A, Kalpokiene J, Jurcys P (2020) Regulating deep fakes: legal and ethical considerations. J Intell Property Law Practice 15(1):24–31. https://doi.org/10.1093/jiplp/jpz167
22. Milmo D (2023) Age checks, trolls and deepfakes: what's in the online safety bill? The Guardian
23. Ministry of Justice, Raab D (2022) New laws to better protect victims from abuse of intimate images. U.K. Government. Available at https://www.gov.uk/government/news/new-laws-to-better-protect-victims-from-abuse-of-intimate-images. Accessed 12 Sept 2023
24. NCMEC (2023) About us. TakeItDown. Available at https://takeitdown.ncmec.org/about-us/. Accessed 12 Sept 2023
25. Papenfuss M (2020) Fake video! Trump tweets creepy 'independence day' spoof starring him. HuffPost UK. Retrieved 30 Apr 2023, from https://www.huffingtonpost.co.uk/entry/independence-day-trump-aliens-tweet-video_n_5ec05fd9c5b641b8b123ba89
26. Pawelec M (2022) Deepfakes and democracy (theory): how synthetic audio-visual media for disinformation and hate speech threaten core democratic functions. Digital Soc 1(2). https://doi.org/10.1007/s44206-022-00010-6
27. Snyder V (2023) How covid-19 has changed social media habits. business.com. Retrieved 2 May 2023, from https://www.business.com/articles/social-media-patterns-during-the-pandemic/
28. StopNCII (2023) About us. About StopNCII.org. Retrieved 30 Apr 2023, from https://stopncii.org/about-us/
29. Stupp C (2019) Fraudsters used AI to mimic Ceo's voice in unusual cybercrime case. Wall St J. Retrieved 28 Apr 2023, from https://www.wsj.com/articles/fraudsters-use-ai-to-mimic-ceos-voice-in-unusual-cybercrime-case-11567157402
30. University of Exeter (2021) Unreliable witness testimony biggest cause of miscarriages of justice over the past 50 years, study suggests. Retrieved 2 May 2023, from https://news-archive.exeter.ac.uk/homepage/title_855937_en.html
31. Venema AE, Geradts ZJ (2020) Digital forensics deepfakes and the legal process. Retrieved 20 Apr 2023, from https://www.essentialresearch.eu/wp-content/uploads/2020/07/Digital-Forensics-Deepfakes-and-the-Legal-Process-Venema-Geradts2020.pdf

32. Wang S, Kim S (2022) Users' emotional and behavioral responses to deepfake videos of K-pop idols. Comput Hum Behav 134:107305. https://doi.org/10.1016/j.chb.2022.107305
33. Watson A (2022) Usage of social media as a news source worldwide 2022. Statista. Retrieved 30 Apr 2023, from https://www.statista.com/statistics/718019/social-media-news-source/
34. Wu P (2015) Impossible to regulate: social media, terrorists, and the role for the U.N. Chicago J Int Law: 281–311

Chapter 13
The Impact of Facial Recognition Technology on the Fundamental Right to Privacy and Associated Recommendations

Abstract The rapid proliferation of biometric technologies, particularly facial recognition, has extended its reach across both public and private spheres, serving the dual objectives of surveillance and security. Amid the ongoing evolution of this technology, its substantial benefits coexist with formidable challenges that necessitate equal scrutiny. This chapter aims to provide a critical analysis of the impact of facial recognition on the fundamental right to privacy. To this end, the chapter seeks to expound upon the multi-dimensional facets of facial recognition technology (FRT), evaluating both its merits and inherent limitations. Acknowledging the manifold benefits of FRT, the chapter emphasises the importance of a balanced perspective. This perspective not only recognises its potential to enhance security but also addresses its potential intrusiveness, thus considering its impact on privacy rights. In the wake of this comprehensive analysis, the chapter offers a set of recommendations intended to safeguard the fundamental right to privacy in the context of facial recognition deployment. The findings reveal that as the landscape of biometric technology continues to evolve, imperatives such as increased accuracy, enhanced transparency, and the establishment of a robust legal framework surface prominently. Furthermore, the findings underscore the necessity for a careful integration of technology and ethics, a convergence that assumes paramount importance in upholding individual privacy.

Keywords Facial recognition technology · FRT · Privacy rights · Biometrics · Artificial intelligence · AI · GDPR · Human rights · Surveillance

13.1 Introduction

FRT operates through the application of biometric methods, enabling the identification and authentication of an individual's identity based on their distinct facial features. The mechanism of FRT involves an initial detection of facial characteristics, followed by a comprehensive analysis of these features. Subsequently, sophisticated algorithms are deployed to conduct a comparison and match the extracted image

© The Author(s), under exclusive license to Springer Nature Switzerland AG 2024   259
R. Montasari, *Cyberspace, Cyberterrorism and the International Security in the Fourth Industrial Revolution*, Advanced Sciences and Technologies for Security Applications, https://doi.org/10.1007/978-3-031-50454-9_13

against preexisting data repositories [16]. The right to privacy is protected under Article 8 of the Human Rights Act 1998, which stipulates the "right to respect for private and family life" [14]. The preservation of this right assumes an indispensable role in upholding democratic societies, as it is inherently intertwined with the notion of individual autonomy. Privacy empowers individuals to demarcate personal boundaries, asserting control over the degree of external interference, whether from fellow citizens or governmental entities. This empowerment, in turn, facilitates the exercise of personal freedom and choice. Nonetheless, the use of FRT can infringe upon the right to privacy under certain circumstances which will be examined within this chapter. Such infringement occurs when the technology is employed devoid of informed consent or when its deployment disproportionately outweighs its public security benefits.

In light of these considerations, this chapter aims to conduct a critical analysis of the multifaceted impact of FRT on the right to privacy. To achieve this aim, the chapter examines the pivotal challenges that arise in the complex balancing act between harnessing FRT's potential and safeguarding privacy rights. The discourse will specifically delve into several key challenges, each examined through a critical lens. These challenges include the security of sensitive data, mass government surveillance, the pressing concern of discrimination embedded within FRT systems, the imperative of enhancing public awareness, and the pressing need for well-defined legal frameworks and regulations. As the discourse advances, the chapter also offers a series of recommendations aimed at mitigating the identified challenges. These recommendations are anticipated to foster greater transparency in the use of FRT, increase its accuracy and efficacy, and, most notably, contribute to the refinement and elucidation of pertinent laws and regulations. In this manner, the chapter envisages contributing to the ongoing dialogue concerning the delicate equilibrium between technological advancement and the upholding of core privacy principles.

The subsequent sections of the chapter are structured as follows: Sect. 13.2 provides essential contextual background information, setting the study within its pertinent framework. Moving forward, Sect. 13.3 rigorously analyses the key challenges that are intrinsically linked with the use of FRT. This exploration seeks to elucidate the multifaceted issues that underscore the discourse on FRT deployment. In tandem, Sect. 13.4 investigates the existing legal frameworks and regulations, aiming to provide an in-depth understanding of the regulatory parameters that govern the application of FRT. Section 13.5 presents a comprehensive set of recommendations that can be implemented to mitigate and/or address the challenges that have been identified and examined earlier in the chapter. Section 13.6 encapsulates a discussion, offering conclusive reflections extending to a pathway for future research pursuits. Finally, the chapter concludes in Sect. 13.7.

13.2 Background

During the mid-2000s, the emergence of FRT garnered significant attention in both public awareness and commercial applications [3, 17, 22]. This marked a notable progression from historical practices such as identification sketches based on eyewitness accounts, which had been commonly employed by law enforcement agencies (LEAs). However, these sketches faced widespread criticism due to their inherent inaccuracies [28]. FRT operates with the primary objective of identifying or verifying an individual's identity by creating a template of their facial image and subsequently comparing it to an existing repository of photographs, videos, or real-time surveillance data [12, 16]. This process relies on complex algorithms that leverage biometric markers to detect facial attributes, including measurements such as the distance between an individual's eyes and dimensions from forehead to chin, as well as other distinguishing features, all aimed at establishing potential matches [27]. FRT exhibits utility in one-to-one applications, where it concurrently enhances convenience and security, such as individuals gaining access to workplaces through facial scans without a need for physical keys or unlocking mobile phones [16]. However, its true potential lies in one-to-many applications, where it effectively assists LEAs in identifying victims, locating missing children, apprehending criminals, and bolstering national security at border crossings and airports. Illustrating this perspective, a study conducted by Jain et al. underscores the efficacy of FRT in heightening security, preventing criminal activities, and amplifying convenience, particularly in real-time scenarios (2016). As technological landscape evolves, so too will its security and protection measures. Thus, it is plausible that facial recognition's multifaceted potential will continue to intersect with broader societal domains, accentuating the intersection of security and convenience within an ethical framework.

One commonly used form of FRT is 'facial characterisation' which involves detecting a person's reactions or emotions based on their facial features and microexpressions. This particular application finds extensive deployment in marketing and advertising contexts, aiming to gather insights for refining targeting strategies towards specific demographics. While facial characterisation does not inherently collect data for identification purposes, many individuals still perceive it as a violation of their right to privacy [24]. FRT consists of a diverse array of applications, spanning both identification-centric and non-identification applications. As discussed, the key categories that encapsulate these applications are detection, characterisation, verification, and identification, each presenting distinct implications for privacy rights. Prominent instances of FRT deployment encompass domains such as airport security and border control, surveillance facilitated by Closed-Circuit Television (CCTV), locating missing children, and identity verification to unlock mobile devices. Furthermore, its purview extends to personalised experiences, targeted advertising, medical diagnoses predicated on facial feature analysis, attendance tracking within educational institutions and workplaces, as well as facial identification for social media activities such as filters, photo tagging, and friend recommendations based on photos.

In summary, FRT emerged in the mid-2000s, replacing unreliable methods such as eyewitness sketches used by LEAs. As a result, this technological shift has introduced complex algorithms and biometric markers used to create facial templates for identification. Furthermore, FRT finds its initial applications in one-to-one scenarios, enhancing convenience and security for activities such as workplace access and mobile phone unlocking. Nevertheless, its full potential is realised in one-to-many applications, where it assists law enforcement in various tasks. Moving beyond its core functions, facial characterisation in FRT serves purposes such as emotion detection for marketing but raises significant privacy concerns. Beyond these considerations, FRT spans a wide array of applications, including security, surveillance, locating missing individuals, identity verification, personalized experiences, medical diagnoses, attendance tracking, and social media activities. Finally, as technology advances, FRT will further develop, with an ongoing effort to strike a balance between security and convenience while respecting ethical boundaries.

13.3   Challenges

Some scholars and professionals advocate for a complete halt in the use of FRT until further research is conducted and a more robust legal framework and regulations are established. The initial notable opposition to the use of FRT manifested in 2019 when San Francisco enacted a prohibition on its governmental and law enforcement use, subsequently triggering similar actions in other jurisdictions [7]. Similarly, others have described the widespread adoption of FRT as the ultimate demise of privacy [3]. However, a predominant stance within academia, while maintaining a degree of scepticism regarding the enduring ramifications for privacy rights and recognising the inadequacies of existing privacy safeguards, concurrently acknowledges the merits of facial recognition in bolstering public security. Others, however, posit that the potential risks linked to FRT outweigh its benefits, contending that its intrinsically biased nature engenders social inequality, exacerbates prejudiced law enforcement practices, and thus advocates for its proscription [20].

A salient apprehension regarding the application of FRT pertains to its implications for the fundamental right to privacy, particularly concerning security considerations. While FRT holds the potential to enhance public security, the inherent susceptibility of the sensitive biometric data amassed renders it vulnerable to breaches or cyberattacks. This, in turn, compromises individual security in the event of hacking or data breaches. Furthermore, the irreversibility of damages stemming from the sale or unauthorised disclosure of data due to negligence, corruption, or foreign intelligence hacking underscores the paramount significance of data, intensifying concerns over information and identity theft. This concern emanates from the immutable nature of biometric identifiers, unlike mutable passwords [22]. Furthermore, unlike other forms of biometric data, such as fingerprints, facial recognition can be obtained without the individual's knowledge, significantly impacting both consent and privacy [24]. As result, several researchers, policymakers and digital activists alike have expressed caution against the increased use and normalisation of technologies such as facial recognition due to concerns about their adverse effects on individual privacy [18]. Additionally, concerns arise regarding discriminatory practices resulting from inaccuracies and biases associated with FRT, which violate individual privacy.

Liu et al. [21] and Big Brother Watch [4] highlight a multitude of concerns regarding the current use of FRT, including high error rates for people of colour and women, leading to discriminatory outcomes. Such practices contravene individuals' fundamental right to privacy, as these demographic groups are disproportionately targeted or marginalised predicated upon their racial or gender attributes. Moreover, these practices can have severe consequences, such as wrongful arrests or denial of access to services and opportunities. The broader repercussions consist of potential instances of wrongful arrests and the unjust denial of access to essential services or opportunities. Amnesty International's "ban the scan" campaign, initiated in 2021, further underscores these concerns [2]. This initiative seeks to proscribe the application of FRT due to its adverse discriminatory impact, particularly on marginalised populations. In this context, Stark [26] aptly employs the metaphor of plutonium to convey the notion that FRT demands acknowledgment as an intrinsically dangerous technology, thereby warranting specialised application rather than normalisation. Concurrently, a prominent concern with regard to the accuracy of FRT pertains to its potential for the abuse of authority and the curtailment or exploitation of individual liberties [6]. Instances such as China's extensive use of FRT for purposes of control serve as illustrative exemplars of the dangers involved. This underscores the necessity for enhanced regulation and oversight to ensure the judicious deployment of FRT while respecting fundamental human rights and freedoms [23].

A 2019 survey conducted by the Ada Lovelace Institute revealed a dearth in knowledge and understanding regarding the technical limitations inherent to FRT [1]. Noteworthy findings indicated that a substantial proportion of respondents held the erroneous belief of the technology's accuracy (18%), and a quarter erroneously surmised that its use by law enforcement exhibited no discrimination on the basis of race or gender (24%) (Ibid.). However, the shortfall in public awareness significantly curtailed its ability to undertake a discerning evaluation of the advantages and perils and to adopt an informed position on the employment of FRT. Furthermore, increasing awareness and education concerning technologies such as FRT would indubitably foster increased public engagement and ultimately strengthen the democratic foundation for the development of such technologies (Ibid.).

In summary, the topic of FRT is a subject of vigorous debate among scholars and professionals alike. Some advocate for an immediate halt in its use until comprehensive research and robust legal regulations are firmly in place, while others are enthusiastic about its potential for bolstering public security. Central to this debate are concerns that revolve around privacy rights, data vulnerability, and inherent biases embedded within FRT systems. Critics contend that FRT can result in discriminatory outcomes, with particular impacts on people of colour and women. For instance, Amnesty International has launched a campaign calling for a complete ban on FRT due to its undeniable discriminatory impact. Furthermore, there are growing apprehensions regarding authoritarian applications of FRT, exemplified by its extensive misuse in China. Consequently, there is a growing consensus that enhanced regulation and oversight are imperative. Furthermore, as discussed above, the 2019 survey by Ada Lovelace Institute (2019) underscored a profound lack of public understanding about FRT, emphasising the pressing need for education and awareness initiatives. This knowledge deficit is critical for individuals to make informed decisions about the use of this technology and navigate the complex landscape of its ethical and societal implications.

13.4   Current Legal Framework and Regulations

In the United Kingdom (UK), a combination of legislative and regulatory approaches is employed to protect the right to privacy in conjunction with the use of FRT. Legislative measures establish a legal framework that law enforcement must adhere to, while regulatory approaches provide additional guidelines and regulations to safeguard privacy rights in the private sector's utilisation of FRT. The legislative aspect of protecting the impact of FRT on the right to privacy in the UK is primarily governed by the Data Protection Act 2018 [8] and the Human Rights Act 1998 [14]. The Data Protection Act 2018 offers a framework for the collection, use, and safeguarding of personal data, including biometric data such as FRT. It requires explicit consent from individuals before their personal data, including biometric information, is collected and processed. Additionally, the law mandates that the processing of such data must be necessary and proportionate. The Human Rights Act 1998, on the other hand, is aimed at safeguarding the fundamental rights and freedoms of individuals, including the right to privacy. For this reason, public authorities are obligated to act in a manner compatible with the European Convention on Human Rights [9], which includes the right to respect for private and family life.

Moreover, the regulatory approaches primarily focus on the private sector's use of FRT and aim to protect the right to privacy from potential violations. These approaches include the Surveillance Camera Code of Practice 2013, as amended in 2021, and the guidance provided by the Information Commissioner's Office [15]. The Surveillance Camera Code of Practice 2013 offers guidance to organisations on the use of surveillance cameras, including FRT. It requires organisations to conduct privacy impact assessments before deploying the technology to ensure that its use is necessary and proportionate, and to provide clear and transparent information to the public (Ibid.). The ICO, as a regulatory agency responsible for enforcing data protection laws in England and Wales, provides guidance and recommendations on obtaining consent, ensuring accuracy and fairness, and implementing appropriate security measures when using FRT. However, it should be noted that these regulatory approaches are non-binding guidelines, and organisations are not legally obliged to follow them.

In summary, in the UK, the protection of privacy in the context of FRT relies on a combination of legislative and regulatory measures. Legislative acts, such as the Data Protection Act 2018 and the Human Rights Act 1998, establish legal frameworks governing the use of FRT, emphasising consent and proportionality. In contrast, the Human Rights Act ensures compliance with the European Convention on Human Rights, including the right to privacy. Additionally, regulatory approaches, encompassing the Surveillance Camera Code of Practice and guidance from the Information Commissioner's Office, focus on safeguarding privacy in the private sector's deployment of FRT. These measures necessitate privacy impact assessments, transparency, and adherence to data protection principles even though they are not legally binding.

13.5 Recommendations

One proposed course of action in response to the concerns surrounding FRT and its potential infringement upon the right to privacy entails ensuring increased transparency from organisations, LEAs, and governmental bodies regarding the deployment methods and locations of stated technology. Furthermore, organisations must be held responsible for any instances of technology misuse, a goal that can be achieved through the implementation of regular audits and evaluations. An additional recommendation involves fostering increased awareness and education concerning the application, accuracy, reliability, and limitations of FRT. This proposed course of action aligns with the findings of Kostka et al.'s [19] research, which indicated a positive and significant correlation between a more advanced level of education and a favourable inclination towards embracing the use of FRT. Furthermore, in addressing concerns pertaining to mass surveillance and data security, numerous proponents contend that the deployment of such technology should be restricted exclusively to legitimate purposes. These purposes comprise national security needs, matters of public interest, and facilitation of law enforcement activities.

As revealed by the [1] survey, a minority of respondents advocated an outright ban on FRT, while the majority aligned with the notion of governmental regulations that circumscribed its application to specific contexts. This survey underscores the acceptance of FRT when deployed to serve the broader interests of society and personal security. However, concurrently, there exists an emerging trepidation among the public regarding the normalisation of surveillance due to the escalating use of FRT. This dichotomy underscores the meticulous equilibrium requisite between harnessing FRT for societal advancement and mitigating the infringement upon individual privacy. Noteworthy is the prevailing support for LEA's incorporation of FRT, contingent upon its perceived contribution to crime reduction and the assurance of appropriate safeguards and limitations, as stipulated by the findings of the [1] survey. This stance underscores the necessity for increased public engagement in order to cultivate a more comprehensive understanding of the concerns associated with the use of FRT in the realm of policing.

Furthermore, the examination of public perspectives concerning the employment of FRT by LEAs has yielded noteworthy insights. Research conducted by Kostka et al. [18] and Bradford et al. [5] collectively underscores a prevailing consensus centred around key apprehensions, which consist of concerns related to the erosion of privacy, potential democratic implications, and a perceptible lack of trust. Consequently, these studies posit that an increased level of trust in governmental entities correlates with a higher propensity to embrace novel technologies, such as FRT, while concurrently diminishing concerns pertaining to individual privacy infringements.

Another significant area of concern revolves around the potential violations of the right to privacy, a matter accentuated particularly in instances of ethical usage by private sectors. The findings of the [1] substantiate that the nature and intent behind the implementation of FRT bear substantial implications for the violation of privacy rights. This investigation underscores the public's inclination to resist the infringement on their privacy rights in exchange for commercial gains facilitated by FRT. However, the level of public awareness concerning the commercial application of facial recognition was notably lower in comparison to its employment within law enforcement contexts [1]. This observation underlines the imperative for policymakers to elucidate the legal frameworks and regulatory paradigms surrounding FRT, with a particular emphasis on its application within the private sector. Therefore, considering the potential for greater ambiguity and reduced stringency in legislative provisions within this sector, the necessity for clarification becomes more pronounced. Furthermore, efforts to establish regulatory mechanisms within the private sector carry the promise of nurturing increased levels of public confidence, thus contributing to the assurance of safeguarding individual privacy rights. Similarly, alternative viewpoints articulated by scholars such as Hartzog and Selinger [13] contend for a complete prohibition on FRT, distinguishing it from other variants of biometric technology. This perspective is premised on the notion that FRT, due to its increased intrusiveness and the limited capacity for obtaining explicit consent, should be proscribed. This stance is particularly pronounced when considering live facial recognition, which entails the real-time capture and subsequent comparison of facial images with those stored in databases [5].

In light of the escalating concerns surrounding privacy infringements attributed to FRT, a prevailing recommendation calls for the implementation of more lucid and stringent regulations, guidelines, and legislative measures [25]. Within this context, a consensus emerges suggesting that governments should institute comprehensive privacy laws and regulations applicable to both public and private entities. This initiative should aim to safeguard individuals' right to privacy while also conferring the autonomy to govern the use of their collected data. Notably, proponents argue for the necessity to reconceptualise the notion of privacy to accommodate the distinctive attributes of the digital milieu [25]. Responding to these concerns, the European Union (EU) has taken proactive steps to introduce regulatory frameworks in the form of the General Data Protection Regulation (GDPR) [11]. This legislative effort seeks to provide individuals with higher control over their personal data and introduces mechanisms to constrain the exploitation of their biometric data, including that of facial recognition (Ibid.).

Despite the substantial impact of FRT on contemporary law enforcement practices, a notable turning point transpired in the UK when the legal challenge of *R (on the application of Bridges) v Chief Constable of South Wales Police* emerged in August 2020 [29]. In this landmark case, the UK's Court of Appeal rendered a verdict that deemed the employment of FRT by the police to be unlawful. Subsequent to this legal ruling, the House of Commons Science and Technology Committee released a consequential report, which recommended the imposition of a moratorium on the use of FRT by both LEAs and public authorities (Ibid.). This proposed suspension aims to facilitate a more extensive investigation into the associated risks. Here, this juncture underscores the dual nature of FRT's impact in that, while it offers transformative potential for enhanced victim and perpetrator identification, it concurrently harbours the capacity to encroach upon privacy rights, potentially culminating in unlawful applications (Ibid.). The Bridges case holds particular significance as it imposed discernible limitations on the proliferation of FRT deployment by law enforcement in the UK. Importantly, the court's ruling acknowledges the technology's legitimacy in principle, contingent upon its alignment with law enforcement objectives (Ibid.).

The *Bridges* ruling, while acknowledging the immediate impact of FRT on the right to privacy, failed to fully comprehend its broader societal implications within a democratic context. Moreover, it understated the imperative of safeguarding against intrusive surveillance in public spaces. Furthermore, while the ruling's initiation of the discourse is laudable, it lacked the mandate for transparency, which could have fostered greater public trust concerning extensive surveillance practices (Ibid.). Therefore, the ruling's commendable short-term effects are overshadowed by an insufficient consideration of its prospective long-term consequences, potentially jeopardising the protection of privacy rights in democratic societies in the future (Ibid.). Preceding the Bridges case, the UK Equality and Human Rights Commission had advocated for a moratorium on the use of FRT in response to concerns regarding its potential discriminatory ramifications [10]. However, the practical implementation of this recommendation appears to have been limited.

Another recommendation in the context of the foregoing discussions is that individuals should be accorded the prerogative to access data acquired through FRT, along with the ability to request its deletion. This, in turn, ensures informed consent and bolstering data security. Furthermore, individuals should be apprised of FRT deployment and offered the choice to opt out. While this proposition appears reasonable in theory, its realisation in practice is hampered by complexities, particularly regarding the law enforcement's use of the technology, where consent might not be feasible, and considering the fragmented legal and policy framework in the UK. This underscores the imperative for a comprehensive review and clarification of the legal framework to align more effectively with evolving technological landscapes and public expectations [1]. Finally, active engagement between governments, organisations, and the public within the decision-making process is imperative. This participatory approach not only ensures the incorporation of concerns and diverse perspectives but also fosters increased trust among the public, governmental bodies, LEAs, and organisations. As a result, the ongoing and substantive nature of public engagement is essential for cultivating a favourable environment that would likely facilitate the positive and ethical application of FRT in the future.

In summary, in response to growing concerns surrounding FRT and its impact on privacy, this section has offered several key recommendations that seek to balance FRT's benefits with privacy considerations. Firstly, enhanced transparency has been highlighted as crucial among organisations, LEAs, and governmental bodies regarding FRT deployment methods and locations. Within this context, regular audits of organisations that deploy FRT could deter misuse and ensure responsibility. Additionally, raising public awareness and education about FRT, including its applications and limitations, has been considered paramount. In the debate over whether to ban or regulate FRT, regulated application has been revealed to be favoured to strike a fair balance. Thus, public engagement in FRT deployment decisions has been highlighted as vital. Similarly, this section has provided recommendations regarding the ethical use of FRT in the private sector. To this end, the section has highlighted the significant importance of clear policies and regulatory mechanisms to build trust. Within this context, stringent regulations and privacy laws, exemplified by the EU's GDPR, have been considered essential for protecting privacy rights. This is supported by legal challenges such as the Bridges case in the UK, underscoring the need for clear legal guidelines. Finally, addressing this is supported by citing legal challenges, such as the Bridges case in the UK.

13.6   Discussion

The paramount findings regarding the influence of FRT on the right to privacy centre on several salient aspects. Foremost among these is the pronounced dearth of education pertaining to its accuracy, which gives rise to concerns of discrimination and bias. In addition, a substantial revelation pertains to the insufficient organisational accountability within the context of the prevailing legal framework and regulatory apparatus. Furthermore, a pertinent apprehension arises in relation to potential security breaches that could lead to the exposure of sensitive information. These insights serve to accentuate the potential necessity for implementing more frequent audits and reviews. This, in turn, enhances accountability mechanisms and proactively mitigates privacy breaches. The findings, in a broader context, suggest that the absence of transparency, coupled with limited education and awareness concerning FRT, contributes to concerns surrounding its deployment within private sectors and its potential expansion. Despite the reservations expressed, a substantial number of scholars and organisations advocate for the adoption of FRT, contingent upon increased regulatory oversight and the implementation of safeguards aimed at preserving the fundamental right to privacy. Therefore, a promising avenue for future research could involve exploring the use of FRT in other countries, such as Australia, where it is more prevalent. Additionally, comparing and contrasting the legal frameworks and regulations in these jurisdictions with those in the UK could yield valuable insights into effective measures for safeguarding individuals from privacy breaches associated with the application of FRT.

13.7 Conclusion

In conclusion, the ramifications of FRT on the fundamental right to privacy are contingent upon the levels of trust vested in both governmental bodies and private organisations. Furthermore, one crucial determinant lies in the extent of a government's rapport with the citizenry and its historical track record concerning the misuse of public surveillance and data. However, an increase in transparency and public awareness, achieved through measures such as access to facial recognition system data and the ability to delete facial recognition data, serves to mitigate the impact of FRT. Moreover, the implementation of lucid and stringent regulations and guidelines governing the use of FRT across both public and private sectors reduces the potential for data misuse and enhances data security. Similarly, substantial enhancements in the technology's accuracy are imperative to prevent discriminatory biases that violate additional rights. Furthermore, while FRT offers prospects of increased efficiency and security, it simultaneously raises concerns about privacy erosion, necessitating its circumscribed application in specialised contexts rather than its unfettered normalisation. Finally, throughout the discourse in the chapter, it has been elucidated that while FRT provides considerable benefits and holds a distinct position in future security paradigms, its adoption should be undertaken cautiously due to its potential ramifications for privacy. Instead, the recommendations elucidated herein warrant heed, with the aim of potentially guiding the judicious use of FRT in a more secure and refined manner. Ultimately, a complete prohibition on FRT might not be the most suitable approach; instead, its deployment should be careful and balanced, limited to specific contexts that enhance law enforcement practices and public security, all while avoiding its normalisation.

References

1. Ada Lovelace Institute (2019) Beyond face value: public attitudes to facial recognition technology. Available at: https://www.adalovelaceinstitute.org/report/beyond-face-value-public-attitudes-to-facial-recognition-technology/. Accessed: 28 Sept 2023
2. Amnesty International (2021) Ban dangerous facial recognition technology that amplifies racist policing. Racist policing
3. Andrejevic M, Selwyn N (2022) Facial recognition. Wiley
4. Big Brother Watch (2018) Face off: the lawless growth of facial recognition in UK policing. https://bigbrotherwatch.org.uk/wp-content/uploads/2018/05/Face-Off-final-digital-1.pdf
5. Bradford B, Yesberg JA, Jackson J, Dawson P (2020) Live facial recognition: trust and legitimacy as predictors of public support for police use of new technology. Br J Criminol 60(6):1502–1522. https://doi.org/10.1093/bjc/azaa032
6. Bragias A, Hine K, Fleet R (2021) 'Only in our best interest, right?' Public perceptions of police use of facial recognition technology. Police Pract Res 22(6):1637–1654
7. Conger K, Fausset R, Kovaleski SF (2019) San Francisco bans facial recognition technology. The New York Times 14(1)
8. Data Protection Act (2018) c. 12. Available at: https://www.legislation.gov.uk/ukpga/2018/12/contents/enacted. Accessed: 22 Sept 2023

9. ECHR (2022) Guide on Article 8 of the European convention on human rights: right to respect for private and family life, home and correspondence. Available at: https://www.echr.coe.int/documents/d/echr/guide_art_8_eng. Accessed: 22 Sept 2023

10. Equality and Human Rights Commission (2020) Facial recognition technology and predictive policing algorithms out-pacing the law. Available at: https://www.equalityhumanrights.com/en/our-work/news/facial-recognition-technology-and-predictive-policing-algorithms-out-pacing-law. Accessed: 22 Sept 2023

11. European Union (2016) Regulation (EU) 2016/679 of the European Parliament and of the Council of 27 April 2016 on the protection of natural persons with regard to the processing of personal data and on the free movement of such data, and repealing Directive 95/46/EC (General Data Protection Regulation). Off J Eur Union L119/1–L119/88

12. Hamann K, Smith R (2019) Facial recognition technology. Crim Justice 34(1):9–13

13. Hartzog W, Selinger E (2019) Why you can no longer get lost in the crowd. The New York Times. Available at: https://www.nytimes.com/2019/04/17/opinion/data-privacy.html. Accessed: 22 Sept 2023

14. Human Right Act (1998) c. 42. Available at: https://www.legislation.gov.uk/ukpga/1998/42/contents. Accessed: 22 Sept 2023

15. ICO (2022) Guidance on video surveillance (including CCTV). Available at: https://ico.org.uk/for-organisations/uk-gdpr-guidance-and-resources/cctv-and-video-surveillance/guidance-on-video-surveillance-including-cctv/about-this-guidance/. Accessed: 22 Sept 2023

16. Jacques L (2021) Facial recognition technology and privacy: race and gender how to ensure the right to privacy is protected. San Diego Int Law J 23(1):111–156

17. Jain AK, Ross A, Nandakumar K (2016) Introduction to biometrics. Springer

18. Kostka G, Steinacker L, Meckel M (2023) Under big brother's watchful eye: cross-country attitudes toward facial recognition technology. Gov Inf Q 40(1):101761. https://doi.org/10.1016/j.giq.2022.101761

19. Kostka G, Steinacker L, Meckel M (2021) Between security and convenience: facial recognition technology in the eyes of citizens in China, Germany, the United Kingdom, and the United States. Public Underst Sci 30(6):671–690

20. Liao TF (2019) The dangers of facial recognition technology. Harvard J Law Technol 33(1):1–55

21. Liu X, Lu L, Shen Z, Lu K (2018) A novel face recognition algorithm via weighted kernel sparse representation. Futur Gener Comput Syst 80:653–663

22. Naker S, Greenbaum D (2017) Now you see me: now you still do: facial recognition technology and the growing lack of privacy. Boston Univ J Sci Technol Law 23:88

23. Nesterova I (2020) Mass data gathering and surveillance: the fight against facial recognition technology in the globalized world. SHS Web Conf 74:3006. https://doi.org/10.1051/shsconf/20207403006

24. Raposo VL (2023) (Do not) Remember my face: uses of facial recognition technology in light of the general data protection regulation. Inf Commun Technol Law 32(1):45–63

25. Selvadurai N (2015) Not just a face in the crowd: addressing the intrusive potential of the online application of face recognition technologies. Int J Law Inf Technol 23(3):187–218

26. Stark L (2019) Facial recognition is the plutonium of AI. XRDS: Crossroads ACM Mag Students 25(3):50–55

27. Symanovich S (2019) How does facial recognition work? https://us.norton.com/internetsecurity-iot-how-facial-recognition-software-works.html

28. Valentine T, Davis JP (2015) Forensic facial identification: theory and practice of identification from eyewitnesses, composites and CCTV. Wiley

29. Zalnieriute M (2020) Burning bridges: the automated facial recognition technology and public space surveillance in the modern state. Columbia Sci Technol Law Rev 22:284

Printed in the United States
by Baker & Taylor Publisher Services